STATISTICAL MODELS FOR NUCLEAR DECAY

Related Titles

Neutrons, Nuclei and Matter
J Byrne

Basic Ideas and Concepts in Nuclear Physics (2nd Edition)
K Heyde

Non-accelerator Particle Physics
H V Klapdor-Kleingrothaus and A Staudt

Nuclear Physics: Energy and Matter
J M Pearson

Nuclear Decay Modes
D N Poenaru

Nuclear Particles in Cancer Treatment
J F Fowler

Linear Accelerators for Radiation Therapy (2nd Edition)
D Greene and P C Williams

Nuclear Methods in Science and Technology
Y M Tsipenyuk

FUNDAMENTAL AND APPLIED NUCLEAR PHYSICS SERIES

Series Editors
R R Betts and W Greiner

STATISTICAL MODELS FOR NUCLEAR DECAY

From Evaporation to Vaporization

A J Cole

Institut des Sciences Nucléaires, Grenoble, France

INSTITUTE OF PHYSICS PUBLISHING
BRISTOL AND PHILADELPHIA

British Library Cataloguing-in-Publication Data

A catalogue record for this book is available from the British Library.

ISBN 0 7503 0512 6

Library of Congress Cataloging-in-Publication Data are available

Commissioning Editor: James Revill
Publisher: Nicki Dennis
Production Editor: Simon Laurenson
Production Control: Sarah Plenty
Cover Design: Victoria Le Billon
Marketing Executive: Colin Fenton

Published by Institute of Physics Publishing, wholly owned by The Institute of Physics, London

Institute of Physics Publishing, Dirac House, Temple Back, Bristol BS1 6BE, UK

US Office: Institute of Physics Publishing, The Public Ledger Building, Suite 1035, 150 South Independence Mall West, Philadelphia, PA 19106, USA

Typeset in TeX using the IOP Bookmaker Macros
Printed in the UK by Bookcraft, Midsomer Norton, Somerset

to Bernard and Margaret Harvey

Contents

Preface

The purpose of this book is to present and discuss statistical models which are used to describe the decay of excited atomic nuclei. The subject dates from about 1937 when, building on the Bohr concept of the compound nucleus as a system in temporal equilibrium, Weisskopf first proposed a quantitatively successful statistical model to describe the 'evaporation' of neutrons from excited compound nuclei.

The book is intended for use by senior undergraduates and postgraduates as well as confirmed experimentalists seeking some global perspective concerning the history and current status of this research domain.

Establishing the validity of the statistical approach is a difficult problem at all excitation energies. Historically, following Bohr, the use of the statistical theory at low excitation energies was justified by supposing that the reaction between two nuclei proceeded via a succession of intermediate states of the compound nucleus (projectile + target) so that beyond conservation laws the input reaction channel had no specific influence on the relative weights of decay channels. We may expect this 'loss of memory' to be more and more effective if the lifetime of the decaying system is much longer than the characteristic time for a transition between compound nuclear states.

With increasing excitation energy the lifetime of the compound system rapidly decreases. The Bohr argument, which is not watertight even at low energies, therefore becomes less and less valid and, in fact, probably cannot be used at excitation energies above about 2–3 MeV/nucleon. This seems to pose a real problem which has been largely ignored in recent research. Indeed, workers concerned with the development of purely statistical models have appealed to comparison of predictions with experimental data to establish the simultaneous validity of the statistical approach and of the particular statistical model involved. My own view, expressed towards the end of chapter 1, is that the justification for application of simple 'geometrical' statistical models to nuclear decay is inextricably associated with the *concept* of equilibrium rather than with arguments concerning timescales.

There are two other important aspects which distinguish the nuclear problem from applications more familiar to students of statistical mechanics. The first is that the systems investigated are typically composed of 10^2 particles (nucleons) and must be considered as isolated from exterior influences. Consequently, the use of thermodynamics to describe such systems is limited. The second is that (contrasting with a thermometer placed in a beaker of water) measurements are made on the decay products long after the system has ceased to exist. The practical application of statistical mechanics is therefore necessarily limited to prediction of asymptotic properties of the corresponding macrostates.

It is fair to say that statistical models of nuclear decay may be classified in terms of increasing excitation energy (and angular momentum) and, simultaneously, increasing complexity. The fact that increasing complexity corresponds to the chronological development of the subject has more or less dictated the organization of the material presented in this work.

The book is divided into eight chapters. The first two chapters are concerned with introductions to the statistical mechanics and nuclear physics which are necessary for understanding the development of statistical models of nuclear decay. These chapters were written in the hope of providing a largely self-contained text. Of course this is never completely achieved and further background reading from the cited references is recommended.

The second part of the book (chapters 3 and 4) describes the statistical models which were created to describe decay processes at low excitation energy. In chapter 3, the topics include the Weisskopf theory for evaporation of neutrons, the Hauser–Feshbach evaporation theory, fusion, the Griffin, Blann–Cline and Harp–Miller–Berne approaches to pre-equilibrium particle emission, and, finally, early statistical theories of low energy fission. The discussion of fission includes Fong's model which assumes equilibrium at scission. The model is usually considered as a failure of the statistical approach emphasizing the need to consider the detailed dynamical progression of the fission process. Nevertheless, variations of this model are still useful especially at higher excitation energies.

Concerning the Hauser–Feshbach theory, one may distinguish two main fields of application. The first, which is *not* dealt with in this book, involves low energy compound nucleus reactions leading to specific final states. Averaging over compound nuclear resonances appears here as a justification for the incoherent summing procedure which forms the basis of the theory. Averaging over intermediate resonances is, of course, also important for decay of the compound nucleus at higher excitation energies in which the residual nucleus is represented by a density of states. However, in this case, it is also possible to consider the Hauser–Feshbach model (the designation 'model' is due to E Vogt) on a lower level as a semiclassical extension of the Weisskopf or Bohr–Wheeler evaporation models.

At this point, it is as well to state that a large body of literature concerning the construction of a detailed quantum mechanical theory of reactions proceeding via the formation of a compound nucleus has not been considered herein. The modern theories, exemplified by the work of Nishioka and collaborators, are technically complex. To provide even a sketch of progress in this field (including a comprehensive description of S-matrix theory and random Hamiltonian matrices) would have considerably lengthened the text. Furthermore, the theoretical developments, associated notably with the names of Feshbach, Moldhauer, Weidenmüller, Mahaux and collaborators, are mainly relevant at energies which are rather low when compared with the range of excitation considered herein.

At the phenomenological model level, once the conceptual relation between the Weisskopf and the Bohr–Wheeler models is established (e.g., for neutron emission) we can almost say that the scene is set. Later developments applied to sequential decay of 'hot' nuclei, even stretching up to the 1990s, all use the same basic idea which is that an emitted particle can be considered as being originally situated somewhere on the surface of the emitting nucleus with a randomly directed velocity. This is even true of Griffin's model for pre-equilibrium emission which, in the context of the temporal equilibrium picture proposed by Bohr, appears as one of the boldest propositions in the history of the subject.

Chapter 4, which begins with an introduction to Monte Carlo simulations, is mainly concerned with applications of the Hauser–Feshbach theory to single and multistep evaporation. The multistep model relies on the assumption that light particles are emitted successively and independently from the excited compound nucleus. The success of the theory in predicting observables is truly remarkable and we illustrate our description with results from several experimental studies. The reader will probably be especially impressed by the accurate prediction of correlations when several particles are emitted from the 'parent' system.

The third and last part of the text is devoted to the description of decay processes which are 'modern' insofar as the experimental work has mainly been carried out over the last 20 years. These applications include the so-called sequential binary decay mechanism (chapter 5), the most popular description of which relies on an extension of the Bohr–Wheeler theory. Multifragmentation is discussed in chapters 6 and 7. The possibility of dissociation of a highly excited nucleus into several large fragments and light particles was first investigated in the late 1970s. The main thrust of experimental work (which is continuing) occurred from about 1985 onwards and was spurred on by the availability of higher and higher energy heavy ion beams, and especially by the construction of charged particle multidetectors.

It is currently believed that compression of the composite system produced in head-on collisions of projectile and target nuclei plays an important role in the multifragmentation process. Statistical models are therefore required to be able to include a collective decompressional energy as well as the interfragment Coulomb repulsion. To complicate matters further the emission of pre-equilibrium nucleons and alpha particles during the initial stages of the collision remove mass (charge) and energy from the multifragmenting system. Thus, the application of statistical models of multifragmentation is conditioned by global variables which depend not only on initial conditions but also on dynamical evolution. This is, indeed, a minor revolution compared with, say, the evaporative process, which is essentially governed by simple conservation laws.

Current theories, which are exposed in chapter 6, again hark back to Fong and to Bohr and Wheeler. On the one hand, one has witnessed the developement of 'multiple scission like' theories which picture the intial state as a set of newly formed fragments enclosed in some small 'freeze-out' volume and interacting via Coulomb forces. On the other hand, the transition state method considers that equilibrium is established close to the multifragmentation barrier (more precisely, at the mimimum of the density of microstates considered as a function of a size parameter).

The second 'multifragmentation' chapter (chapter 7) is mainly concerned with the investigation of multifragmentation considered as a change of phase. The original impetus for the search for phase transition 'signals' is due to the observation that the distribution of the size of fragments produced in multifragmentation decay followed, approximately, a power law, as expected for a liquid–vapour transition at the critical temperature. The subjects treated include a study of the rather detailed analogy between percolation and multifragmentation. The use of the simple cubic bond breaking percolation model in small systems not only succeeds in reproducing multifragmentation data but also allows us to test schemes designed to look for a second order phase change. Experimental investigations based both on the fragment size distribution and on nuclear calorimetry are also discussed. The current status of this latter field is a little confused because of difficulties in selecting a reliable 'thermometer' which can be used to measure nuclear temperatures.

At very high excitation energies there is a measurable probability that the nucleus explodes into nucleons and alpha particles. No larger fragments remain. The statistical description of this 'vaporization' process is still in its infancy although interesting progress has recently been made by members of the INDRA collaboration. In the last part of chapter 7 we will provide a description of the mechanism and show that the equilibrium concept provides the key to understanding the relative importance of the various disintegration channels.

Finally, in chapter 8, we attempt the difficult task of drawing the threads together in order to build a coherent picture of applications of statistical mechanics to nuclear decay. In this chapter we also indicate a few areas of research where further work may be useful.

So much for the global organization of the material presented in this book. I shall now say a few words concerning personal preferences and style. The choice of topics undoubtedly reflects the point of view of the author who, for the last 20 years, has worked on the successive phases of this subject being involved both with experiments and with the task of building and applying statistical models. This bias is partly responsible for some omissions, notably that of the careful work carried out at low energies concerning the application of the Hauser–Feshbach formula to elastic and inelastic scattering. In my defence, besides invoking space limitations, may I say that a rather complete review of the relevant literature is given in the recent book of Gadioli and Hodgson and that I felt that I could hardly improve on their treatment.

Under the heading 'mea culpa', I should also say that I have voluntarily simplified the presentation of certain models in order to convey the principles while avoiding the (sometimes tedious) details concering concrete applications presented in the original research papers. The alternative would probably have been to leave out the account altogether. I ask to be forgiven for any distortions which may have been introduced by this procedure. In all cases I have provided references to the original publications.

Throughout the second and third parts of the text, I have also made frequent reference to published experimental results and techniques which have been instrumental (no pun intended) in providing a fairly comprehensive understanding of the formation and decay of highly excited atomic nuclei. However, given the scope of the subject, the reference list is necessarily incomplete and I apologize to colleagues whose contributions have not been explicitly cited. I would ask them to understand that, in a work of this nature, one selects material from the specialized journals in order to illustrate particular points. There is thus no intention of providing a comprehensive survey of the literature.

The index refers only to page numbers where something interesting concerning keywords is to be found. I have always disliked searching through page numbers only to find that many cited pages contain little more than the keywords themselves.

Concerning the main body of the text, it is important to state that I have been at some pains to avoid the 'review' style which allows an author to summarize a large body of literature all the while presenting equations, the justification and detailed understanding of which can only be aquired by careful study of the cited research articles. Rather, I have tried to present a largely self-explanatory text in which the overwhelming majority of results from theory are explicitly derived. The relatively few exceptions correspond to results for which the derivations are available in standard texts.

I have also tried to emphasize the physical content of the statistical models and thereby to provide an informative and readable book. Of course, I have not been able to avoid expressing developments in mathematical terms. However, with the exception of a few simple results from quantum mechanics the mathematics involved has been known for at least a century and should pose no problems for a senior undergraduate or postgraduate student. No prior knowledge of statistical mechanics is assumed although some familiarity with the arguments concerning the foundations of this subject would help the reader to understand parts of chapter 1 and perhaps to form a personal opinion concerning the justification for the use of statistical methods.

Finally, wherever possible I have followed the modern trend in nuclear heavy ion physics of using classical statistical mechanics. I am currently aware of only one minor difficulty (concerning dimensions of microstate densities) which results from this choice. One exception to the classical emphasis has to do with a model for vaporization where the Boson or Fermion character of light fragments is explicitly taken into account. It seems however that this feature is not very important except at low energies.

Let me terminate this preface with a brief word of warning concerning multifragmentation and vaporization. Most of the research concerning these topics is quite recent. Indeed, borrowing the terminology employed most instructively by John Ziman, one could say that while there exists a common and well-understood (*consensible*) language concerning the physics of these processes it cannot be said that there is a global *concensus* concerning the mechanism. It is therefore probable that some of the material relating to these subjects will require some degree of revision over the next ten years. Nevertheless, I believe that the present treatment should provide a fairly comprehensive background for understanding the context of current research and an appropriate springboard towards future discoveries.

ISN Grenoble, January 2000

Acknowledgements

Several friends and colleagues were kind enough to read and criticize all or parts of the manuscript. Special thanks are due to Pierre Désesquelles, Igor Gontchar and Maria-Ester Brandan for painstaking work resulting in correction of organizational problems, changes in notation and correction of a few errors in equations. Improvements were also suggested by Marie-Laure Andrieux, Chris Balding, Cyrilla Cole, Sylvain David, Jean-Pierre Longequeue, Hervé Ńifenecker, Pascale Pham and Jean-Marc Richard. I thank them all for their generous assistance. During the course of writing I also benefited from discussions with Bernard Sylvestre-Brac, Jaume Carbonell, Mireille Durand, Alain Giorni and Claude Gignoux. Their advice was most welcome.

Clarifications concerning important models were discussed in e-mail or telephone conversations with Bernard Borderie, Bob Charity, Bill Friedman, Stephan Hilaire, Jörgen Randrup, Jean Richert and Pierre Wagner, and further advice on notation and 'last minute' corrections were forthcoming from IOPP referees Nobby Clarke and Russell Betts. Their participation is gratefully acknowledged.

A large proportion of the drawings were painstakingly achieved by Christian Favro. I do not know how I would have managed without his good-humoured collaboration.

Our librarian, Claudine Tur, contributed most efficiently by quickly obtaining many 'ancient' articles which I wished to include as references. I also thank Claudine for turning a conspiratorial blind eye to my somewhat unorthodox book-borrowing habits especially during the early part of the work.

Thanks, also, to my wife, Annie Müller, for putting up with 'weekend working' but also for insisting on pauses and short holidays throughout the 18 months during which the work was accomplished. Some of my best ideas for improvements somehow appeared during these breaks.

Welcome encouragement came from the series editors, Walter Greiner, Dennis Hamilton (now deceased) and Russell Betts, and help at all stages of the project was available from IOPP editor Jim Revill. I thank them most heartily for their support.

Finally, let me express my gratitude to the IN2P3(CNRS) and particularly to Hubert Flocard, Directeur Adjoint Scientifique at the IN2P3 in Paris, and Joël

Chauvin, Director of the Institut des Sciences Nucléaires in Grenoble, for supporting and encouraging me to go through with the 'adventure' of book writing.

A J Cole
Grenoble
January 2000

Useful Constants

Velocity of light (c) *in vacuo*	$2.99792458 \times 10^{23}$ fm s^{-1}
(1 fm $= 10^{-15}$ m)	
Mega electronvolt (MeV)	$1.60217733 \times 10^{-13}$ Joules
(1 Joule (J) = work done by force of 1 Newton moving through 1 m)	
Planck's constant (h)	$6.6260755 \times 10^{-34}$ J s
	$= 4.1356692 \times 10^{-21}$ MeV s
$\hbar$ (hbar, h/2π)	$6.5821220 \times 10^{-22}$ MeV s
$\hbar$ c	197.327053 MeV fm
Pascal (Pa)	1 Pa $= 10^{-4}$ Newton cm^{-2}
(1 bar $= 10^5$ Pa, 1 cm of water at 4°C exerts a pressure of 98.0637 Pa)	
Avogadro's number	6.0221367×10^{23} mol^{-1}
(atoms or molecules per mole)	
(1 mole of a substance contains the molecular weight in grammes, e.g., 1 mole of Carbon contains 12 g)	
NTP normal temperature and pressure:	
Normal temperature	273.15 K
Normal pressure—1 atmosphere	1.01325×10^5 Pa
Loschmidt's number	2.686763×10^{19}
(number of atoms or molecules per cm^3 at normal temperature and pressure)	
Boltzmann's constant k_B	8.617385×10^{-11} MeV K^{-1}
Atomic mass unit (u)	1.660540×10^{-24} g
	=931.49432 MeV c^{-2}
Mass of proton	1.007276470 u
	=938.27231 MeV c^{-2}
Mass of neutron	1.008664904 u
	=939.56563 MeV c^{-2}
Mass of electron	$5.48579903 \times 10^{-4}$ u
	=0.51099906 MeV c^{-2}
Mass excess (Δ) neutron	8.071 MeV

Mass excess hydrogen atom (mass $\times\, c^2$ of neutral atom with mass number A is $A \times uc^2 + \Delta$)	7.289 MeV
Barn (b, used for cross-sections)	10^{-28} m^2 = 100 fm^2
Millibarn (mb)	10^{-3} b = 10^{-27} cm^2
1 pint (U.K.)	0.56825 litres

PART 1

STATISTICAL MECHANICS AND NUCLEAR PHYSICS

Chapter 1

Elements of equilibrium statistical mechanics

And thick and fast, they came at last,
And more and more and more,
All hopping through the frothy waves,
And scrambling to the shore.

The Walrus and the Carpenter, Lewis Caroll

1.1 Introduction

This chapter is devoted to the presentation of elements of equilibrium statistical mechanics which are later applied to the decay of excited atomic nuclei. The selection of topics is therefore oriented towards practical applications in small isolated systems. In spite of this restriction, it is hoped that the subject matter will be useful (without necessary reference to nuclear physics) for those wishing to complement introductory studies in thermodynamics and statistical mechanics.

Most people who follow some basic course in equilibrium statistical mechanics are led to consider the subject matter and techniques of this branch of physics as a well-defined and understood body of knowledge. However, in 1985, Ma [1] wrote: 'I think that statistical mechanics, as it is at present, is an ill-proportioned subject, with many successful applications but relatively little understanding of the basic principles. Yet most textbooks treat statistical mechanics as a complete science'.

Whether or not one agrees with this opinion, it is certainly true that various books approach the subject from distinct standpoints which are often related to a particular field of application (this chapter is no exception). One such distinction concerns the use of the so-called Boltzmann weight for states of systems in contact with an 'infinite' heat reservoir. Feynman [2] refers to this weight as the 'summit of statistical mechanics'. Yet, the Boltzmann weight and the corrresponding canonical partition function have to be used with considerable caution in nuclear physics applications. Another more conceptual distinction which is of relevance to the nuclear decay problem is concerned with the 'ergodic' hypothesis which is used in connection with single systems. At any moment in time, a system can be considered to be in one of a very large number of possible microscopic configurations, usually referred to as microstates. The essential idea behind the ergodic hypothesis is that a system in equilibrium evolves through a representative set of all accessible microstates (possibly the complete set) over the time interval associated with a physical measurement. Thus, a single measurement is an average value, the average being taken over the sample of the full set of microscopic configurations. Furthermore, equilibrium is maintained over a time interval which is much longer than the characteristic measurement time, so that repeated measurements give essentially the same results. It follows that, for 'ergodic' systems, a theoretical treatment of equilibrium can be constructed either in terms of the properties of a single system which are measured over a finite time interval or, usually more conveniently, in terms of the properties of a fictive ensemble of identically constrained systems which provides a representative sample of all attainable configurations. The truth of the ergodic hypothesis is notoriously difficult to establish for most applications. Perhaps this difficulty lies behind Ma's criticism.

In most nuclear physics experiments measurements are carried out using a foil target which is bombarded by a stream (beam) of incident projectile nuclei. Data are accumulated as a succession of independent events. Each event corresponds to a collision between a projectile nucleus and a target nucleus and the products of each collision are detected in suitable detectors placed around the target. Thus, in contrast with the vast majority of applications, most nuclear physics experiments provide us with an ensemble of events. To the extent that each event can be associated with a particular microscopic configuration of the system under study one might therefore consider the ergodic hypothesis to be unnecessary because experiment directly accesses an ensemble of microstates. The crucial question is obviously to decide whether or not the observed sample 'represents' the full set. In practice, this means that we must decide whether the set of microscopic configurations corresponding to the observed set of events can be considered as a random selection from the set of all such configurations (see section 1.8).

This being said, it is a matter of fact that the conceptual basis provided by the ergodic hypothesis has dominated statistical models of nuclear collisions until quite recently. This is, to some extent, a legacy from early experimental investigations involving the production of a weakly excited intermediate (compound) system composed of projectile and target nucleons. The corresponding theoretical models suppose that each excited compound system behaves as an ergodic system prior to decay. In this picture, decay is a rare event in the sense that in a small time interval the vast majority of compound system microstates do not evolve towards decay but rather to other compound nuclear states. The set of microscopic configurations corresponding to observed events referred to in the preceding paragraph is identified, in the compound nucleus picture, with the set of decaying states. These 'cataclysmic' decay states represent only a minute fraction of the full microstate population so that the system seldom finds itself in such a state. One consequence of this scenario is that decay probabilities do not depend on the way in which the compound system was formed. Ergodic behaviour implies that a representative sample of compound nuclear states, which are consistent with conservation laws, are visited before the system finds itself in a 'no-return' decaying state, whatever the initial conditions (e.g. projectile and target mass and charge) characterizing the collision.

For high excitations the preceding picture is difficult to maintain because decay of the intermediate system takes place very rapidly. However, it is still usually possible to specify a small time interval, Δt, such that each event may be associated with a particular characteristic 'decaying' microstate existing within Δt. The essential difference is that the set of decaying microstates leading to the various types of observable events now represent a significant fraction of all microstates. Under these circumstances, the relevance of the compound nucleus concept is questionable. It is therefore less evident that the probabilties associated with the various modes of decay can be assumed to be consistent with the concept of equilibrium, in particular with regard to the possible influence of the initial conditions.

Whether or not decaying microstates are considered to represent a small fraction of the total, the difficulty arises as to how these states are to be identified and how to describe their temporal evolution. This problem lies at the heart of statistical theories of nuclear decay over the full range of excitation energy considered herein. This will first become apparent in chapter 3 where we deal, in more detail, with the historical development of the subject.

Moving away from conceptual problems, it is important to note that we will be principally concerned with statistical mechanics rather than with thermodynamics. The aim of thermodynamics is to describe macroscopic properties of systems in equilibrium taken in the limit of infinite volume and infinite particle number. An excited nucleus, however, is a *small isolated* system so that the usual arsenal of thermodynamics applies only in a rather restricted sense. One consequence is that we discuss the canonical (Boltzmann) distribution, which may be thought of as one cornerstone of statistical mechanics and thermodynamics, mainly as an approximation which facilitates the evaluation of sums of microstates. It is obvious that Feynman's remark, cited earlier, applies only to 'thermodynamic' systems.

In the next section we introduce the microcanonical ensemble which has provided a suprisingly successful description of the disintegration of nuclei over a wide range of excitation energy. In subsequent sections (1.3 and 1.4) we discuss the basic convolution problem, the canonical distribution, and thereby the relation between statistical mechanics and thermodynamics (section 1.5). A short section (1.6) is also devoted to the grand canonical distribution which pertains when the system under study may be considered to share not only energy (as in the canonical distribution) but also particles with some large 'reservoir'.

Considerable importance is currently attached to the possibility of observing a phase transition in studies of nuclear multifragmentation. For this reason we have included (section 1.7) an introduction to equations of state and the liquid–gas phase transition. We also present and examine the concept of 'pseudo-equilibrium' (section 1.8) which is concerned with deciding whether or not a given set of measured events implies equilibrium in the underlying ensemble of microstates. We conclude our presentation of equilibrium statistical mechanics with a short overview of the problems which arise in studies of nuclear decay processes as the energy in the system is increased.

There are, of course, many excellent books which provide a much broader coverage of thermodynamics and equilibrium statistical mechanics than can possibly be attempted in this short introductory chapter. Two personal favourites are the works of Greiner, Neise and Stöcker [3] (translated from the original version in German) and the English edition of the Ecole Polytechnique lectures by Balian [4]. Both works not only provide required mathematical developments but also dwell on concepts drawing examples from many different fields of research. We also recommend to the interested reader the work of Feynman [2] and that of Thompson [5] and, especially for thermodynamics, that of Callen [6]. One should also cite the book of Klimontovich [7] (translated from the original Russian version) which emphasizes dynamical evolution and, in this sense, presents the subject from a viewpoint which is complementary to that adopted in most other works. For readers of French, the translation of the book by Vassiliev [8] provides a most useful introduction to statistical mechanics and its relation to thermodynamics. Finally we would draw the reader's attention to the discussion of probability and entropy in the first part of the book by Baierlin [9].

1.2 Microstates and macrostates

The most complete description which is necessary to specify any system in a given physical context is referred to as a microstate. The context is usually formed by imposed macroscopic constraints which are conserved quantities such as total energy together with physical constraints (volume, shape etc). The definition of a microstate is facilitated by introducing the notion of a degree of freedom. In mechanics, it is customary to use a semantically natural definition of a degree of freedom as any independant coordinate which may be used to characterize the motion of a set of particles so that a system of N particles in ordinary space has $3N$ degrees of freedom. However, in classical statistical mechanics it is more useful to define a degree of freedom of the system as any independent coordinate used to specify the microscopic state. Thus, for example, the degrees of freedom for a set of N structureless particles in ordinary space consist of the three position and three momentum coordinates for each particle so that there are $6N$ degrees of freedom. The total energy of the system may, in principle, be distributed over all degrees of freedom. However, this is not always the case. For example, if our

N particles do not interact and there are no external forces then all the energy is necessarily concentrated in the momentum degrees of freedom.

One may imagine that a microscopic description necessarily involves all possible degrees of freedom that one can think of to describe the system. This is not the case however. The context is important. The first excited state of the nucleon is the Δ resonance which occurs at an excitation energy of 293 MeV (the nucleon mass is approximately 939 MeV/c^2 and the mass of the delta is 1232 MeV/c^2) so that for a nucleus excited to say 50 MeV it makes no sense to invoke internal nucleonic degrees of freedom. Even if the total excitation energy of a nucleus made up of a few tens of nucleons is of the order of a few hundred MeV, it is extremely unlikely that a large fraction of this energy would be concentrated in just one nucleon.

Once the variables required to specify the relevant degrees of freedom have been selected, one can define a microstate of the system in terms of a set of particular numerical values for these variables. One could, in principle, model the system by attempting, with the help of a computer, to generate and examine all microstates which are consistent with the macroscopic constraints. However, this is almost always impossible due simply to the enormous number of such states.

We can now state the fundamental property which defines equilibrium. In equilibrium, each and every microstate occurs with the same probability. Thus if $\Omega(\boldsymbol{c})$ is the number of microstates consistent with the set of constraints $\boldsymbol{c}$, then the probablity of occurrence of any given microstate is

$$P(\boldsymbol{c}) = \frac{1}{\Omega(\boldsymbol{c})}. \tag{1.2.1}$$

This brings us immediately to the definition of *macrostates*. We used the word 'occurrence' rather than 'observation' above because naturally occurring microstates cannot normally be observed. This may be due to the complexity of the microscopic configuration or/and to the impossibility of making measurements on the relevant microscopic scale. For example, the (classical) description of the microscopic state of one cubic centimetre of gas molecules at a temperature of 0°C (273.15 K) and a pressure of 1 atmosphere (1.01325×10^5 Pa) requires the instantaneous measurement of the six position and momentum coordinates of each molecule. The number of molecules is Loschmitt's number, $L = 2.69 \times 10^{19}$ [10], so that the specification of any one microscopic configuration involves 1.61×10^{20} coordinates. Such a specification is, of course, completely impossible in real experiments. In nuclear physics the number of position–momentum coordinates is much smaller. Even taking into account internal excitation of nuclear fragments, the total number of variables required to describe multifragmentation microstates (chapter 6) is only of the order of a few hundred. However, in addition to the impossibility of making measurements on a length scale of 1 fm (10^{-15} m) a new difficulty arises because the lifetime of the microstate (typically of the order of 10^{-21} s) is out of reach of experiment. Indeed, decay products which result from the microscopic state formed during the collision of projectile and target nuclei are measured at typical times of 10^{-9} s, long after the microstate has ceased to exist.

What *can* be observed therefore usually amounts to a partial or incomplete specification of the system. Furthermore, this specification is usually made in terms of (macroscopic) variables which may not be relevant for the microscopic description. By definition, such a partial description is referred to as a macrostate. Given that this description is incomplete, we can immediately see that distinct microstates must correspond to the same macrostate. Put in a slightly different way, we can state that to each macrostate there corresponds a subset of the total set of microstates. In addition, the set of all macrostates is exhaustive. Thus if we denote by $\boldsymbol{m}$ the set of additional conditions required to single out a given macrostate then we have

$$\sum_{\boldsymbol{m}} \Omega(\boldsymbol{m}; \boldsymbol{c}) = \Omega(\boldsymbol{c}), \tag{1.2.2}$$

in which the notation $X(\boldsymbol{m}; \boldsymbol{c})$ refers to the property X for the macrostate $\boldsymbol{m}$ under the constraints, $\boldsymbol{c}$. Now the equilibrium property implies that the probability

$$P(\boldsymbol{m}; \boldsymbol{c}) = \frac{\Omega(\boldsymbol{m}; \boldsymbol{c})}{\Omega(\boldsymbol{c})} = \frac{\Omega(\boldsymbol{m}; \boldsymbol{c})}{\sum_{\boldsymbol{m}'} \Omega(\boldsymbol{m}'; \boldsymbol{c})}. \tag{1.2.3}$$

This innocent looking equation has far-reaching consequences. It implies that predictions of experimental observables can be made on the basis of the equilibrium assumption if we can evaluate the *number* of microstates associated with each and every macrostate. We have no need to estimate the statistical weight (probability of occurrence) to be associated with each microstate because, in equilibrium, they all have the same weight. We have only to count them.

In some cases the distinction between microstates and macrostates is not useful. Throwing dice provides an example. If the dice are thrown on a tray then the full microscopic description requires at least the two-dimensional positional and three-dimensional rotational coordinates of each dice. The macroscopic state involves just two *discrete* rotational coordinates for each dice. However, in this particular case, integration over the continuous variables provides a number which is the same for all macrostates. The consequence is that we can, in practice, neglect this number and set up a description in which microstates and macrostates are identical.

In most cases, this kind of simplification is impossible. Recall the example of a gas. A microstate is technically very complicated especially if the gas molecules interact. On the other hand, the experiment may measure, say, the pressure on the walls of the container or the density of gas molecules.

As another example we may consider the decay of a highly excited nucleus which leads to the formation of a set of fragments. These products may be detected, and identified using suitable detectors. We thus obtain, for each measured event, a macrostate specified by the number of fragments and by the mass, charge and vector velocity of each fragment. Of course, the precise definition of the macrostate is at the disposition of the investigator. Thus, for example, one could choose to analyse the data uniquely in terms of the number of fragments (usually referred to as the multiplicity) produced in the decay process. The corresponding microstates are not so easy to define. The fragments were all formed during the projectile–target collision, i.e., in some small region of space with perhaps some initial velocities. Furthermore, the fragments, which are, of course, nuclei in their own right, could certainly be excited at the moment of formation and may interact amongst themselves via Coulomb and nuclear forces. It will be clear that the microstate definition involves both technical and conceptual difficulties and thus presents quite a challenge to the would-be model maker.

The quantity $\Omega(\boldsymbol{c})$ defined in (1.2.1) is usually referred to as the microcanonical sum of states. The natural logarithm of this sum is the microcanonical entropy, $S_{\rm M}(\boldsymbol{c})$, i.e.,

$$S_{\rm M}(\boldsymbol{c}) = k_{\rm B} \ln[\Omega(\boldsymbol{c})], \tag{1.2.4}$$

where $k_{\rm B}$ is Boltzmann's constant ($k_{\rm B} = 8.617385 \times 10^{-11}$ MeV per degree Kelvin [10]). In thermodynamics, the dimension of $S_{\rm M}$ is identical with that of $k_{\rm B}$ and the quantity $k_{\rm B}T$ (T is temperature) has the dimension of an energy. Furthermore, it is customary to make use of the quantity $\beta = 1/(k_{\rm B}T)$ which has the dimension of an inverse energy. In nuclear physics, we retain the definition of β but define temperature as $T = 1/\beta$ so that the constant $k_{\rm B}$ is suppressed and the temperature is measured in units of energy. A temperature of 1 MeV is thus equivalent to a 'thermodynamic' temperature of 1.16×10^{10} K.

All physical observables are obtained from macrostate probabilities which are ratios of partial and total sums such as $\Omega(\boldsymbol{m}; \boldsymbol{c})$ and $\Omega(\boldsymbol{c})$. 'Physics' is therefore insensitive to arbitrary renormalization of $\Omega(\boldsymbol{m}; \boldsymbol{c})$, i.e., to the insertion of an arbitrary additive constant in the entropy. Usually however, in accordance with quantum statistical mechanics, one defines the entropy so that a system which possesses only one microstate has $S_{\rm M} = 0$.

There is a much used approximation for the entropy which involves the notion of a density of microstates. Indeed, the idea of a smoothed density is one of the most universally applied concepts of statistical physics. Consider a system which exhibits a discrete spectrum of energies. As the energy, E, increases from $E = 0$ one finds precise values of E for which the number of microstates $\Omega(E) > 0$ and other 'in between' values for which $\Omega(E) = 0$. This situation is highly inconvenient for calculations. Moreover, in many circumstances (e.g., in the case of highly excited nuclei), it is practically impossible to calculate or measure the precise energy of all states.

If a precise energy *can* be assigned to each state then the number of states up to energy E can be written

$$N(E) = \sum_n g_n \int_0^E \delta(E' - E_n)\,{\rm d}E', \tag{1.2.5}$$

where g_n is the number of states at the energy E_n and is referred to as the degeneracy. The exact *density of states* which we label $g(E)$ can now be defined from (1.2.5). Thus,

$$N(E) = \int_0^E g(E')\,\mathrm{d}E', \tag{1.2.6}$$

so that

$$g(E') = \sum_n g_n \delta(E' - E_n) = \sum_i \delta(E' - E_i), \tag{1.2.7}$$

where the first sum is over all energy levels of the system and the second sum is performed over all individual states of the system irrespective of their energies. As an example, we can cite the case of atomic nuclei where, at energy, E_n, and angular momentum, J_n, there are $g_n = 2J_n + 1$ degenerate states with angular momentum projection quantum numbers, M, from $-J_n$ to $+J_n$ in unit steps.

The definition (1.2.7) is useful for formal manipulation (the delta function is not a proper function and normally appears only as part of an integrand). However, for analytical developments, it is more interesting to construct a 'smoothed' microstate density which is obtained by writing

$$\rho(E) = \frac{\Delta N}{\Delta E} = \frac{[N(E + \Delta E) - N(E)]}{\Delta E} = \frac{1}{\Delta E}\int_E^{E+\Delta E} g(E')\,\mathrm{d}E', \tag{1.2.8}$$

which is the average number of states per unit energy, the average being performed over the small energy interval $(E, E + \Delta E)$. Equation (1.2.8) may not be very useful if the real microstates occur at well-separated intervals. This may be seen if we compare the smooth density approximation (1.2.8) with the exact expression (1.2.7) for the particular case of a particle of mass m, enclosed in a box with sides of length a, b and c. Classically, this system is characterized by the position and momentum of the particle and the number of states corresponding to these continuous variables is thus infinite. In the quantum mechanical description the system is represented by a wavefunction which must go to zero at the walls of the container. Possible states are therefore characterized by a wavefunction of the form

$$\Psi(\boldsymbol{r}) = A \sin\frac{\pi n_x x}{a} \sin\frac{\pi n_y y}{b} \sin\frac{\pi n_z z}{c}$$

and are specified by three integers $\{n_x, n_y, n_z\}$ in terms of which the energy is obtained from the time independent Schrödinger equation

$$\frac{-\hbar^2}{2m}\nabla^2\Psi(\boldsymbol{r}) = E_n\Psi(\boldsymbol{r}) = \frac{(\pi\hbar)^2}{2m}\left[\frac{n_x^2}{a^2} + \frac{n_y^2}{b^2} + \frac{n_z^2}{c^2}\right]\Psi(\boldsymbol{r}). \tag{1.2.9}$$

In a cubic box the energy is thus proportional to $n^2 = n_x^2 + n_y^2 + n_z^2$. We see, in figure 1.1, which represents state densities in such a box, that the smoothed density may sometimes be a rather drastic approximation. However, such an approximation may be justified if the 'smoothing width', ΔE, is no larger than the energy resolution which characterizes a measurement.

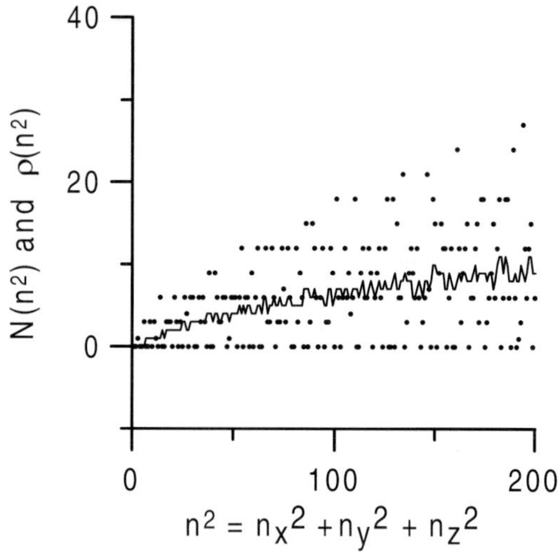

Figure 1.1. Number of states $N(n^2)$ in a cubic box as a function of the rescaled energy n^2. The continuous line which represents the density of states is obtained by averaging the original distribution. In this case we used nine-point smoothing, i.e., $\rho(n^2) = \langle N(n^2)\rangle = \frac{1}{9}\sum_{k=n^2-4}^{n^2+4} N(k)$.

The idea of smoothing out a discrete density leads to a new way of defining the sum of microstates, i.e., we equate $\Omega(E)$ with the number of states ΔN in the interval $(E, E + \Delta E)$. Thus

$$\Omega(E) = \int_E^{E+\Delta E} g(E')\,\mathrm{d}E' = \rho(E)\Delta E. \tag{1.2.10}$$

Then, from (1.2.4), the entropy (from here on we drop Boltzmann's constant) in this approximation is

$$S_{\mathrm{M}}(E) = \ln[\rho(E)] + \ln[\Delta E]. \tag{1.2.11}$$

Now for practical comparisons of probabilities of different macrostates the ΔE term cancels. Thus for two macrostates $\boldsymbol{m}$ and $\boldsymbol{m}'$,

$$\frac{P(\boldsymbol{m}; E)}{P(\boldsymbol{m}'; E)} = \frac{\rho(\boldsymbol{m}, E)\Delta E}{\rho(\boldsymbol{m}', E)\Delta E} = \frac{\rho(\boldsymbol{m}, E)}{\rho(\boldsymbol{m}', E)}. \tag{1.2.12}$$

Because of (1.2.12), one can drop the (constant) $\ln[\Delta E]$ term and redefine the entropy so that $S_{\mathrm{M}}(E)$ is given numerically by

$$S_{\mathrm{M}}(E) = \ln[\rho(E)]. \tag{1.2.13}$$

One could also proceed by choosing the energy scale so that $\ln[\Delta E] = 0$. For example, in a nucleus at excitation energy E^* which possesses a significant number of states in a 1 MeV slice of energy, by defining $\rho(E^*)$ as the number of states per MeV so that $\Delta E = 1$ MeV we eliminate the $\ln[\Delta E]$ term.

Despite the appearance of figure 1.1 it is fair to say that the density of states approximation constitutes an essential element of most applications. It should be remembered however that each physical situation has its own characteristic difficulties. For example, in nuclear physics, the quantitative specification of the density of microstates in highly excited nuclei remains somewhat speculative, principally due to the fact that it is not directly accessible to experimental measurement.

The quantum mechanical treatment of a particle in a box not only shows that the state space is not continuous in finite systems, but also provides the required correction to the density. Classically a measure of the position–momentum space for such a particle would be provided by the generalized volume element

$$\mathrm{d}\Gamma \equiv \mathrm{d}^3r\,\mathrm{d}^3p \equiv \mathrm{d}r_x\,\mathrm{d}r_y\,\mathrm{d}r_z\,\mathrm{d}p_x\,\mathrm{d}p_y\,\mathrm{d}p_z. \tag{1.2.14}$$

To understand the modification introduced by the quantum mechanical treatment we may return to equation (1.2.9). If we set $a = b = c = L$ in this equation then the number of states in an octant of 'radius' n ($n^2 = n_x^2 + n_y^2 + n_z^2$) is approximately given by $\frac{1}{8}\frac{4}{3}\pi n^3$. The number of states between n^2 and $n^2 + \mathrm{d}n^2$ is thus $\rho(n^2)\,\mathrm{d}n^2 = \frac{1}{8}2\pi n\;\mathrm{d}n^2$ and is *finite* because the energy of the particle is quantized. We can exhibit more clearly the consequence of quantization if we express the wavefunction in terms of the wavenumber components $\{k_x, k_y, k_z\}$

$$\Psi(\boldsymbol{r}) = A\sin(k_x x)\sin(k_y y)\sin(k_z z), \tag{1.2.15}$$

where $k_x = \pi n_x/L$ with similar definitions for k_y and k_z. Defining the wavenumber $k = \sqrt{k_x^2 + k_y^2 + k_z^2}$, the number of states enclosed in a k-sphere (of radius $k = \pi n/L$) is then expressed in terms of k rather than in terms of n as $\left(\frac{L}{2\pi}\right)^3\frac{4}{3}\pi k^3$ and the number of states between k and $k + \mathrm{d}k$ is $\rho(k)\,\mathrm{d}k = \left(\frac{L}{2\pi}\right)^3 4\pi k^2\,\mathrm{d}k$. The quantum mechanical momentum operator is given, in terms of the unit vector $(\hat{u}_x, \hat{u}_y, \hat{u}_z)$ as $-\mathrm{i}\hbar\left(\hat{u}_x\frac{\partial}{\partial x} + \hat{u}_y\frac{\partial}{\partial y} + \hat{u}_z\frac{\partial}{\partial z}\right)$. Applying this operator to the wavefunction in (1.2.15) we find the momentum $p = hk/2\pi$ where h is Planck's constant ($h = 4.1357 \times 10^{-21}$ MeV s). Expressed in terms of momentum we therefore find $\rho(p)\,\mathrm{d}p = \frac{V}{h^3}4\pi p^2\,\mathrm{d}p$, where the volume $V = L^3$. This quantity is simply the $\mathrm{d}\Gamma$ of equation (1.2.14) expressed in polar coordinates, integrated over spatial coordinates and divided by h^3. Thus the effect of quantization is to introduce a *finite* measure of position–momentum space. The argument may be extended to an arbitrary number of particles so that a dimensionless quantity, which may be thought of as the number of states in a volume $\mathrm{d}\Gamma$ corresponding to N particles, is obtained by dividing the N particle analogue of (1.2.14) by h^{3N}. The total number of microstates in some finite region, Γ', of N particle position–momentum space is then given by the integral

$$\Omega(\Gamma') = \frac{1}{h^{3N}} \int_{\Gamma'} \mathrm{d}\Gamma = \frac{1}{h^{3N}} \int_{\Gamma'} \mathrm{d}^3 r_1\, \mathrm{d}^3 r_2 \ldots \mathrm{d}^3 r_N\, \mathrm{d}^3 p_1\, \mathrm{d}^3 p_2 \ldots \mathrm{d}^3 p_N. \tag{1.2.16}$$

(Note the oft-used notational shortcut achieved by writing a multidimensional integral using a single integral symbol.) It is customary, following the American physicist J W Gibbs, to refer to the multidimensional space defined by the integrand in (1.2.16) as 'phase space' so that $\mathrm{d}\Gamma$ is a volume element in phase space. This usage will be adopted throughout this book. It appears, however, that this terminology has been generalized to refer to the space of all microstates so that a 'phase space' model may be identified with an equilibrium statistical model.

As an example of the application of (1.2.16), which will be useful later, we can calculate the density of phase space states which should be associated with two structureless non-interacting fragments of masses m_1 and m_2 and momenta $\boldsymbol{p}_1$ and $\boldsymbol{p}_2$ enclosed in a volume V such that the total energy

$$E = E_1 + E_2 = \frac{p_1^2}{2m_1} + \frac{p_2^2}{2m_2} \tag{1.2.17}$$

is fixed. We write

$$\rho_{12}(E, V) = \frac{\int \mathrm{d}^3 r_1\, \mathrm{d}^3 r_2 \int \delta\left(\frac{p_1^2}{2m_1} + \frac{p_2^2}{2m_2} - E\right) \mathrm{d}^3 p_1\, \mathrm{d}^3 p_2}{h^6}, \tag{1.2.18}$$

and note that ρ_{12} is, by definition, a continuous function of energy and does indeed represent a number of states per unit energy. Performing the integration over the spatial coordinates and writing the double integration over momentum (energy) explicitly, we obtain (with $\mathrm{d}^3 p = 2\pi p\, \mathrm{d}p^2 = 2\pi (2m)^{3/2} E^{1/2}\, \mathrm{d}E$)

$$\begin{aligned}
\rho_{12}(E, V) &= \left(\frac{2\pi V}{h^3}\right)^2 (2m_1)^{3/2} (2m_2)^{3/2} \int_0^E \sqrt{E_1} \\
&\quad \times \left[\int_0^E \delta(E_1 + E_2 - E) \sqrt{E_2}\, \mathrm{d}E_2\right] \mathrm{d}E_1 \\
&= \left(\frac{2\pi V}{h^3}\right)^2 (2m_1)^{3/2} (2m_2)^{3/2} E^2 \int_0^1 \sqrt{(1-x)x}\, \mathrm{d}x \\
&= \left(\frac{V}{h^3}\right)^2 (2\pi)^3 m_1^{3/2} m_2^{3/2} \frac{E^2}{2}.
\end{aligned} \tag{1.2.19}$$

In the next section we shall show how (1.2.19) may be simply generalized to the case of N fragments. While on this subject, however, it seems appropriate to discuss the two-body phase space in a slightly different manner. In many cases of interest we are concerned with particles (molecules, nuclear fragments, etc) which interact pairwise so that the interaction energy can be calculated as a sum over particle pairs. In this case it is useful to carry out a coordinate transformation in order to exhibit the relative coordinates explicitly. For more than a few particles this transformation is rather complicated [11]. For just two particles with masses m_1 and m_2, and vector positions and momenta $\boldsymbol{r}_1, \boldsymbol{r}_2$ and $\boldsymbol{p}_1, \boldsymbol{p}_2$, respectively, the transformation is defined by

$$\boldsymbol{R} = \frac{m_1\boldsymbol{r}_1 + m_2\boldsymbol{r}_2}{m_1 + m_2}, \tag{1.2.20}$$

$$\boldsymbol{P} = \boldsymbol{p}_1 + \boldsymbol{p}_2, \tag{1.2.21}$$

$$M = m_1 + m_2, \tag{1.2.22}$$

$$\boldsymbol{r} = \boldsymbol{r}_1 - \boldsymbol{r}_2, \tag{1.2.23}$$

$$\boldsymbol{p} = \frac{m_2\boldsymbol{p}_1 - m_1\boldsymbol{p}_2}{m_1 + m_2}, \tag{1.2.24}$$

$$\mu = \frac{m_1 m_2}{m_1 + m_2}, \tag{1.2.25}$$

where equations (1.2.20)–(1.2.22) define the position, momentum and mass which characterizes the overall motion of the centre-of-mass and the remaining equations refer to relative motion of the two particles in the centre-of-mass system. It is easy to verify that the total energy of the system can be written as

$$\frac{p_1^2}{2m_1} + \frac{p_2^2}{2m_2} = \frac{P^2}{2M} + \frac{p^2}{2\mu} = E_{\mathrm{CM}} + E_{\mathrm{cm}}, \tag{1.2.26}$$

where E_{CM} is the kinetic energy associated with the motion of the centre of mass and E_{cm} is the kinetic energy associated with relative motion of the two particles.

We shall be applying this transformation to nuclear collisions in chapter 2. An important feature which we need immediately, however, is that two-body interactions affect only the momentum, $\boldsymbol{p}$, in the centre of mass so that, in the absence of external forces, the total momentum, $\boldsymbol{P}$, is a constant of the motion. Under these circumstances it is possible to perform a Galilean transformation on the whole system in order to set $\boldsymbol{P}$ to any desired value. For notational consistency we will say that, whether such a transformation is performed or not, the energy which characterizes the motion of the centre of mass is $E_{\mathrm{CM}} = E_0$ and the momentum, $\boldsymbol{P} = \boldsymbol{P}_0$. The position of the centre of mass may be also arbitrarily fixed (e.g., $\boldsymbol{R} = \boldsymbol{R}_0 = 0$). We may now re-evaluate the phase space volume element that appears in (1.2.18), i.e.,

$$\mathrm{d}^3p_1\,\mathrm{d}^3p_2\,\mathrm{d}^3r_1\,\mathrm{d}^3r_2 = (\mathrm{d}p_{1,x}\,\mathrm{d}r_{1,x}\,\mathrm{d}p_{2,x}\,\mathrm{d}r_{2,x})(\mathrm{d}p_{1,y}\,\mathrm{d}r_{1,y}\,\mathrm{d}p_{2,y}\,\mathrm{d}r_{2,y})\\ (\mathrm{d}p_{1,z}\,\mathrm{d}r_{1,z}\,\mathrm{d}p_{2,z}\,\mathrm{d}r_{2,z}), \tag{1.2.27}$$

incorporating these physical constraints. The first step is to express the volume element using the coordinate transformation (1.2.20)–(1.2.25)

$$(\mathrm{d}p_{1,x}\,\mathrm{d}r_{1,x}\,\mathrm{d}p_{2,x}\,\mathrm{d}r_{2,x})(\mathrm{d}p_{1,y}\,\mathrm{d}r_{1,y}\,\mathrm{d}p_{2,y}\,\mathrm{d}r_{2,y})(\mathrm{d}p_{1,z}\,\mathrm{d}r_{1,z}\,\mathrm{d}p_{2,z}\,\mathrm{d}r_{2,z})$$
$$= J_x(\mathrm{d}P_x\,\mathrm{d}R_x\,\mathrm{d}p_x\,\mathrm{d}r_x)J_y(\mathrm{d}P_y\,\mathrm{d}R_y\,\mathrm{d}p_y\,\mathrm{d}r_y)J_z(\mathrm{d}P_z\,\mathrm{d}R_z\,\mathrm{d}p_z\,\mathrm{d}r_z). \tag{1.2.28}$$

Equation (1.2.28) results from the fact that the Jacobian, J, for the full transformation is the product of separate Jacobians for each Cartesian coordinate. If equations (1.2.20), (1.2.21) and (1.2.23) are written in Cartesian component form we obviously get three identical sets of equations. Therefore J_x, J_y and J_z must have the same functional dependence on their respective arguments. The appropriate function can be thus obtained using just one component, e.g. J_x. In fact it is quickly seen that the Jacobian factors are all unity. Position and momentum factors can be separated so that we only have to calculate the two corresponding determinants. For example, using (1.2.20)–(1.2.25),

$$J_x = \left[\frac{\partial r_{1,x}}{\partial R_x}\frac{\partial r_{2,x}}{\partial r_x} - \frac{\partial r_{1,x}}{\partial r_x}\frac{\partial r_{2,x}}{\partial R_x}\right]\left[\frac{\partial p_{1,x}}{\partial P_x}\frac{\partial p_{2,x}}{\partial p_x} - \frac{\partial p_{1,x}}{\partial p_x}\frac{\partial p_{2,x}}{\partial P_x}\right] = 1, \tag{1.2.29}$$

so that we can conclude

$$\mathrm{d}^3p_1\,\mathrm{d}^3p_2\,\mathrm{d}^3r_1\,\mathrm{d}^3r_2 = \mathrm{d}^3P\,\mathrm{d}^3p\,\mathrm{d}^3R\,\mathrm{d}^3r. \tag{1.2.30}$$

Armed with (1.2.30) we calculate the modified phase space density straightforwardly. Because we are restricting the phase space by fixing $\boldsymbol{R}$ and $\boldsymbol{P}$ (to $\boldsymbol{R}_0$ and $\boldsymbol{P}_0$ respectively) these coordinates make no contribution. For this reason we divide by h^3 rather than by h^6. In fact one should always introduce position and momentum restrictions in pairs (although this is not always done in practice). Thus a one-dimensional constraint would be represented by inserting a factor of the form $h\delta(r - r_0)\delta(p - p_0)$. The remaining coordinates can be considered as those of a ficticious single particle with mass μ and energy E_{cm} in a volume V. The phase space density (you may care to verify the dimension) corresponding to (1.2.18) is thus

$$\rho_{12} = \frac{\int \delta^3(\boldsymbol{R} - \boldsymbol{R}_0)\,\mathrm{d}^3R\,\mathrm{d}^3r \int \delta^3(\boldsymbol{P} - \boldsymbol{P}_0)\delta\left(\frac{P^2}{2M} + \frac{p^2}{2\mu} - E\right)\mathrm{d}^3p\,\mathrm{d}^3P}{h^3}, \tag{1.2.31}$$

where, following Balian [4], we have used the symbol $\delta^3(\boldsymbol{A})$ to designate $\delta(A_x)$ $\delta(A_y)\delta(A_z)$. The spatial integration is reduced to that over $\boldsymbol{r}$ and thus yields a single factor of V. The integration over $\boldsymbol{P}$ (recall that $P_0^2/2M = E_0 = E - E_{\mathrm{cm}}$)

then yields

$$\rho_{12}(E, V, \boldsymbol{R} = \boldsymbol{R}_0, \boldsymbol{P} = \boldsymbol{P}_0) = \frac{2\pi V}{h^3} \int \delta\left(\frac{p^2}{2\mu} - (E - E_0)\right) p \, \mathrm{d}p^2$$

$$= \frac{2\pi V}{h^3} \int \delta\left(\frac{p^2}{2\mu} - E_{\mathrm{cm}}\right) p \, \mathrm{d}p^2 = \frac{2\pi V}{h^3} (2\mu)^{3/2} \int \delta(E' - E_{\mathrm{cm}}) \sqrt{E'} \, \mathrm{d}E'$$

$$= \frac{2\pi V}{h^3} (2\mu)^{3/2} E_{\mathrm{cm}}^{1/2}. \tag{1.2.32}$$

The reader will have no difficulty in showing that, as expected, this equation is identical with that for a single particle of mass μ in a volume V constrained only by the 'total' energy E_{cm}.

1.3 Subsystems and convolution

In many cases of practical interest the system under investigation can be divided into distinct parts or components for each of which it is possible to calculate a density of microstates. Each part may involve one or more degrees of freedom. The problem then arises as to how the full microstate density is to be calculated if the component densities are known. The simplest case is that of a dual component system, with total energy E. If we distribute the total energy by putting energy E_1 in the first subsystem and E_2 in the second, then the total number of states for this particular energy division is simply

$$\Omega_{12}(E_1, E_2; E) = \Omega_1(E_1)\Omega_2(E_2), \tag{1.3.1}$$

and the microcanonical sum is obtained by summing all possible combinations such as (1.3.1), i.e.

$$\Omega_{12}(E) = \sum \Omega_1(E_1)\Omega_2(E_2), \tag{1.3.2}$$

where the sum is carried out over all partitions of the energy E such that $E_1 + E_2 = E$. If we replace the terms in (1.3.2) by their continuous approximations, or if the state space is continuous, we can rewrite the analogue of this equation as

$$\rho_{12}(E) = \int_0^E \int_0^E \rho_1(E_1)\rho_2(E_2)\delta(E_1 + E_2 - E) \, \mathrm{d}E_1 \, \mathrm{d}E_2$$

$$= \int_0^E \rho_1(E_1)\rho_2(E - E_1) \, \mathrm{d}E_1. \tag{1.3.3}$$

Equation (1.3.3) is referred to as a convolution integral. The important property of such integrals is that the Laplace transform of $\rho_{12}(E)$, which is defined by

$$L_\beta[\rho_{12}] = \int_0^\infty \rho_{12}(E) \, \mathrm{e}^{-\beta E} \, \mathrm{d}E, \tag{1.3.4}$$

is the product of the transforms of ρ_1 and ρ_2. This theorem may be extended to include an arbitrary number of components. We can see this immediately by considering a three-component system for which the extension of (1.3.3) (slightly rearranged) is

$$\rho_{123}(E) = \int_0^E \rho_1(E_1)\left[\int_0^E \int_0^E \rho_2(E_2)\rho_3(E_3)\delta(E_2 + E_3 - (E - E_1))\,\mathrm{d}E_2\,\mathrm{d}E_3\right]\mathrm{d}E_1. \tag{1.3.5}$$

The double integral in square brackets is a two density convolution which we can write as $\rho_{23}(E - E_1)$. Then (1.3.5) becomes

$$\rho_{123}(E) = \int_0^E \rho_1(E_1)\rho_{23}(E - E_1)\,\mathrm{d}E_1. \tag{1.3.6}$$

It follows that an arbitrary number of components can be included in integrals such as (1.3.5) and that the Laplace transform of the full density of states is the product of the Laplace transforms of each of the components. (The usefulness of the density of states approximation will probably now be apparent.)

The density of states for a single particle with mass, m_1, and kinetic energy, E_1, enclosed in a volume, V, is (see (1.2.32))

$$\rho_1(E_1) = \frac{V}{h^3}(2\pi m_1)^{3/2}\frac{E_1^{1/2}}{\Gamma(3/2)}, \tag{1.3.7}$$

where $\Gamma(x)$ is the Gamma function $(\Gamma(3/2) = \sqrt{\pi}/2)$. The Laplace transform of (1.3.7) is

$$L_\beta[\rho_1] = \frac{V}{h^3}\left(\frac{2\pi m_1}{\beta}\right)^{3/2}. \tag{1.3.8}$$

The product of N such transforms for fragments with masses $m_1 \ldots m_N$ is thus

$$L_\beta[\rho_N] = \left(\frac{V}{h^3}\right)^N (2\pi)^{3N/2}\left[\prod_i m_i^{3/2}\right]\beta^{-3N/2}. \tag{1.3.9}$$

where the symbol $\prod_i$ designates the product of the N terms. The inverse transform now provides the expression for the density of states for N fragments with total energy $E(= \Sigma E_i)$ enclosed in a volume V, i.e.,

$$\rho_N(E) = \left(\frac{V}{h^3}\right)^N (2\pi)^{3N/2}\left[\prod_i m_i^{3/2}\right]\frac{E^{3N/2-1}}{\Gamma(3N/2)}, \tag{1.3.10}$$

which, with $N = 2$, gives back equation (1.2.19). We shall be investigating state densities such as (1.3.10) in more detail when we come to the topic of multifragmentation.

There is yet another property of the convolution integral which turns out to be very useful in nuclear physics for approximating integrals such as (1.3.3) and (1.3.5). Suppose that the integrand in (1.3.3) has a single maximum at $E_1 > 0$ (not obvious, but often the case) and that we wish to calculate the value at the maximum. To this end, we define the differential $f' = \frac{\mathrm{d}}{\mathrm{d}E_1}[(\rho_1(E_1)\rho_2(E - E_1)]$. Given that $E - E_1 = E_2$, the maximum of the integrand then occurs when

$$f' = \rho_2(E_2)\frac{\mathrm{d}\rho_1(E_1)}{\mathrm{d}E_1} - \rho_1(E_1)\frac{\mathrm{d}\rho_2(E_2)}{\mathrm{d}E_2} = 0, \qquad (1.3.11)$$

or

$$\frac{\mathrm{d}}{\mathrm{d}E_1}\ln[\rho_1(E_1)] = \frac{\mathrm{d}}{\mathrm{d}E_2}\ln[\rho_2(E_2)]. \qquad (1.3.12)$$

The practical significance of the exercise is that it may be possible to obtain a numerical estimate of the integral (1.3.3) by using the maximum value. Thus, if a maximum occurs at $E_1 = E_1^{\max}$ we may approximate the integral by the value of the integrand at this point, i.e.,

$$\int_0^E \rho_1(E_1)\rho_2(E - E_1)\,\mathrm{d}E_1 \approx \int_0^E \rho_1(E_1)\rho_2(E - E_1)\delta(E_1 - E_1^{\max})\,\mathrm{d}E_1$$
$$= \rho_1(E_1^{\max})\rho_2(E - E_1^{\max}). \qquad (1.3.13)$$

Clearly (1.3.13) provides only a numerical estimate because it does not have the right dimension. However, we can use an argument similar to that used after equation (1.2.11) to justify its use in practical calculations. Of course, the precise value, $E_1^{\max}$, can only be found if we specify the energy dependence of the densities ρ_1 and ρ_2. An example is shown in figure 1.2 which represents the convolution integrand of the densities corresponding to internal excitation of two nuclei whose mass numbers add to 200. Two cases are shown: one (on the right of the figure) in which the total number of nucleons is divided into two equal parts, i.e., $A_1 = A_2 = 100$, and one (on the left) with $A_1 = 30$ and $A_2 = 170$. For reasons which will become apparent in the following section, the maximum of the integrand in each case is equal to e^{100} which is approximately 2.7×10^{43}. On the other hand the full integrals (calculated numerically) are respectively 6.6×10^{44} and 4.8×10^{44} so that we may conclude that the entropy is reasonably well estimated using only the maximum values. This being said, we would be unwise to use the maximum values to estimate the ratio of the probabilities of the (100, 100) and (30, 170) splits. If we did, we would of course conclude that the two probabilities are equal whereas the actual ratio is $6.6/4.8 = 1.375$. Another feature of figure 1.2 which will be useful later is that the integrands are, in both cases, bell shaped and reasonably well described by Gaussian curves.

Equations (1.3.12) and (1.3.13) may be extended to include any number of subsystems. However, it seems evident that (1.3.13) represents a rather drastic approximation and its use will need to be justified in any particular situation.

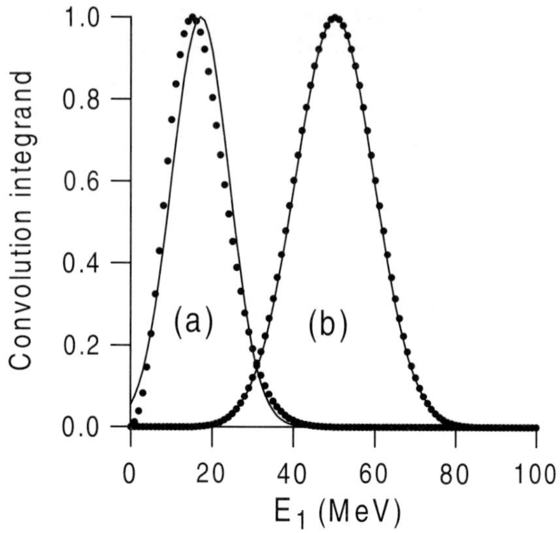

Figure 1.2. Convolution integrands. The points represent the integrand of the convolution of two densities corresponding to excited states of nuclei with mass numbers A_1 and A_2, (*a*) $A_1 = 30$, $A_2 = 170$, (*b*) $A_1 = A_2 = 100$. In both cases the convolution integrand corresponds to $E = E_1 + E_2 = 100$ MeV. The densities are each of the form $\rho(E^*) = \mathrm{e}^{2\sqrt{aE^*}}$ where $a = A/8$. The integrand is displayed divided by the maximum $\mathrm{e}^{2\sqrt{(a_1+a_2)E}}$ as a function of the energy, E_1, in the first fragment. The solid curves represent Gaussian fits to the integrands.

We shall see in chapter 3 (in connection with nuclear fission) that the approximation can be improved by exploiting the characteristic 'Gaussian' form of the integrands.

Equation (1.3.12) provides a link with thermodynamics. In thermodynamic equilibrium a multicomponent system is composed of subsystems which all have the same temperature. The microcanonical temperature for a subsystem with density of states $\rho(E)$ is defined as

$$\frac{1}{T} = \frac{\mathrm{d}}{\mathrm{d}E} \ln[\rho(E)]. \tag{1.3.14}$$

We see that the 'microcanonical' notion of uniform temperature is based on the approximation of integrals such as that depicted in (1.3.13) and is useful provided that the most probable division of the total energy among different subsystems is 'overwhelmingly' the most probable. This gives rise, in some texts, to the misleading definition of the equilibrium state as that macrostate (subset of microstates) which is characterized by the most probable energy division. At the risk of repetition, we say again that the equilibrium state is that for which all microstates have the same probability.

The importance of a proper command of convolution techniques and approximations cannot be overstated. This is particularly true in multicomponent systems where choosing the order in which the various densities are to be convoluted may result in considerable simplification of the problem of estimating total or partial sums of microstates.

1.4 The Boltzmann distribution

The representation of the two-system convolution by the maximum value of the integrand suggests that we write the densities in the form

$$\rho(E) = \rho(E^{\max})\, \mathrm{e}^{\beta(E - E^{\max})} \tag{1.4.1}$$

(which implies $\beta = \mathrm{d}\ln[\rho]/\,\mathrm{d}E$ at $E = E^{\max}$) in which case the maximum of the integrand occurs when each of the two components has the same value of β. Obviously, β is the inverse of the microcanonical temperature. One may object that it should not be necessary to introduce both the expansion parameter of a microstate density, β, which is a statistical mechanical concept, and the temperature, T, which is a thermodynamical concept. In the context of statistical mechanics, this is true. However, many applications are discussed in terms of temperature so that the concept is difficult to avoid. It is important to understand that we have not proved that the temperature defined in (1.3.14) is equivalent to the *thermodynamical* temperature although this is in fact the case.

The expansion (1.4.1) can be used to improve the estimation of the convolution integral provided by equation (1.3.13) (try it!) but that is not our objective in this section. Rather we wish to consider a two-system convolution in which just one of the densities (say that of system 1) is represented as in equation (1.4.1) which we now write more directly as

$$\rho(E + \Delta E) = \rho(E)\, \mathrm{e}^{\beta \Delta E}. \tag{1.4.2}$$

Whether or not equation (1.4.2) can be considered as a useful approximation depends on the application. In nuclear physics the density of states in a nucleus composed of A nucleons at excitation energy, E^*, can be quite well represented in terms of the so-called 'level density parameter', $a \approx \frac{A}{8}$, by

$$\rho(E^*)\,\mathrm{d}E = c\,\mathrm{e}^{2\sqrt{aE^*}}\,\mathrm{d}E, \tag{1.4.3}$$

where c is a constant. (In fact figure 1.2 was constructed using (1.4.3) with $c = 1$.) Thus we immediately find

$$\beta = \frac{\mathrm{d}(2\sqrt{aE^*})}{\mathrm{d}E} = \sqrt{\frac{a}{E^*}}. \tag{1.4.4}$$

(You may now go back and check that the integrands represented in figure 1.2 do actually peak at e^{100}.) Equation (1.4.4) implies, of course, that it is impossible to ascribe an energy independent temperature to an excited nucleus. Given that β varies only as the square root of the energy, however, one may consider using a constant value of β over a restricted region of energy. This point is illustrated in figure 1.3. As one may verify by differentiating equation (1.4.4), the 'constant β' approximation improves with increasing energy.

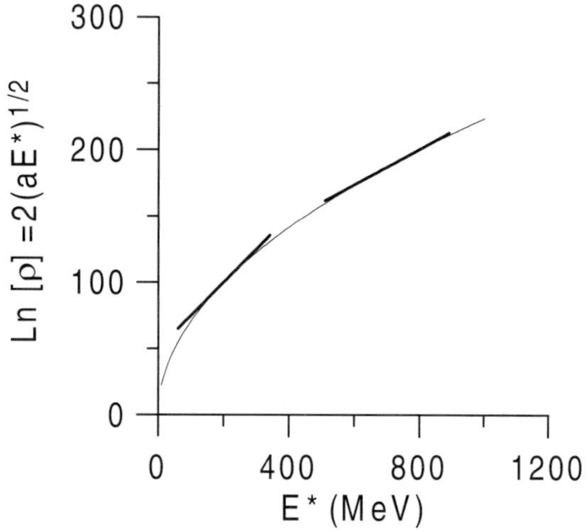

Figure 1.3. The density of states in a nucleus with mass number $A = 100$. The density is represented by the approximate expression $\rho(E^*) = e^{2\sqrt{aE^*}}$. The straight lines shown on the logarithmic plot are constant temperature approximations at $E^* = 200$ and 700 MeV.

We are now ready to understand the concept of a heat bath. The conventional treatment refers to a large 'heat' reservoir which somehow is in contact with the 'small' system under investigation. It has its origins in macroscopic thermodynamics which has limited relevance for small systems. We will try to avoid such references. The notion of 'contact' and energy flow between subsystems seem to imply a time evolution and thus be rooted in the ergodic picture of equilibrium (for a description of this picture the reader is referred to the work of Klimontovitch [7]). In the ensemble approach we require only the equilibrium characteristic of uniform population of microstates, providing we consider microstates of the full system and not just of one of its component subsystems.

Consider again our two-component system with total energy E and imagine that system 2 is characterized by a discrete level scheme which, for the moment, we *do not* approximate by a continuous density function. Let us enumerate the energy levels in system 2 using the index n and suppose the lowest level has $\varepsilon_0 = 0$. At each energy level we may find one or more microstates. We thus denote the degeneracy of the nth level as g_n. We will also need to refer to the set of microstates as opposed to the set of energy levels. This will be done using the index 'i' so that ε_i is the energy of the ith microstate.

The two systems are depicted in figure 1.4. Suppose that system 1 is represented by a microstate density function. We begin by placing the second system in its lowest energy level. If there is just one state at energy ε_0, the number of microscopic configurations which contributes to the sum of microstates is simply

$$(\rho_1(E) \times 1)\Delta E. \tag{1.4.5}$$

Successive levels in system 2 are separated in energy so that we can move up to the next level (situated at say ε_1 with, say, g_1 degenerate states). Each of the g_1 states contributes

$$(\rho_1(E - \varepsilon_1) \times 1)\Delta E, \tag{1.4.6}$$

so that the contribution from all states at ε_1 is simply $(\rho_1(E - \varepsilon_1) \times g_1)\Delta E$. We see that the contribution from the energy level at ε_n is $(\rho_1(E - \varepsilon_n) \times g_n)\Delta E$. By adopting the expansion (1.4.2), this contribution can be written

$$(\rho_1(E)\,\mathrm{e}^{-\beta_1\varepsilon_n} \times g_n)\Delta E, \tag{1.4.7}$$

where we denote the expansion parameter as β_1 to emphasize that it characterizes system 1. This equation has the status of an approximation (involving a first order development of $\ln[\rho_1]$). We can see, however, that it is generally valid, either if the change in energy is small enough so that β_1 can be considered to be constant, or if we allow for an energy dependence of β_1 such that (1.4.2) is strictly valid at all energies. In either case, if we adopt (1.4.2), the *relative* number of microstates corresponding to any energy level in system 2 is given by (we drop the constant factor $\rho_1(E)\Delta E$)

$$\Omega(\varepsilon_n) = g_n\,\mathrm{e}^{-\beta_1\varepsilon_n}. \tag{1.4.8}$$

It follows that the weight to be attached to any single microstate at energy ε_i is

$$w_i = \mathrm{e}^{-\beta_1\varepsilon_i}. \tag{1.4.9}$$

(Note again the use of the subscript 'i' to refer to a single microstate.) If β_1 can be considered *constant* for all system 2 states, equation (1.4.9) is referred to as the Boltzmann distribution. It appears that the alternative term 'canonical distribution' was coined by Gibbs (*Elementary Principles in Statistical Mechanics*, 1902) because of its overriding importance in statistical mechanics (a canon is a general rule or principle).

We can now resume the concept of a heat bath by saying that the weight to be assigned to any single 'system 2' microstate is given by the canonical distribution (equation (1.4.9)). Furthermore, the 'heat bath' system (system 1) must have a state density which behaves exponentially (i.e., with constant β_1) for all system 2 energies. Finally, we require that the sum of microstates converges to a fixed value as the system 2 energy increases.

The energy spectrum in the combined system is, of course, given by (1.4.8). If it is possible to define a level density $\rho_2(\varepsilon)$ in system 2, then the energy

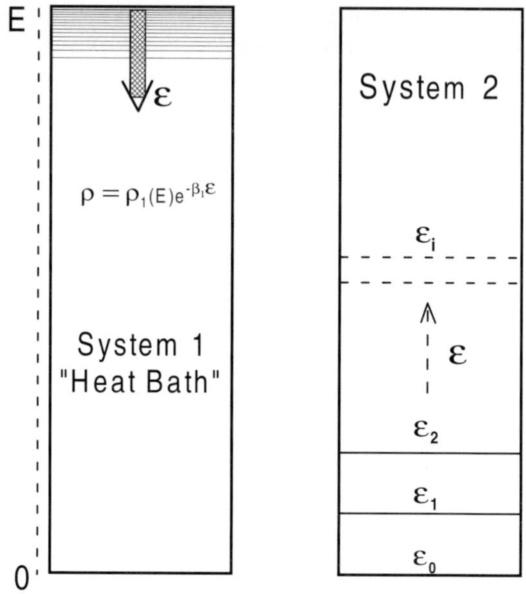

Figure 1.4. Image of heat bath system 1 associated with a subsystem 2. The total energy is E and system 2 is shown as having discrete, non-degenerate, states with energies $\varepsilon_0, \varepsilon_1, \varepsilon_2, \ldots \varepsilon_i$. It may also be that energy *levels* in system 2 are degenerate in which case each level, ε_n, has an associated integer degeneracy $g_n \geq 1$.

spectrum is straightforwardly given (replacing g_n by $\rho_2(\varepsilon)\,\mathrm{d}\varepsilon$) as the analogue of (1.4.8), i.e.,

$$\rho_{12}(\varepsilon)\,\mathrm{d}\varepsilon = \rho_2(\varepsilon)\,\mathrm{e}^{-\beta_1\varepsilon}\,\mathrm{d}\varepsilon. \tag{1.4.10}$$

The spectrum obviously exhibits both the characteristics of the heat bath and of the 'system 2' under study. We can, of course, expand the logarithm of $\rho_2(\varepsilon)$ in the same way as we expanded the heat bath density. However, we have no guarantee that the corresponding β_2 will be constant and indeed we would expect, in general, that $\beta_2 = \beta_2(\varepsilon)$. We then quickly find (as expected) that the maximum of the integrand represented in (1.4.10) occurs when $\beta_2(\varepsilon) = \beta_1$, and thus that the maximum microstate density (the maximum probability) occurs when the subsystem 2 has the same microcanonical temperature as the heat bath.

If system 2 can be represented by a continuous density of states as in (1.4.10) the convergence property implies that the integral of (1.4.10) converges for some finite value of ε which is less than the total energy, E, of the combined system and thus can be approximated as the Laplace transform of $\rho_2(\varepsilon)$.

$$\int_0^E \rho_2(\varepsilon)\,\mathrm{e}^{-\beta_1\varepsilon}\,\mathrm{d}\varepsilon \approx \int_0^\infty \rho_2(\varepsilon)\,\mathrm{e}^{-\beta_1\varepsilon}\,\mathrm{d}\varepsilon. \tag{1.4.11}$$

Indeed, to within the constant $\rho_1(E)\Delta E$ this integral which is referred to as the canonical partition sum (Z_{can}) represents the full microcanonical sum of states of the combined system.

Let us now suppose we have several subsystems (say $s = 2$ to N) together with a system ($s = 1$) which behaves as a heat bath and that we wish to calculate the microstate density for a total energy E. If the energy in system s is labelled E_s, the microstate density convolution is written straightforwardly as the extension of equation (1.3.5)

$$\int_0^E \rho_2(E_2) \int_0^E \rho_3(E_3) \ldots \left[\int_0^E \rho_1(E_1)\delta[E - E_1 - (E_2 + E_3 + \ldots)] \,\mathrm{d}E_1 \right] \mathrm{d}E_2 \,\mathrm{d}E_3 \ldots . \tag{1.4.12}$$

Now writing $\rho_1(E_1) = \rho_1(E)\,\mathrm{e}^{-\beta_1(E-E_1)}$, equation (1.4.12) becomes

$$\rho_1(E) \int_0^E \rho_2(E_2) \int_0^E \rho_3(E_3) \ldots \left[\int_0^E \mathrm{e}^{-\beta_1(E-E_1)}\delta[E - E_1 - (E_2 + E_3 + \ldots)] \,\mathrm{d}E_1 \right] \mathrm{d}E_2 \,\mathrm{d}E_3 \ldots, \tag{1.4.13}$$

so that by carrying out the integral over E_1 we obtain the product

$$\begin{aligned} \rho_{123\ldots}(E) &= \rho_1(E) \left[\int_0^E \rho_2(E_2)\,\mathrm{e}^{-\beta_1 E_2} \,\mathrm{d}E_2 \right] \\ &\quad \times \left[\int_0^E \rho_3(E_3)\,\mathrm{e}^{-\beta_1 E_3} \,\mathrm{d}E_3 \right] \ldots \times \ldots \\ &= \rho_1(E) \prod_{s=2}^{N} \left[\int_0^E \rho_s(E_s)\,\mathrm{e}^{-\beta_1 E_s} \,\mathrm{d}E_s \right]. \end{aligned} \tag{1.4.14}$$

Notice that as for the single system 2, the most probable distribution of the energy occurs again when all subsystems have the same temperature (i.e., for all s, $\beta_s(E_s) = \beta_1$). If we now drop the constant factor $\rho_1(E)$ and replace the upper limit in each of the integrals by infinity (see (1.4.11)) we can write the sum of states (canonical partition sum) as the product of the sums for each system

$$Z_{\text{can}}(\beta_1) = \prod_{s=2}^{N} Z_{\text{can}}^{(s)}(\beta_1). \tag{1.4.15}$$

Equation (1.4.15) represents a product of Laplace transforms and is thus the transform of the convolution of the densities of systems 2 to N. Notice that the product form for the partition sum (1.4.15) does not occur in the microcanonical ensemble. Therefore, there is no heat bath and the best we can do to simplify the evaluation of the partition sum is to apply the convolution theorem directly to the set of subsystems (assuming that subsystems with known state densities can be identified). For subsystems best described in terms of discrete states we find, by inserting equation (1.2.7) in (1.4.14), that

$$Z_{\text{can}}(\beta_1) = \prod_{s=2}^{N} Z_{\text{can}}^{(s)}(\beta_1) = \prod_{s=2}^{N} \left[\sum_i e^{-\beta_1 E_i^{(s)}} \right], \qquad (1.4.16)$$

where $E_i^{(s)}$ is the energy of the ith microstate in the system s. We again remind the reader that the sums on the right-hand side of equation (1.4.16) are over individual microstates for the subsystem s. Expressed as a sum over energy levels we would write the equivalent expression

$$Z_{\text{can}}(\beta_1) = \prod_{s=2}^{N} \left[\sum_{n(s)} g_n^{(s)} e^{-\beta_1 E_n^{(s)}} \right], \qquad (1.4.17)$$

where $E_n^{(s)}$ the energy of the nth energy level in system s and $g_n^{(s)}$ is the number of microstates to be found at that energy.

In nuclear physics the use of the canonical distribution is rather restricted. The best choice for the heat bath is usually provided by the degrees of freedom associated with internal excitation of a single fragment or a group of fragments. However, as indicated in equation (1.4.4) and figure 1.3, the quantity β_1 corresponding to this choice is normally energy dependent. We shall discuss this matter further when we come to the subject of multifragmentation.

Equation (1.4.15) provides us with an opportunity to make a simple introduction to the problem of identical particles or, more generally, of identical subsystems. In evaluating any sum of microstates we must, of course, be sure to include each microscopic configuration only once. Any two microscopic configurations are considered to be identical (one and the same microstate) if they are indistiguishable. Now the actual examination of all configurations in order to sort out indistinguishable microstates is practically impossible (because the number of microstates is usually a huge number). Thus, in many cases, this problem has no simple solution. A possible, and much used, strategy is to evaluate sums of microstates without worrying about indistinguishable configurations and then to include a correction factor in the sum. We will consider two examples for which the appropriate corrections are simply based on combinatorial arguments. Let us, however, emphasize from the start that our discussion concerns classical indistinguishability and takes no account of quantum mechanical permutation symmetry constraints which apply to states or wavefunctions and which have no classical counterpart. As a first example, consider a set of N subsystems (which we now label from 1 to N) supplemented by a 'heat bath' system. The full microcanonical sum of states is, to within a constant (equations (1.4.15), (1.4.16)), a product of the partition sums corresponding to each subsystem,

$$Z_{\text{can}} = \prod_{s=1}^{N} Z_s. \tag{1.4.18}$$

Suppose that, of the N subsystems, three systems (1, 2 and 3) are identical. Let us also assume for simplicity that all energy levels are non-degenerate so that there is only one microstate at each level. If the heat bath system is characterized by an expansion parameter, β, the corresponding part of the partition sum is

$$Z' = \sum_i e^{-\beta E_i^{(1)}} \sum_j e^{-\beta E_j^{(2)}} \sum_k e^{-\beta E_k^{(3)}}. \tag{1.4.19}$$

Now let us take a typical term in the sum, $e^{-\beta[E_i^{(1)}+E_j^{(2)}+E_k^{(3)}]}$ and note that it corresponds to system 1 being found in the state at energy E_i, system 2 in the state at energy E_j and system 3 in a state at energy E_k. The partition sum, Z', will also contain a term $e^{-\beta[E_i^{(2)}+E_j^{(1)}+E_k^{(3)}]}$ with system 1 in the state at E_j, system 2 in the state at E_i and system 3 again in the state at E_k. The set of energy levels in each subsystem is identical so that, as we cannot tell which system is which, the second cited contribution is not only equal but *identical* with the first, i.e., it refers to the *same* microstate. By including all such states explicitly in the sum we are thus overcounting the number of distinct configurations. In this particular case it is easy to see that we are, in fact, overcounting by the number of ways we can permute the three energies which, of course, is simply 3!. We can easily generalize this argument. If the product of N partition functions for N subsystems associated with a heat bath system includes identical (indistinguishable) subsystems and we label successive sets of identical systems using the index k so that the set k contains n_k members we can conclude that the partition sum is

$$Z_{\text{can}} = \prod_k \frac{Z_k^{n_k}}{n_k!}, \tag{1.4.20}$$

where the product is now over distinguishable systems and $\sum n_k = N$.

Although the above argument was developed for non-degenerate states it is easy to see that degeneracy causes no problem because each microstate at a given energy, which forms one member of a group of degenerate states, must have a microscopic structure which distinguishes it from any of its 'companion' states. Otherwise we would, of course, be referring to one and the same state!

The second example concerns a linear chain of s atoms (referred to in percolation theory as sites) connected by bonds (figure 1.5). We suppose that N fragments are formed by breaking $N - 1$ bonds and we seek the total number of microstates as a function of N. We might initially guess that this number is $N!$ which is the number of ways of permuting N fragments. However, permutations of chains (fragments) of equal length produce indistinguishable configurations. The number of permutations of n_k entities is, as in the previous example, $n_k!$ so that the (partial) sum of microstates is

$$\Omega(N, s) = \sum_{n_1, n_2 \ldots n_s} \frac{N!}{n_1! n_2! \ldots n_s!}, \qquad (1.4.21)$$

in which n_k is the number of fragments of length k and the sums are to be performed under the constraints of the fixed total number of sites, $s = \sum_k k n_k$, and multiplicity, $N = \sum_k n_k$. This expression is not easy to evaluate. Furthermore, the problem becomes much more difficult if we go to more than one dimension. It is a useful exercise to, at least, try to appreciate the difficulty by making a sketch of the percolation problem in two dimensions (the two-dimensional equivalent of figure 1.5). We will be discussing percolation again in the context of nuclear multifragmentation.

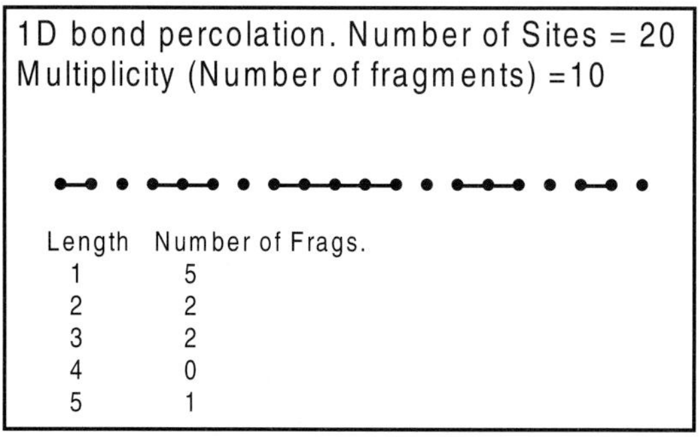

Figure 1.5. Bond breaking percolation in one dimension.

1.5 Statistical mechanics and thermodynamics

In this and the following two sections we assume the validity of the heat bath approximation, i.e., the constancy of β_1. Thus we consider our system '2' to be characterized by a constant temperature. Given this condition, much useful physics can be obtained from the partition sum. We begin this section by examining some properties of this function which for discrete microstates we write as

$$Z_{\text{can}} = \sum_{n(\text{levels})} g_n \, \mathrm{e}^{-\beta\varepsilon_n} = \sum_{i(\text{states})} \mathrm{e}^{-\beta\varepsilon_i}, \qquad (1.5.1)$$

where we have explicitly indicated that the first sum on the right is over energies and the second over all microstates considered individually. At the risk of further repetition (see after (1.4.7)), we again emphasize that, to within the constant $\rho_1(E)\Delta E$, which we dropped in the previous section, Z_{can} represents the *microcanonical* sum of states for the system composed of the heat bath + system 2.

Now following on from (1.4.9) the probability to be associated with any microstate is

$$p_i = \frac{e^{-\beta\varepsilon_i}}{\sum_i e^{-\beta\varepsilon_i}} = \frac{e^{-\beta\varepsilon_i}}{Z_{\text{can}}}. \tag{1.5.2}$$

For example, the average value of the energy (referred to as the internal energy) is given as

$$U = \langle\varepsilon\rangle = \sum_i p_i\varepsilon_i = \frac{\sum_i \varepsilon_i e^{-\beta\varepsilon_i}}{\sum_i e^{-\beta\varepsilon_i}}. \tag{1.5.3}$$

This quantity is a function only of the spectrum of energy levels of the system (which, as can be seen from equation (1.2.9), may be affected by constraints such as the volume) and of the inverse temperature associated with the heat bath. It can be directly expressed using the partition function as

$$U = -\frac{\partial \ln Z_{\text{can}}}{\partial\beta}. \tag{1.5.4}$$

Equation (1.5.4) has been written rather loosely. Although more cumbersome, we could (should) have written

$$U(V, \beta) = -\left.\frac{\partial \ln Z_{\text{can}}(\beta)}{\partial\beta}\right|_{V=\text{const}}$$

in order to emphasize that the derivative is evaluated at constant volume and, indeed, that U is a function of V and β. Usually, no confusion arises in such expressions so that the more compact notation can be retained. Nevertheless, we shall sometimes use the notation $U(V, \beta)$ or $U(V, T)$.

Probably the most important quantity in statistical thermodynamics is the (Helmholtz) free energy which is defined via the relation

$$F = -T \ln[Z_{\text{can}}]. \tag{1.5.5}$$

Using the free energy we can rewrite the probabilities, p_i, as

$$p_i = e^{\beta(F-\varepsilon_i)}. \tag{1.5.6}$$

Following their definition, the sum of these probabilities is unity for all values of β (in phase space this is expressed as $\int e^{\beta(F-\varepsilon)}\, d\Gamma = 1$). The differential of this sum with respect to β is therefore equal to zero so that

$$\begin{aligned} 0 &= \sum_i e^{\beta(F-\varepsilon_i)} \frac{\partial[\beta(F-\varepsilon_i)]}{\partial\beta} \\ &= \sum_i e^{\beta(F-\varepsilon_i)}\left(F + \beta\frac{\partial F}{\partial\beta} - \varepsilon_i\right). \end{aligned} \tag{1.5.7}$$

Given that $\beta\partial/\partial\beta = -T\partial/\partial T$ and that the free energy, F, at constant β and V, is a constant, we can write $\sum_i F\,\mathrm{e}^{\beta(F-\varepsilon_i)} = \sum_i F p_i = F$, and thus obtain, with the help of (1.5.3), the relation known as the Gibbs–Helmholtz equation, i.e.,

$$U(V,T) = \left(1 - \frac{\partial}{\partial \ln[T]}\right)F = F - T\frac{\partial F}{\partial T}. \tag{1.5.8}$$

We can give a meaning to the free energy which is most useful in thermodynamics [8]. Consider N atoms (we will henceforth use the word 'particles') enclosed in a container of volume V and let us calculate the pressure on the walls of the container at fixed temperature T. The word 'enclosed' indicates that the potential energy of any given particle increases rapidly as it approaches any wall of the container. If the coordinate of particle i along the x-axis is x_i and we consider a portion of the container wall at (positive) x this means that the potential $\phi(x-x_i)$ decreases rapidly with increasing x. Now suppose that a portion of the container wall with surface s at coordinate x is displaced by an infinitesimal amount, $\mathrm{d}x$, due to the pressure exerted by the particles. The corresponding increase in volume of the container is $\mathrm{d}V = s\,\mathrm{d}x$ and can be thought of as resulting from a force, f, exerted by the particles on the container wall. The force is, of course, the negative of the derivative of the potential energy with respect to x (remember, ϕ *decreases* with increasing x for fixed x_i). Thus

$$f = \sum_{i=1}^{N} f_i = -\frac{\mathrm{d}}{\mathrm{d}x}\sum_{i=1}^{N}\phi(x-x_i). \tag{1.5.9}$$

The pressure is defined as the average force per unit area. In order to calculate the derivative of the potential energy we note that the total energy, E, is

$$E = \sum_{i=1}^{N}\phi(x-x_i) + \sum_{i=1}^{N}\frac{p_i^2}{2m_i} + V_{\mathrm{int}} + V_{\mathrm{ext}}, \tag{1.5.10}$$

where the second term is the total kinetic energy and the third and fourth terms are, respectively, the energy of interaction between particles and the energy due to any external fields. For infinitesimal displacements the latter three terms are independent of x so that

$$\frac{\mathrm{d}}{\mathrm{d}x}\left[\sum_{i=1}^{N}\phi(x-x_i)\right] = \frac{\mathrm{d}E}{\mathrm{d}x}. \tag{1.5.11}$$

The pressure can now be written by substituting (1.5.11) in (1.5.9), dividing by s and using the Gibbs probability $\mathrm{e}^{(F-E)/T}$ to perform the weighted average over the coordinates of the particles in the system. Thus,

$$P = -\frac{1}{s}\int\frac{\mathrm{d}E}{\mathrm{d}x}\,\mathrm{e}^{-(F-E)/T}\,\mathrm{d}\Gamma = -\int\frac{\mathrm{d}E}{\mathrm{d}V}\,\mathrm{e}^{-(F-E)/T}\,\mathrm{d}\Gamma, \tag{1.5.12}$$

where the integral is over position and momentum coordinates of all particles. Equation (1.5.12) may be developed using the normalization condition (implied by the definition of F), i.e., $\int \mathrm{e}^{-(F-E)/T}\,\mathrm{d}\Gamma = 1$. Differentiation of this equation with respect to the volume V obviously produces zero on the the right-hand side and, consequently, the result

$$\frac{1}{T}\int \frac{\partial E}{\partial V}\,\mathrm{e}^{-(F-E)/T}\,\mathrm{d}\Gamma = \frac{1}{T}\int \frac{\partial F}{\partial V}\,\mathrm{e}^{-(F-E)/T}\,\mathrm{d}\Gamma. \tag{1.5.13}$$

Since the free energy does not depend on the coordinates and momenta of the particles we arrive at the important result (applicable to a system at given temperature and volume)

$$P(V,T) = -\int \frac{\partial F}{\partial V}\,\mathrm{e}^{-(F-E)/T}\,\mathrm{d}\Gamma = -\frac{\partial F}{\partial V}. \tag{1.5.14}$$

Equation (1.5.14) is referred to as an equation of state. It describes the macroscopic behaviour of the system. We shall develop this subject further in section 1.7.

Another fundamental result concerns the so-called Gibbs entropy. Again by definition (1.5.5) the differential on the right-hand side of (1.5.8) is

$$\begin{aligned}
-\frac{\partial F}{\partial T} = \frac{\partial}{\partial T}\left(T\ln\left[\sum_i \mathrm{e}^{-\varepsilon_i/T}\right]\right) &= T\frac{\partial}{\partial T}\ln\left[\sum_i \mathrm{e}^{-\varepsilon_i/T}\right] + \ln\left[\sum_i \mathrm{e}^{-\varepsilon_i/T}\right] \\
&= \frac{T\sum_i \frac{\varepsilon_i}{T^2}\,\mathrm{e}^{-\varepsilon_i/T}}{\sum_i \mathrm{e}^{-\varepsilon_i/T}} + \ln\left[\sum_i \mathrm{e}^{-\varepsilon_i/T}\right] \\
&= -\sum_i p_i\left(\frac{-\varepsilon_i}{T} - \ln\left[\sum_{i'} \mathrm{e}^{\varepsilon_{i'}/T}\right]\right) \\
&= -\sum_i p_i \ln[p_i].
\end{aligned} \tag{1.5.15}$$

Equation (1.5.15) defines the Gibbs entropy,

$$S_G = -\frac{\partial F}{\partial T} = -\sum_i p_i \ln[p_i]. \tag{1.5.16}$$

(If the temperature is measured in degrees kelvin rather than in energy units the Gibbs entropy is written as

$$S_G = -k_\mathrm{B}\sum_i p_i \ln[p_i],$$

see (1.2.4).) The first law of thermodynamics can be expressed by stating that an increase in energy of a system may be brought about either by adding an amount of heat energy (say ΔQ) to the system or by doing work on the system (e.g., by pushing on a piston which works against the internal pressure so as to decrease the volume). The change in energy, for a system characterized by its temperature and presssure, is thus expressed as

$$\Delta U = -P\Delta V + \Delta Q. \tag{1.5.17}$$

The significance of the Gibbs entropy can be understood by differentiating the Gibbs–Helmholtz equation (taking $F = F(V, T)$) and comparing the result with (1.5.17)

$$\begin{aligned} \mathrm{d}U &= \mathrm{d}F - \mathrm{d}\left(T\frac{\partial F}{\partial T}\right) = \frac{\partial F}{\partial V}\,\mathrm{d}V + \frac{\partial F}{\partial T}\,\mathrm{d}T - \frac{\partial F}{\partial T}\,\mathrm{d}T - T\,\mathrm{d}\left(\frac{\partial F}{\partial T}\right) \\ &= -\,P\,\mathrm{d}V + T\,\mathrm{d}S_G. \end{aligned} \tag{1.5.18}$$

We see that, at constant temperature and volume, the introduction of a small quantity of heat energy, ΔQ, corresponds to an increase in entropy, $\Delta S_G = \Delta Q/T$ (we use the word 'small' to indicate that the system remains practically in equilibrium although, of course, it will eventually 'settle down' at a slightly higher temperature). Equation (1.5.18) also indicates that a change in volume (or, more generally, the act of performing mechanical work on the system) which involves no heat exchange with the surroundings (adiabatic process) is isentropic. This observation is important for thermodynamical models of expanding nuclei which must be considered as isolated systems.

The Gibbs entropy has another interpretation which is extremely important irrespective of whether systems are described in terms of microstates or macrostates. The entropy represents the degree of disorder or randomness in the system [9]. Given an arbitrary set of events and no further information the 'reasonable man' would assume the events to be equi-probable (e.g., the six faces of a dice). This, in fact, corresponds to maximizing the Gibbs entropy. To see why, let us assume that each macrostate has an associated probability p_i. We now proceed to maximize the entropy, taking into account the only known constraint, $\sum_i p_i = 1$. Using the Lagrange method of undetermined multipliers, we find, for all i,

$$\frac{\mathrm{d}}{\mathrm{d}p_i}\left[-\sum_i p_i \ln[p_i] + \lambda\left(\sum_i p_i - 1\right)\right] = 0, \tag{1.5.19}$$

which yields $\ln[p_i] = \lambda - 1 = \text{const.}$ We thus obtain the microcanonical equilibrium distribution. Furthermore, if all p_i are equal and we have Ω states (so that $p_i = 1/\Omega$) we quickly find that $S_G = \ln[\Omega]$ and is therefore identical with the microcanonical entropy S_{M}.

This method is, in fact, generally applicable. As a second example, we may consider the canonical distribution. Our derivation at the beginning of section 1.4 was based uniquely on the supposed behaviour of the density of microstates corresponding to the heat bath system (system 1), i.e., we supposed the density to vary as $\rho(E + \Delta E) = \rho(E)\,\mathrm{e}^{\beta\Delta E}$ (equation (1.4.2)). We subsequently found that the *most probable* energy division between system 1 and system 2 was such that the logarithmic dervatives of the two densities were equal. This allowed us to identify the most probable macrostate as the state for which system 2 has the same microcanonical temperature as the reservoir, although, strictly speaking we had little use for the concept of 'temperature'.

We can consider an alternative procedure in which the characteristics of the density of states in the heat bath are hardly mentioned. The only role attributed to the heat bath is that, in equilibrium, it fixes not the most probable, but the average energy in system 2. Apart from this additional constraint, which we can write $\sum_i \varepsilon_i p_i = U$, we forget about the heat bath and maximize the Gibbs entropy in system 2. We write

$$\frac{\mathrm{d}}{\mathrm{d}p_i}\left[-\sum_i p_i \ln[p_i] - \lambda_1\left(\sum_i p_i - 1\right) - \lambda_2\left(\sum_i \varepsilon_i p_i - U\right)\right] = 0, \quad (1.5.20)$$

where the two constraints are introduced with the undetermined multipliers $-\lambda_1$ and $-\lambda_2$. Equation (1.5.20) yields

$$-1 - \ln[p_i] - \lambda_1 - \lambda_2\varepsilon_i = 0, \quad (1.5.21)$$

so that by defining $\ln[c] = -(1 + \lambda_1)$ we obtain

$$p_i = c\,\mathrm{e}^{-\lambda_2\varepsilon_i}. \quad (1.5.22)$$

Now, the constraint $\sum_i p_i = 1$ obviously gives $c = \left(\sum_i \mathrm{e}^{-\lambda_2\varepsilon_i}\right)^{-1}$. The second constraint then takes the form

$$\frac{\sum_i \varepsilon_i\,\mathrm{e}^{-\lambda_2\varepsilon_i}}{\sum_i \mathrm{e}^{-\lambda_2\varepsilon_i}} = U, \quad (1.5.23)$$

which is identical with equation (1.5.3) providing we identify λ_2 with β. We have thus shown that maximizing the Gibbs entropy in system 2, assuming that the heat bath fixes the average energy, U, is completely equivalent to our previous assumption concerning the density of states in system 1 (the heat bath). Otherwise (1.5.23) would not hold.

As a conceptual basis, it appears that the specific consideration of the density of states in the heat bath is better adapted to the nuclear physics situation due to the fact that a heat bath, if introduced at all, corresponds to some part of the system which behaves only approximately as (1.4.2) (see figure 1.3).

A comparison of microcanonical and Gibbs entropies follows from the Gibbs–Helmholtz equation which can be written

$$S_G = \beta U + \ln[Z_{\mathrm{can}}], \quad (1.5.24)$$

so that

$$\frac{\partial S_G}{\partial U} = \beta. \quad (1.5.25)$$

This result is to be compared with the microcanonical definition, which from (1.3.14) is

$$\frac{\mathrm{d}S_{\mathrm{M}}}{\mathrm{d}E} = \frac{1}{T}. \tag{1.5.26}$$

It is thus the *average* energy in the canonical ensemble which plays the role of the energy in the microcanonical ensemble. Indeed, the Gibbs–Helmholtz equation relates the *average* energy and the *average value* of $\ln[p_i]$. The averaging is carried out over all states of the combined system. This kind of relation is typical of thermodynamics and, indeed, is appropriate for many large systems for which fluctuations away from the average values can be neglected at least to first order (for a study of fluctuations see Klimontovitch [7]). In such systems it is often true that average values and most probable values are essentially identical, since the construction of the average is dominated by the most probable macrostate. In small systems, while average values may have well-understood physical significance, fluctuations are often so important that one is usually well advised to study the full probability distributions rather than just the first few moments (the average value is the moment of order 1 divided by the moment of order 0). This is particularly true of the Gibbs entropy whose distribution may be strongly influenced by macrostates with small probabilities. Furthermore, in small systems, average values and most probable values may be quite different.

1.6 The grand canonical ensemble

In this book we will mainly be concerned with statistical models which involve fixing the number of particles, the volume (or density) and the energy (or possibly the temperature). However, it will occasionally be useful to consider varying the number of particles in the system.

Variation of particle number can be taken into account by an extension of the reasoning leading to the canonical distribution. The extended distribution is referred to as the grand canonical distribution (or grand canonical ensemble). The density of microstates now depends not only on the energy but also on the number of particles and is written $\rho(E, N)$. For discrete states, in the same way as we defined the canonical partition function, we can define the grand canonical partition function as

$$\mathbb{Z}(\alpha, \beta) = \sum_{\nu=0}^{\infty} \mathrm{e}^{\alpha\nu} \sum_{i} \mathrm{e}^{-\beta E_i(\nu)} = \sum_{\nu=0}^{\infty} \mathrm{e}^{\beta\mu\nu} \sum_{i} \mathrm{e}^{-\beta E_i(\nu)}, \tag{1.6.1}$$

where the sum over i refers to all microstates of a system with ν particles and $\mu = \alpha/\beta$ is referred to as the chemical potential. Just as thermodynamic systems in thermal equilibrium have the same temperature (β) so thermodynamic systems which are simultaneously in chemical and thermal equilibrium (the situation almost always encountered) have the same value of μ. However, as emphasized above in connection with the notion of temperature, the notion 'same chemical potential' refers to the maximum of an integrand of a convolution integral and thus must be used with caution in small systems.

It is possible also to extend the definition of the density of states given in equation (1.2.7). We write

$$g(E, N) = \sum_{\nu,i} \delta(\nu - N)\delta(E - E_i(\nu)), \tag{1.6.2}$$

so that, just as the canonical partition function is the Laplace transform of the one-dimensional density $g(E) = \sum_i \delta(E - E_i)$, $\mathbb{Z}(\mu, \beta)$ (or $\mathbb{Z}(\alpha, \beta)$) is the (double) Laplace transform of $g(E, N)$ i.e.,

$$\begin{aligned}
\mathbb{Z}(\mu, \beta) &= \int_0^\infty \int_0^\infty \sum_{\nu,i} \delta(\nu - N)\delta(E - E_i(\nu))\,\mathrm{e}^{-\beta(E-\mu N)}\,\mathrm{d}E\,\mathrm{d}N \\
&= \sum_\nu \int_0^\infty \mathrm{e}^{\beta\mu N}\delta(\nu - N)\left[\sum_i \int_0^\infty \mathrm{e}^{-\beta E}\delta(E - E_i(\nu))\,\mathrm{d}E\right]\mathrm{d}N \\
&= \sum_{\nu,i} \mathrm{e}^{-\beta(E_i(\nu)-\mu\nu)},
\end{aligned} \tag{1.6.3}$$

which is the same as (1.6.1). In the same way as we introduced the free energy in the canonical distribution we may introduce a constant referred to as the grand potential which is usually designated by the symbol Ω. To avoid confusion with the microcanonical sum of states we will refer to this quantity as Ω^*. Thus with $\Omega^* = -T\ln[\mathbb{Z}(\mu, \beta)]$ we have

$$1 = \sum_{\nu,i} \mathrm{e}^{\beta(\Omega^* - E_i(\nu)+\mu\nu)}, \tag{1.6.4}$$

and differentiation with respect to β yields

$$\begin{aligned}
0 &= \sum_{\nu,i}\left(\Omega^* + \beta\frac{\partial\Omega^*}{\partial\beta} - E_i(\nu) + \mu\nu\right)\mathrm{e}^{\beta(\Omega^* - E_i(\nu)+\mu\nu)} \\
&= \Omega^* + \beta\frac{\partial\Omega^*}{\partial\beta} - \sum_{\nu,i}(E_i(\nu) - \mu\nu)\,\mathrm{e}^{\beta(\Omega^* - E_i(\nu)+\mu\nu)},
\end{aligned} \tag{1.6.5}$$

which, by taking the average energy $U = \langle E_i(\nu)\rangle$ to the left-hand side and noting again that $\partial/\partial(\ln[\beta]) = -\partial/\partial(\ln[T])$, becomes

$$\sum_{\nu,i} E_i(\nu)\,\mathrm{e}^{\beta(\Omega^* - E_i(\nu)+\mu\nu)} = U = \Omega^* - T\frac{\partial\Omega^*}{\partial T} + \mu\langle\nu\rangle. \tag{1.6.6}$$

From the definition of Ω^* it follows straightforwardly that $\langle\nu\rangle = -\partial\Omega^*/\partial\mu$, so that finally

$$U = \Omega^* - T\frac{\partial\Omega^*}{\partial T} - \mu\frac{\partial\Omega^*}{\partial\mu}. \tag{1.6.7}$$

Equation (1.6.5) (or (1.6.6)) can be thought of as the analogue of the Gibbs–Helmholtz equation. In the canonical distribution we identified the entropy as $-\sum p_i \ln[p_i]$. The same definition, using the probability associated with a state in the grand canonical ensemble, leads to

$$S^* = -\frac{\partial\Omega^*}{\partial T}. \tag{1.6.8}$$

A most important relation in a system described in terms of volume, temperature and chemical potential concerns the differential of the internal (average) energy

$$\begin{aligned} \mathrm{d}U &= \frac{\partial\Omega^*}{\partial V}\,\mathrm{d}V + \frac{\partial\Omega^*}{\partial T}\,\mathrm{d}T + \frac{\partial\Omega^*}{\partial\mu}\,\mathrm{d}\mu - \mathrm{d}\left[T\frac{\partial\Omega^*}{\partial T}\right] - \mathrm{d}\left[\mu\frac{\partial\Omega^*}{\partial\mu}\right] \\ &= \frac{\partial\Omega^*}{\partial V}\,\mathrm{d}V + \frac{\partial\Omega^*}{\partial T}\,\mathrm{d}T + \frac{\partial\Omega^*}{\partial\mu}\,\mathrm{d}\mu - \frac{\partial\Omega^*}{\partial T}\,\mathrm{d}T - T\,\mathrm{d}\left[\frac{\partial\Omega^*}{\partial T}\right] \\ &\quad - \frac{\partial\Omega^*}{\partial\mu}\,\mathrm{d}\mu - \mu\,\mathrm{d}\left[\frac{\partial\Omega^*}{\partial\mu}\right], \end{aligned} \tag{1.6.9}$$

which with (1.6.7) yields

$$\mathrm{d}U = \frac{\partial\Omega^*}{\partial V}\,\mathrm{d}V + T\,\mathrm{d}S^* + \mu\,\mathrm{d}\langle\nu\rangle. \tag{1.6.10}$$

This equation shows that in a system with constant volume and entropy, the increase in the the internal energy when a particle is added to the system is equal to μ (hence the name, chemical *potential*). In thermodynamics we consider the average value of the particle number to be the actual particle number N. In an isolated vessel which may contain more than one type of particle, equation (1.6.9) thus generalizes to

$$\mathrm{d}U = -P\,\mathrm{d}V + T\,\mathrm{d}S^* + \sum_k \mu_k\,\mathrm{d}N_k. \tag{1.6.11}$$

(The identification of the pressure, P, with $-\partial\Omega^*/\partial V$ is left as an exercise for the reader.) Chemical equilibrium which implies $\sum_k \mu_k\,\mathrm{d}N_k = 0$ will be further considered in chapter 7 in connection with the subject of caloric curves.

We will be using the grand canonical partition function in chapter 2 when we consider the density of states in an excited nucleus. Its use there depends on the fact that knowing the density $\rho(E, N)$ is completely equivalent to knowing the grand partition function $\mathbb{Z}$. By modelling $\mathbb{Z}$ and approximating the inverse transform we can obtain a 'smoothed' density and thus avoid the inconvenience of manipulating the delta functions.

There is one more topic which concerns particle numbers. In quantum mechanics, particles are divided into two categories, bosons and fermions. Bosons, which are said to obey Bose–Einstein statistics, are particles with integral spin such as deuterons or alpha particles, whereas fermions (electrons, nucleons) have half integral spin and obey Fermi–Dirac statistics. Any given quantum (energy) state may be occupied by an arbitrary number of bosons. However, the Pauli principle states that not more than one fermion may occupy a particular state. The underlying reason is that a wavefunction which represents a multiparticle state of identical particles must have a definite symmetry under interchange of particle coordinates. Under permutation of the spatial and spin coordinates of any two particles, boson wavefunctions are unchanged whereas fermion wavefunctions show only a change of sign. Thus, for example, if we consider just two particles, each of which may occupy a set of one-particle states, and we confine our analysis to combined spin symmetric states we can define products of orthogonal normalized (orthonormal) wavefunctions $\phi_1(r_1)\phi_2(r_2)$ in which it is to be understood that particle 1 occupies state 1 and particle 2 occupies state 2. Such simple products have no definite symmetry under permutation of particle coordinates ($\phi_1(r_1)\phi_2(r_2)$ and $\phi_1(r_2)\phi_2(r_1)$ may be quite different functions) but can be used to construct an antisymmetric wavefunction as a linear combination. Thus,

$$\Phi_A(r_1, r_2) = \frac{1}{\sqrt{2}}[\phi_1(r_1)\phi_2(r_2) - \phi_1(r_2)\phi_2(r_1)] \tag{1.6.12}$$

has the property $\Phi_A(r_1, r_2) = -\Phi_A(r_2, r_1)$ and obviously vanishes when the particles are in the same state ($\phi_1 = \phi_2$). The normalization factor $1/\sqrt{2}$, which ensures that the integral $\int \mathrm{d}r_1\,\mathrm{d}r_2 \Phi_A^*(r_1, r_2)\Phi_A(r_1, r_2) = 1$ generalizes to $1/\sqrt{N!}$ for N particles because there are $N!$ possible permutations of coordinates (N terms in the generalization of (1.6.12)).

The symmetric (Bose–Einstein) state is a little more complicated. It may be constructed as the linear combination (orthogonal to (1.6.12))

$$\Phi_S(r_1, r_2) = \frac{1}{\sqrt{S}}[\phi_1(r_1)\phi_2(r_2) + \phi_1(r_2)\phi_2(r_1)]. \tag{1.6.13}$$

The norm is given by

$$\begin{aligned} S &= \int \mathrm{d}r_1\,\mathrm{d}r_2 \Phi_S^*(r_1, r_2)\Phi_S(r_1, r_2) \\ &= \int \mathrm{d}r_1\,\mathrm{d}r_2 [\phi_1^*(r_1)\phi_2^*(r_2)\phi_1(r_1)\phi_2(r_2) + \phi_1^*(r_2)\phi_2^*(r_1)\phi_1(r_2)\phi_2(r_1) \\ &\quad + \phi_1^*(r_1)\phi_2^*(r_2)\phi_1(r_2)\phi_2(r_1) + \phi_1^*(r_2)\phi_2^*(r_1)\phi_1(r_1)\phi_2(r_2)], \end{aligned} \tag{1.6.14}$$

which is two if the particles are in different states (the second two 'cross' terms are zero) and four if the two particles are in the same state (the four terms are identical). Thus, the Bose–Einstein normalization depends on the occupation numbers. In fact the norm, $\sqrt{S}$, generalizes to $\sqrt{N!n_1!n_2!\ldots}$ where for two particles, either $n_1 = n_2 = 1$ or $n_1 = 2$, $n_2 = 0$, or $n_1 = 0$, $n_2 = 2$. The implication is that permutations between particles in the *same* one-particle state (which cannot occur in the Fermi–Dirac case) do not lead to a new (micro) state.

Thus, the presence of the factor $\sqrt{n_1!n_2!\ldots}$ ensures that, as in the Fermi–Dirac case, the wavefunction is completely specified by the occupation numbers n_i.

The requirement of definite symmetry gives rise to quite distinct partition functions. The two cases can be conveniently examined using the grand canonical ensemble associated with the occupation number representation. Consider a set of non-degenerate energy levels E_i which are available to each member of a set of N non-interacting particles (identical systems) and suppose that n_i particles are to be found in the state with energy E_i. The total energy is then the sum of the energies of the particles $E = \sum_i n_i E_i$ and the total number of particles $N = \sum_i n_i$. The contribution of any N particle state of energy, E, to the partition sum is, of course, weighted by the Boltzmann factor $\mathrm{e}^{-\beta E}$ and by the factor $\mathrm{e}^{\beta\mu N}$. Because the particles are indistinguishable any particular configuration of occupation numbers represents just one microscopic state. Interchanging particles in such a way that the occupation numbers n_i remain unchanged does not produce a new microstate. The full set of microscopic configurations may therefore be explored by examining all possible combinations of the occupation numbers, n_i, which are consistent with the constraints. The grand partition sum can thus be written as a sum over occupation numbers,

$$\begin{aligned}\mathbb{Z} &= \int_0^\infty \int_0^\infty \mathrm{e}^{\beta\mu N} \sum_{n_1,n_2\ldots} \mathrm{e}^{-\beta E} \delta(\Sigma n_i - N)\delta(\Sigma n_i E_i - E)\,\mathrm{d}E\,\mathrm{d}N \\ &= \sum_{n_1,n_2\ldots} \mathrm{e}^{-\beta\Sigma n_i E_i}\,\mathrm{e}^{\beta\mu\Sigma n_i} = \sum_{n_1,n_2\ldots} \mathrm{e}^{-\beta(n_1E_1-\mu n_1)-\beta(n_2E_2-\mu n_2)}\ldots \\ &= \sum_{n_1}[\mathrm{e}^{-\beta(E_1-\mu)}]^{n_1} \times \sum_{n_2}[\mathrm{e}^{-\beta(E_2-\mu)}]^{n_2}\cdots\times\cdots. \end{aligned} \tag{1.6.15}$$

Note in this equation that the second line which represents an unconstrained sum follows because the integral over N frees the constraint on allowed values of the n_k. Each term in (1.6.15) is a geometric progression. For bosons the values of n_i are unrestricted so that recalling that the sum of the geometric progression $(1 + x + x^2 + \cdots) = 1/(1 - x)$ we obtain

$$\sum_{n_i}[\mathrm{e}^{-\beta(E_i-\mu)}]^{n_i} = \frac{1}{1 - \mathrm{e}^{-\beta(E_i-\mu)}} \tag{1.6.16}$$

and the Bose–Einstein partition sum is obtained as the product

$$\mathbb{Z}_{\mathrm{BE}} = \prod_i \frac{1}{1 - \mathrm{e}^{-\beta(E_i-\mu)}}. \tag{1.6.17}$$

The mean number of particles in each state can be obtained from the grand potential

$$\Omega^* = -T\ln[\mathbb{Z}_{\mathrm{BE}}] = \frac{1}{\beta}\sum_i \ln[1 - \mathrm{e}^{-\beta(E_i-\mu)}] = \sum_i \Omega_i^*, \tag{1.6.18}$$

by differentiating Ω_i^* with respect to μ,

$$\langle n_i\rangle_{\rm BE} = -\frac{\partial\Omega_i^*}{\partial\mu} = -\frac{1}{\beta}\frac{\partial}{\partial\mu}(\ln[1 - \mathrm{e}^{-\beta(E_i-\mu)}])$$
$$= \left[\frac{\mathrm{e}^{-\beta(E_i-\mu)}}{1-\mathrm{e}^{-\beta(E_i-\mu)}}\right] = \frac{1}{\mathrm{e}^{\beta(E_i-\mu)}-1}. \tag{1.6.19}$$

For a set of fermions, possible values of n_i are restricted to 0 or 1 so that equation (1.6.15) gives the Fermi–Dirac sum of microstates as

$$\mathbb{Z}_{\rm FD} = \prod_i [1 + \mathrm{e}^{-\beta(E_i-\mu)}], \tag{1.6.20}$$

and the grand potential

$$\Omega^* = \sum_i \Omega_i^* = -T\ln[\mathbb{Z}_{\rm FD}] = -\frac{1}{\beta}\sum_i \ln[1+\mathrm{e}^{-\beta(E_i-\mu)}]. \tag{1.6.21}$$

In this case

$$\langle n_i\rangle_{\rm FD} = -\frac{\partial\Omega_i^*}{\partial\mu} = \frac{1}{\mathrm{e}^{\beta(E_i-\mu)}+1}. \tag{1.6.22}$$

We can use (1.6.19) and (1.6.22) to comment on the difference between quantal and classical systems although a complete understanding would necessitate a proper development of quantum statistical mechanics (see Chapters 11 and 12 in Greiner, Neise and Stöcker [3]). To this end, consider a classical system for which the canonical partition function for N identical constituents (particles, systems) can be written as $Z^N/N!$ where Z is the one particle partition function. Furthermore, let us stipulate that the energy states available to any one particle are not affected by the number of particles (i.e., the particles do not interact). Then the grand canonical partition function for a given N is ('MB' for Maxwell–Boltzmann)

$$\mathbb{Z}_{\rm MB}(N) = \frac{1}{N!}\left(\sum_i \mathrm{e}^{-\beta(E_i-\mu)}\right)^N. \tag{1.6.23}$$

We can recover the occupation number representation by using the multinomial expansion

$$\left(\sum_i x_i\right)^N = N!\sum_{n_1,n_2\ldots}\delta\left(\sum_i n_i - N\right)\prod_i \frac{x_i^{n_i}}{n_i!}, \tag{1.6.24}$$

with the help of which (1.6.23) becomes

$$\mathbb{Z}_{\rm MB}(N) = \frac{1}{N!}\sum_{n_1,n_2\ldots}\left(\frac{N!}{n_1!n_2!\ldots}\mathrm{e}^{-\beta n_1(E_i-\mu)-\beta n_2(E_2-\mu)\ldots}\right). \tag{1.6.25}$$

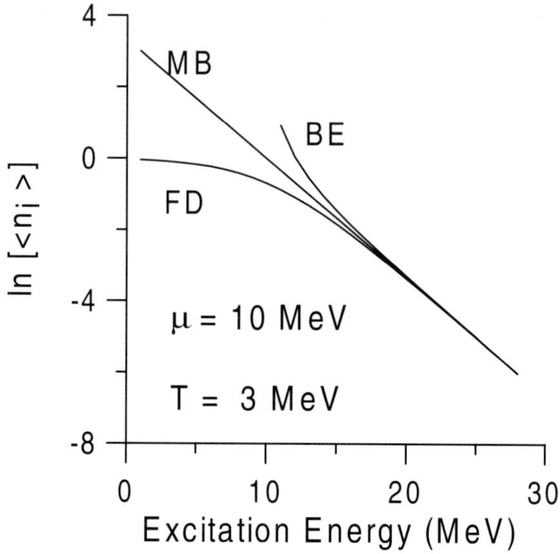

Figure 1.6. Continuous representation of the mean occupation numbers as a function of energy for Fermi–Dirac (FD), Maxwell–Boltzmann (MB) and Bose–Einstein (BE) statistics. Note that FD tends to 1 (0 on log plot) at $E^* = 0$, MB is a straight line with slope $1/T$ and that the BE distribution is singular at $E^* = \mu$. At high energies, where $(E^* - \mu)/T \gg 1$, the three distributions are almost identical.

The similarity of equations (1.6.25) and (1.6.15) is evident except for the presence of the factor $1/(n_1!n_2!\ldots)$. Thus the classical canonical or grand canonical partition sums, when expressed in the occupation number representation, both contain an additional factor $1/(n_1!n_2!\ldots)$ which weights any configuration $\{n_1, n_2 \ldots\}$ of occupation numbers when compared with the Bose–Einstein sum. In contrast, both the Bose–Einstein and the Fermi–Dirac sums imply that a particular set of occupation numbers specifies just one microstate and that the only weighting comes from the associated reservoir.

Without using the occupation number representation the grand canonical sum of states for the classical system can be obtained directly from equation (1.6.23)

$$\mathbb{Z}_{\text{MB}} = \sum_{N=0}^{\infty} \frac{Z^N}{N!} \, \mathrm{e}^{\beta\mu N} = \exp(Z \, \mathrm{e}^{\beta\mu}), \tag{1.6.26}$$

which, if we write the canonical sum for a single system as $Z = \sum_i \mathrm{e}^{-\beta E_i}$, gives us the grand potential

$$\Omega^* = -\frac{1}{\beta} \sum_i \mathrm{e}^{-\beta(E_i - \mu)}. \tag{1.6.27}$$

The mean number of particles in the ith energy state follows from the Poisson distribution (the sum over N in (1.6.26)) or from differentiating the ith term in (1.6.27) with respect to μ. It is simply

$$\langle n_i \rangle_{\text{MB}} = Z\,\mathrm{e}^{\beta\mu} = \mathrm{e}^{-\beta(E_i-\mu)}. \tag{1.6.28}$$

It is instructive to compare this result with that given in the Bose–Einstein and Fermi–Dirac cases. If the energy is much greater than the chemical potential then both in the Bose–Einstein and in Fermi–Dirac cases the exponential term dominates the denominator. We then find that $\langle n_i \rangle \approx \mathrm{e}^{-\beta(E_i-\mu)}$ which, of course, is the result obtained in (1.6.28). In other words, it is the most probable or average energy relative to the chemical potential measured in units of temperature which determines whether or not a given problem can be treated classically. A value of $\beta(U-\mu) \gg 1$ implies small values of $\langle n_i \rangle$ so that the factorial factors such as $1/(n_1!n_2!\ldots)$ and limitations such as $n_i \leq 1$ have little effect. It is therefore not surprising that, in this circumstance, both the the classical and the two quantal ensembles give the same results. A comparison of the Fermi–Dirac, Bose–Einstein and Maxwell–Boltzmann (classical) distributions is shown in figure 1.6.

1.7 Equations of state for ideal and real gases

Arguments based on thermodynamics applied to isolated short-lived systems made up of a small number of particles are certainly open to criticism. Nevertheless, because such arguments are often used to discuss nuclear multifragmentation we present, in this section, the main features of equations of state for ideal (I) and real (R) gases in the hope that the presentation will be useful for understanding the considerable literature concerning this subject.

We begin with the general form for the equation of state which is applicable to a classical system characterized by a temperature T (i.e., associated with a heat bath system). We thus use the canonical ensemble and ignore the fermion or boson character of the constituent particles. The analysis is expected to have some relevance for the nuclear problem provided that the difference between the excitation energy and the chemical potential is much greater than the temperature so that the system behaviour generally conforms quite closely to the classical picture.

Equation (1.5.14) ($P = -\partial F/\partial V$) is, in fact, a general form for the equation of state. Thus, for an ideal gas with constant particle number, N, the free energy is obtained from the partition sum $Z^N/N!$ where Z is the one body partition function. Using equation (1.3.8) we find

$$F_{\text{I}} = -NT\left(\ln V + \tfrac{3}{2}\ln[2\pi mT] - \ln[h^3] - (\ln[N] - 1)\right), \tag{1.7.1}$$

where the last term occurs because, with Stirling's approximation, $\ln N! \approx N$

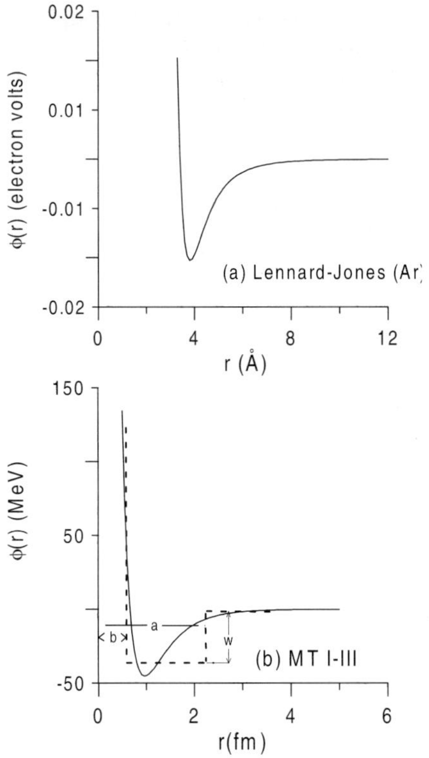

Figure 1.7. (*a*) Lennard–Jones potential representing the interaction between two Ar atoms as a function of the distance (in ångströms) between their mass centres. (*b*) Approximate form of the nucleon–nucleon potential. Compared with the atomic interaction, the length scale is 10^5 times smaller and the energy scale 10^6 times larger.

$(\ln[N]-1)$. From (1.7.1) we immediately obtain the equation of state

$$P_{\rm I} = -\frac{\partial F_{\rm I}}{\partial V} = \frac{NT}{V}. \tag{1.7.2}$$

The entropy of the ideal gas is provided by the definition $S = -\partial F/\partial T$. We find

$$S_{\rm I} = N\left(\ln V + \tfrac{3}{2}\ln[2\pi mT] - \ln[h^3] - (\ln[N]-1) + \tfrac{3}{2}\right). \tag{1.7.3}$$

We can also use the Gibbs–Helmholtz relation, $U = F + TS$ to obtain the internal energy

$$U_{\rm I} = \frac{3NT}{2}. \tag{1.7.4}$$

From equation (1.7.3) we see that isentropic expansion (i.e., expansion at constant entropy) implies that the temperature falls with increasing volume so that $VT^{3/2}$

is constant. Under these circumstances for fixed entropy S, the pressure decreases monotonically with increasing volume as T/V, i.e. as $S^{2/3}/V^{5/3}$, and the internal energy per particle as $(S/V)^{2/3}$. In this case the internal energy per particle and per unit volume behaves in exactly the same way as the pressure.

We now consider a real gas. The characteristic of a real gas is of course that particles interact. Consider, for the moment, a gas composed of just two particles enclosed in a volume V. If the first particle is at a distance r from the second particle, the potential energy corresponding to the interaction, $\phi(r)$, enters into the Boltzmann weight so that, in the centre-of-mass system, the effective one particle partition sum contains an additional term. If we simplify matters by assuming the volume, V, to be a sphere of radius r_V and stipulate that $\phi(r)$ tends to 0 for $r \ll r_V$ (long range forces require special attention, see section 6.7), we can write the integral over coordinate space of the Boltzmann weight corresponding to the interaction of our test particle with the second particle as

$$\int_0^{r_V} 4\pi r^2 \, \mathrm{e}^{-\phi(r)/T} \, \mathrm{d}r. \tag{1.7.5}$$

In the absence of interaction, $\phi(r)$ is everywhere 0 so that (1.7.5) gives the volume V. Typical forms of the potential energy which characterizes the interaction between two atoms and between two nucleons are sketched in figure 1.7. The potential shown in the atomic case (figure 1.7(a)) is the well-known Lennard–Jones potential [1] and the nucleon–nucleon interaction shown in figure 1.7(b) is the S-wave part of the Malfliet–Tjon I–III potential [12]. Both interactions are characterized by a long range attraction and a short range repulsion. In fact the potential energy is often approximated as the sum of a square well attractive potential and a 'hard-core' infinite repulsion (dashed lines in figure 1.7(b)) which is infinite at distances smaller than b. Consider, first, only the repulsion. If we use the hard core approximation then equation (1.7.5) shows that the effective volume available to the first particle is reduced by $V_b = \frac{4}{3}\pi b^3$. Let us now add $(N-2)$ particles to form an N-particle system. The reduction in the volume available to our test particle is now a sum over each of the $n = 2 \ldots N$ other particles and thus globally $(N-1)V_b$. However, if we now select a second particle, and note that we have already taken into account its loss of volume due to interaction with particle 1, we are led to assign an additional loss of volume $(N-2)V_b$ for this particle. The average effective loss of volume per particle for a gas of N particles is thus

$$\Delta V = V_b \frac{1}{N} \sum_{i=1}^{N} (N-i) = \frac{V_b}{N}\left[N^2 - \frac{N}{2}(N+1)\right] = V_b \frac{(N-1)}{2}, \tag{1.7.6}$$

which is, not unexpectedly (and quite generally), half the value for the first particle. This effect will be present for any two-body interaction so that, for an N-particle system, if the potential can be considered to vanish beyond r_ϕ (with

$V_\phi = \frac{4\pi}{3} r_\phi^3$), we can use (1.7.5) and (1.7.6) to define an effective average volume

$$V' = V + \frac{N-1}{2}\left[\int_0^{r_\phi} 4\pi r^2\,\mathrm{e}^{-\phi(r)/T}\,\mathrm{d}r - \int_0^{r_\phi} 4\pi r^2\,\mathrm{d}r\right], \tag{1.7.7}$$

which may be assigned to any member of the set of N particles. Note that in (1.7.7), if we consider only the hard-core potential so that $r_\phi = b$ and $V_\phi = V_b$ we recover (1.7.6), and if we consider the square well approximation to the potential of figure 1.7(b) and split the integral into two parts we get

$$\begin{aligned} V' = V &+ \frac{N-1}{2}\left[\int_0^{b} 4\pi r^2\,\mathrm{e}^{-\infty/T}\,\mathrm{d}r - V_b\right] \\ &+ \frac{N-1}{2}\left[\int_b^{a} 4\pi r^2\,\mathrm{e}^{w/T}\,\mathrm{d}r - (V_a - V_b)\right] \\ = V &+ \frac{N-1}{2}[(V_a - V_b)\,\mathrm{e}^{w/T} - V_a], \end{aligned} \tag{1.7.8}$$

where w is the (constant) well depth. Defining $\delta(T) = [(V_a - V_b)\,\mathrm{e}^{w/T} - V_a]$, and noting that, because the potential energy is momentum independent, the momentum part of the one-body partition function is the same as in the ideal gas, we can write down the N-body partition function

$$Z_\mathrm{R} = \frac{1}{N!}\left[\frac{(2\pi m T)^{3/2} V'}{h^3}\right]^N. \tag{1.7.9}$$

Now again using Stirling's approximation and setting $(N-1)\delta(T) \approx N\delta(T)$ we can immediately write down the free energy

$$F_\mathrm{R} = -T\ln[Z_\mathrm{R}] = -NT\left[\ln\left[V + \frac{N\delta(T)}{2}\right] + \tfrac{3}{2}\ln[2\pi m T] - \ln[h^3] - (\ln[N] - 1)\right], \tag{1.7.10}$$

which, by expanding $\ln[V']$

$$\ln\left[V + \frac{N\delta(T)}{2}\right] = \ln[V] + \ln\left[1 + \frac{N\delta(T)}{2V}\right] \approx \ln[V] + \frac{N\delta(T)}{2V} \tag{1.7.11}$$

becomes

$$\begin{aligned} F_\mathrm{R} &= -NT\left[\ln[V] + \frac{N\delta(T)}{2V} + \frac{3}{2}\ln[2\pi m T] - \ln[h^3] - (\ln[N] - 1)\right] \\ &= F_\mathrm{I} - \frac{N^2 T \delta(T)}{2V}. \end{aligned} \tag{1.7.12}$$

Writing $\delta'(T) = \partial\delta(T)/\partial T$, the entropy is thus

$$S_{\mathrm{R}} = -\frac{\partial F_{\mathrm{R}}}{\partial T} = S_{\mathrm{I}} + \frac{N^2\delta(T)}{2V} + \frac{N^2 T\delta'(T)}{2V}$$

$$= N\Bigg(\ln[V] + \ln[2\pi m T]^{3/2} - \ln[h^3] - (\ln[N] - 1) + \frac{3}{2} + \frac{N\delta(T)}{2V} + \frac{NT\delta'(T)}{2V}\Bigg) \qquad (1.7.13)$$

and the internal energy ($F_{\mathrm{R}} + TS_{\mathrm{R}}$) is

$$U_{\mathrm{R}} = U_{\mathrm{I}} + \frac{N^2T^2\delta'(T)}{2V} = \frac{3NT}{2} + \frac{N^2T^2\delta'(T)}{2V} \qquad (1.7.14)$$

or

$$\frac{U_{\mathrm{R}}}{V} = \frac{3T}{2}\left[\frac{N}{V} + \frac{T\delta'(T)}{3}\left(\frac{N}{V}\right)^2\right]. \qquad (1.7.15)$$

It is also worth noting that, defining the particle density $\rho_{\mathrm{p}} = N/V$, the internal energy per unit volume $U_{\mathrm{R}}/V = U_{\mathrm{R}}\rho_{\mathrm{p}}/N$ so that (1.7.15) can be written

$$\frac{U_{\mathrm{R}}}{N} = \frac{3T}{2}\left[1 + \rho_{\mathrm{p}}\frac{T\delta'(T)}{3}\right]. \qquad (1.7.16)$$

Isentropic expansion (S_{R} = const.) is not as easy to understand in the present case since it depends on the details of the particle–particle interaction. On the other hand, the equation of state is modified by an additional term which goes as the square of the density N/V. We find

$$P_{\mathrm{R}} = -\frac{\partial F_{\mathrm{R}}}{\partial V} = -\frac{\partial F_{\mathrm{I}}}{\partial V} - \frac{N^2T\delta(T)}{2V^2} = T\left[\rho_{\mathrm{p}} - \frac{\delta(T)}{2}\rho_{\mathrm{p}}^2\right], \qquad (1.7.17)$$

which is usually written in terms of the specific volume, $v = 1/\rho_{\mathrm{p}} = V/N$, as

$$\left[P_{\mathrm{R}} + \frac{T\delta(T)}{2v^2}\right]v = T. \qquad (1.7.18)$$

Notice that both equations (1.7.15) and (1.7.17) can be thought of as developments to second order in the particle density. Indeed the similarity between these equations (we remind the reader that $\delta'(T)$ is a negative quantity) has prompted many authors to use (1.7.15) or (1.7.16) as 'equations of state'. From (1.7.17) and (1.7.2) we learn that the pressure of a real gas can be written in terms of the ideal gas pressure ($P_{\mathrm{I}} = \rho_{\mathrm{p}}T$), i.e.

$$\frac{P_{\mathrm{R}}}{P_{\mathrm{I}}} = \left[1 - \frac{\delta(T)}{2}\rho_{\mathrm{p}}\right]. \qquad (1.7.19)$$

Now comes the bad news. It is impossible to obtain a liquid–gas phase transition using our simple theory. This is obvious if we plot the pressure versus the density for various values of the temperature (isotherms). Equation (1.7.17) is quadratic in the density whereas, experimentally, a typical (van der Waals) real gas displays a cubic dependence. The reason has to do with the fact that, at high densities, the particles will tend to be more and more closely packed together so that correlated configurations in which clusters of three or more particles are formed become important. Our simple theory takes no account of this. However, one may think that the pressure would increase rapidly when the particle density, N/V, tends towards the close packing density $\rho_{\mathrm{p}}^{(0)}$ since the particles would tend to 'lock in' together. In the absence of a solid (liquid?) theoretical analysis we may therefore try to correct the problem by inserting an extra term in the equation of state. We thus 'adjust' (1.7.18) to ensure that as the specific volume $1/\rho_{\mathrm{p}} = v \to v_0$, the pressure becomes infinite (it is easily seen that $v_0 \approx V_b$). We then obtain

$$\left[P_{\mathrm{R}} + \frac{T\delta(T)}{2v^2}\right](v - V_b) = T, \tag{1.7.20}$$

which in terms of the particle density is written (with $\rho_{\mathrm{p}}^{(0)} = 1/V_b$)

$$P_{\mathrm{R}} = T\left(\frac{\rho_{\mathrm{p}}^{(0)}\rho_{\mathrm{p}}}{\rho_{\mathrm{p}}^{(0)} - \rho_{\mathrm{p}}} - \frac{\delta(T)\rho_{\mathrm{p}}^2}{2}\right). \tag{1.7.21}$$

Equations (1.7.20) and (1.7.21) are forms of the famous van der Waals equation of state (1873). Typical isotherms (equations of state in the P–V plane for several fixed values of the temperature) are shown in figure 1.8. The lightly shaded area defines the region of phase coexistence (e.g. a liquid–gas mixture). This region includes a spinodal region, i.e., a range of values of the volume for which $\mathrm{d}P/\mathrm{d}V > 0$ thus violating the intuitively obvious requirement that increase in volume leads to a decrease in pressure. This requirement, which is more usefully expressed by stating that the (isothermal) free energy should be a convex function of the volume ($\partial P/\partial V = -\partial^2 F/\partial V^2$), is, in fact, a necessary condition for the existence of the thermodynamic limit [5]. A remedy for the non-physical behaviour which was proposed by Maxwell (Maxwell construction) is to replace the isotherm in the region of the phase change by a horizontal line at constant pressure, P_{M}, such that the integral $\int_{V_1}^{V_2}(P - P_{\mathrm{M}})\,\mathrm{d}V$ vanishes (figure 1.8). The resulting loops above and below this line represent supercooled or supersaturated states which cannot be explained in terms of equilibrium statistical mechanics but must, all the same, be physically meaningful insofar as they are observed experimentally. With the Maxwell construction the isothermal compressibility, $\kappa_T = -\frac{1}{V}\frac{\mathrm{d}V}{\mathrm{d}P}$, diverges at the pressure which characterizes the phase change. In physical terms, change of phase (from a gas to a liquid) is associated with the close-packing of molecules which abruptly leads to severe restrictions on both position and momentum coordinates.

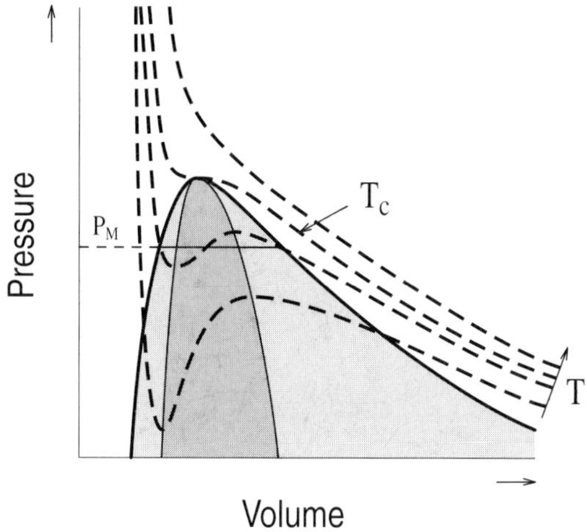

Figure 1.8. Equation of state for a typical van der Waals fluid represented as a set of isotherms in the P–V plane. In this representation, the Maxwell construction is a horizontal line $P = P_M$ (shown on one isotherm). The light shaded area enclosed by the phase boundary indicates the region of phase coexistence. Within this region the darker shaded area indicates the spinodal region where the van der Waals isotherms violate the stability criterion $dP/dV \leq 0$. Above the critical isotherm ($T = T_C$) the phase boundary disappears together with the distinction between liquid and gas phases.

As the temperature is increased, the three intersections of the cubic equation of state with the Maxwell construction merge and finally become identical at the so-called 'critical' temperature. For higher temperatures there is no real solution and, consequently, no boundary between the gas and the liquid phases so that the distinction gas–liquid loses meaning. We shall see later that this idea is employed in one of the models of nuclear multifragmentation in which fragments are considered as heated liquid drops.

In the nuclear case the classical treatment presented above, while providing an introduction to the behaviour of real fluids, is not entirely satisfactory. For one thing, one should somehow incorporate the long range Coulomb repulsion between protons. Secondly, our presentation takes no account of the fact that nucleons are fermions for which the classical gas is an approximation which is useful only at high excitation energies. Finally, it may be difficult to justify the use of a fixed temperature. Despite these difficulties, it is fair to say that a nucleon gas displays many of the features of a classical fluid. A calculation [13] of an equation of state for infinite nuclear matter is displayed in figure 1.9 in which is plotted the pressure (in MeV fm^{-3}) versus the nucleon *density* (rather than volume) for different temperatures. The curves are qualitatively similar to those characteristic of the van der Waals fluid. For $T = 0$ MeV the density at the mechanical equilibrium point (pressure $= 0$) is similar to that found in naturally

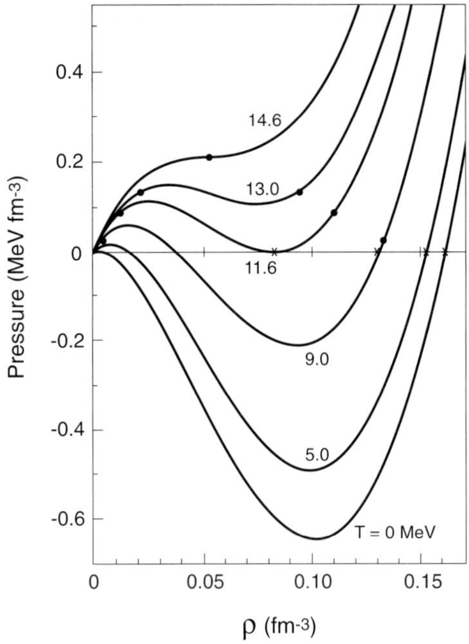

Figure 1.9. Equation of state (plotted as pressure versus nucleon density) at various temperatures (expressed in MeV) for nuclear matter taken from [13]. On each isotherm, the black dots indicate the constant pressure corresponding to the phase transition. At the left-hand point the state is a vapour and at the right-hand point, a liquid. The line joining the two dots on any isotherm is equivalent to the Maxwell construction in the P–V plane. Its length diminishes as the temperature is increased and the line collapses to a single point at the critical temperature, T_C (14.6 MeV). The crosses indicate the equilibrium densities (pressure $= 0$) for the various isotherms. For temperatures above 11.6 MeV, the pressure is positive for all densities. (Reprinted from Brack M, Guet C and Hakansson H-B 1985 *Phys. Rep.* **123** 275, with permission from Elsevier Science.)

occurring ground state nuclei (0.16–0.17 nucleons/fm^3).

The calculation represented in figure 1.9 is probably not the presentation best adapted to the nuclear problem. In finite nuclei there is no possibilty of heat exchange with the 'surroundings' so that the expansion of a highly excited nucleus is adiabatic, and thus takes place at constant entropy rather than at constant temperature (see after (1.5.17)). The (spinodal) region of mechanical instability ($\mathrm{d}P/\mathrm{d}\rho < 0$) which can be taken to indicate the phase change or, at least, a breakup density may be used as the starting point for both statistical and dynamical calculations of multifragmentation.

Identifying and characterizing changes of phase is no easy matter and the difficulty is naturally increased in short-lived nuclear systems. A general feature of phase change in fluids is that they are signalled by some 'non-smooth' behaviour in the variation of the free energy or one of its derivatives (pressure, entropy) as a function of density or temperature (this is evidently true for the van der Waals gas). In isolated systems we may look for non-smooth behaviour of the entropy (or directly the number of microstates) as a function of the total energy of the system. The search for such behaviour lies at the base of the investigation of so-called 'caloric curves' which will be discussed in chapter 7 in connection with multifragmentation.

1.8 Pseudo-equilibrium

The principle, which underlies statistical mechanics and which was discussed at the beginning of this chapter, is that all microstates are *equi-probable*. That is all well and good but, given the huge number of microstates which characterize most problems, one can never make an experimental test of this hypothesis. In many applications (e.g. nuclear physics) the best that can be done is to base the criterion for equilibrium on the notion of a random sample. Thus, if some subset of microstates can be considered to represent a random sample of the full set, we assert that the consequent distribution of macrostate probabilities should correspond to the equilibrium distribution calculated by some suitable model.

The converse of this statement is not necessarily true. Thus, consider the following simple simulation. We define a microscopic space of positive integers between 1 and, say, 1000. We also define a macroscopic variable on this space. The simplest case would be to define, say, ten 'macrostates', the first corresponding to numbers between 1 and 100, the second to numbers between 101 and 200, and so on. The ten probabilities corresponding to each of the ten macrostates may be reasonably established (to within one per cent) in a simulation which generates 10^5 events. Let us now suppose that, by divine intervention, all even numbers are suppressed. This is equivalent to destroying the equilibrium character of the collection of microstates. Nevertheless, a new simulation would yield, to within statistical errors, the same results for the estimates of the macroscopic probabilities.

As a second elementary example, the reader might consider a simulation in which a random selection of 20 microstates (the 20 faces of a special dice) represented by the numbers 0.5, 1.5, 2.5 ... is made for various numbers of Monte Carlo trials. Two simulations are carried out: one in which all states are equally accessible and one in which one of the microstates (the number 3.5) is rendered inaccessible. Of course the direct examination of the microstate population would reveal the defect in the second case immediately. However, let us suppose that our experiment measures only the mean and the standard deviation of the microstate population. Figure 1.10, which exhibits results of an actual simulation, indicates that the difference in mean values shows up only after about 10^4 trials (the expected value for a strictly uniform population is 10.0). However, even after 10^4 trials the standard deviations of the full and truncated populations show no stable

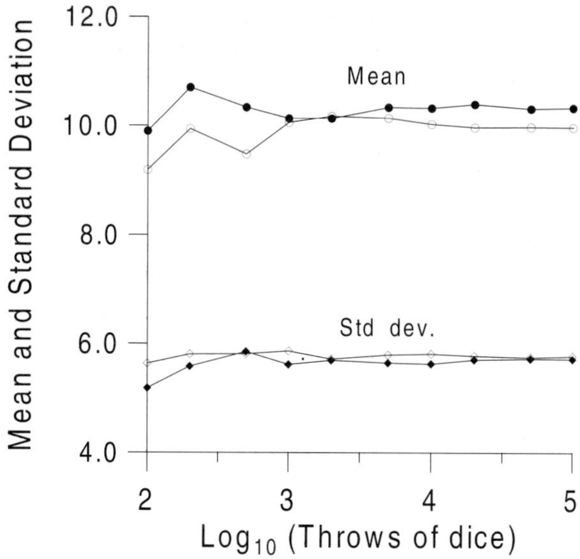

Figure 1.10. Mean values and standard deviations of the numbers (set of microstates) turning up on a 20-sided dice (see text) plotted as a function of the number of throws. The open circles and diamonds refer to the untruncated population whereas the data shown as filled circles or diamonds correspond to the suppression of the outcome represented by the number 3.5.

difference (the asymptotic difference in standard deviations is about one per cent). The example is almost trivial but illustrates quite well that 'defects' in microstate populations, may go undetected if the experiment or simulation does not provide sufficient statistical accuracy (which on the scale of the number of microstates is almost always the case).

There are two consequences which are illustrated by our two examples. The first is that a non-equilibrium situation created by a bias of the microstate population which is *uncorrelated* with the chosen set of macroscopic variables is undetectable. It is just as if we discard randomly selected microstates. The second is that a bias of the microstate population which is *correlated* with the chosen set of macrostates may go undetected if the recorded number of events is too small.

These two observations are most important for reactions in nuclear physics. Let us suppose that we generate a set of events, one by one, by direct *dynamical* simulation of the collision between projectile and target nuclei. Furthermore, let us define microstates of the system by specifying the microscopic configuration at some particular instant in time. This time should be chosen in such a way that the microscopic configurations so-defined correspond one to one with measurable macrostates (i.e., that subsequent time evolution does not induce a one-to-many correspondence between microstates and macrostates). The set of *attainable* microstates in the simulation is restricted by initial conditions but also by the dynamical evolution of the collision which is governed by the physical characteristics of the interaction between projectile and target. Then one may certainly suspect that the microstate population is biased. The question is, with respect to the chosen set of macroscopic variables, how many events are required to establish the existence of a microscopic bias. If this number is far superior to the number of generated (or measured) events, the results of the simulation (or experiment) will appear to be consistent with a uniform microstate population. Then, even though there is no notion of equilibrium as a macroscopic state which is unchanging in time and no certitude concerning 'true' statistical equilibrium, the observed weights for the resulting macrostates will appear (to within statistical errors) to be identical to those which would be observed if all possible microstates were included in their construction.

This perhaps suprising result may be taken as the definition of pseudo-equilibrium [14].

1.9 Statistical models of nuclear decay

We conclude this chapter with an outline of the application of models based on equilibrium statistical mechanics to decay of excited atomic nuclei. We remind the reader that equilibrium is characterized by a unique probability which may be assigned to all attainable microstates. Constraints which determine the set of microstates to be included are obviously provided by conserved quantities such as energy and angular momentum but also by external parameters such as volume.

We take the term 'statistical model' to encompass not only the model structure but also the various ingredients which form the essential input. Confrontation with experiment usually leads to refinements either of the structure or of the physical input. Alternatively, one could consider that differences of predictions with experimental observations are due to defects of the particular model employed and may even call into question the validity of the assumption of equilibrium. However, the fact that statistical models have been largely successful in account ing for measured cross-sections has tended to favourize adaptation of models to fit the data.

Early experiments studied the decay of compound nuclei formed by the complete fusion of projectile and target nuclei. In this situation, the compound nucleus mass, charge and excitation energy can be considered to be known (the latter to within the projectile beam energy resolution). The angular momentum is not known but a probability distribution for this quantity can be modelled quite successfully using theories for fusion.

At low excitation energy and angular momentum only one non-excited light particle (essentially n, p, α) is emitted. Experiments can thus be used to test the microcanonical density of states appropriate for a two-component system composed of the phase space associated with the motion of the emitted particle and the residual nucleus (the compound nucleus minus the emitted particle), and the internal state density of the residual nucleus. As will be seen in chapter 3, this is essentially equivalent to the conception advocated by theorists and put to the test by experimentalists. If the assumed decay mechanism is correct, the simplest observable which can be considered as a crucial test of the model is the relative probability for emission of a given particle. More detailed comparisons involve the emitted particle energy spectra which are mainly determined by the energy dependence of microstate densities associated with the various residual nuclei. The angular distribution of emitted particles is also interesting because it is strongly influenced by the relative orbital angular momentum of the emitted particles and residual nucleus.

At higher excitation energies two or more particles can be emitted sequentially. Experimentally, the situation becomes a little confused because the order of emission in any given event cannot usually be determined and indeed several different sequences involving the same emitted particles may contribute significantly. The measured information in any event is limited to the number and type of emitted particles and the corresponding spectrum of kinetic energies and emission angles for each type of detected particle and for the final residual nucleus. Such information may be gathered by simultaneous measurement of two or more reaction products (exclusive measurements), in which case we obtain 'correlated' spectra, or by measurement of single products (inclusive measurement). In this latter case the correlations are unobserved.

Contrasting with experiment, multistep sequential emission presents no difficulty, in principle, for statistical models. We have to deal with a succession of two component systems rather than the few systems appropriate for the emission of a single particle. If the population of microstates in the residual nucleus after emission of one or more light particles can be considered to be consistent with the equilibrium picture, the standard formalism can be applied to its decay. Thus, multistep emission processes can be modelled with no extension of the basic formalism. However, the 'book-keeping' becomes a little complicated. The solution usually adopted consists in the construction of (Monte Carlo) event generators based on the theoretical statistical models. These implementations have the advantage of automatically taking into account the convolutions and correlations which are inherent to multistep processes.

With increasing excitation energy (of the order of 2–3 MeV/nucleon) decay channels involving the emission of heavier nuclei (i.e. $Z > 2$) enter into competition with the emission of $Z \leq 2$ particles. In this situation the emitted nuclei may themselves be excited and may decay further so that the distinction between emitted particle and residual nucleus is no longer useful. This process is referred to as sequential binary decay and is characterized by a rather complex topology. In many cases, one has to deal with competition between sequential binary and multistep light particle emission processes. In this domain, natural extensions of statistical models applied at lower energies which are quite straightforward in principle, are computationally cumbersome so that, rightly or wrongly, considerable recourse to time saving approximations has been made.

At still higher excitation energies (about 5 MeV/nucleon), the decay process is considered as a prompt disintegration of the parent nucleus in which several light particles and heavier fragments are produced simultaneously. One then refers to 'multifragmentation'. Models for this process are distinct from models for sequential decay because of the many body character of the microstates. If the parent nucleus is partitioned into N fragments, one can imagine a model involving the N fragment phase space and the internal state densities in the fragments. Current statistical models of multifragmentation, which have been successfully applied to the analysis of experimental data, are based on the evaluation of the corresponding density of states in the microcanonical ensemble. Macrostates are usually defined as the partition of the parent nucleus (as in percolation) together with the associated vector momenta of each fragment in the partition.

Of course, for very high excitation energies, heavy fragments are produced more and more in unstable excited states and thus decay. One is left with a large number of neutrons, hydrogen and helium nuclei. The difficulty then is to decide whether the process should be modelled as a prompt 'vaporization' or as a multifragmentation process in which the heavier products decay outside the spatial region in which they were produced. Vaporization is the subject of much on-going investigation and one can expect new results even during the course of production of this book. We will try to present the current situation in chapter 7.

This concludes the discussion of equilibrium statistical mechanics as applied to small systems. In the following chapter we shall present what might be termed as essential background nuclear physics. Chapter 2 contains little reference to statistical mechanics. In contrast, in chapter 3, we will explore the early historical development of our understanding of statistical decay of excited nuclei.

References

[1] Ma S K 1985 *Introduction to Statistical Mechanics* (Philadelphia: World Scientific)
[2] Feynman R P 1972 *Statistical Mechanics* (New York: Addison-Wesley)
[3] Greiner W, Neise L and Stöcker H *Thermodynamics and Statistical Mechanics* (Berlin: Springer)
[4] Balian R 1991 *From Microphysics to Macrophysics* (Berlin: Springer)
[5] Thompson C J 1972 *Mathematical Statistical Mechanics* (Princeton: Princeton University Press)
[6] Callen H B 1985 *Thermodynamics and an Introduction to Thermostatistics* (New York: Wiley)
[7] Klimontovich Y L 1986 *Statistical Physics* (Chur, Switzterland: Harwood)
[8] Vassiliev A 1985 *Introduction à la Physique Statistique* (Moscow: Editions Mir)
[9] Baierlin R 1971 *Atoms and Information Theory* (San Francisco: W H Freeman)
[10] Cohen E R 1996 *The Physics Quick Reference Guide* (Woodbury, New York: AIP)
[11] Messiah A 1961 *Quantum Mechanics* (Amsterdam: North Holland) Ch IX
[12] Carbonell J, Gignoux C and Mekuriev S P 1993 *Few Body Systems* **15** 152; Malfliet R A and Tjon J A 1969 *Nucl. Phys.* A **127** 161

[13] Brack M, Guet C and Hakansson H-B 1985 *Phys. Rep.* **123** 275
[14] Cole A J, Chabane A, Charvet M, Désesquelles P, Giorni A, Heuer D, Lleres A and Viano J B 1997 *J. Phys. G: Nucl. Part. Phys.* **23** 457

Chapter 2

Nuclear physics background

"The time has come," the Walrus said,
"To talk of many things;"

The Walrus and the Carpenter, Lewis Caroll

2.1 Introduction

It is appropriate, at this stage, to include a short revision of those elements of nuclear reaction theory and nuclear properties which are indispensable for understanding the formation and decay of excited nuclei. This chapter has therefore been written with the aim of facilitating the study of the ensuing historical account and, indeed, together with chapter 1, of the rest of this book.

Nuclear reaction theory has been presented in many works. Two examples are the classical text of Messiah [1] and a recent book by Fröbrich and Lipperheide [2]. The reader is referred to these works for a detailed treatment of this subject. Other topics dealt with in this chapter are typical of a basic nuclear physics course. An exception is the discussion of nuclear state densities, the latter part of which was largely inspired by the work of Ericson [3].

Elements of nuclear reaction theory are presented in the following section. Special attention is paid to the notion of cross-section. The succeeding section, 2.3, deals with elastic scattering theory which is essential for discussion of transmission coefficients. These quantities are the basic ingredients of the Hauser–Feshbach evaporation theory. A description of radioactive decay which seems to be absent in many elementary texts is provided in section 2.4, and in section 2.5 we discuss two complementary approaches used to derive microstate densities in excited nuclei. Finally, in section 2.6, we give an introduction to quantum mechanical basis states for angular momentum and to the Clebsch–Gordon (or Wigner) coupling coefficients which are used to transform the basis states for coupled angular momenta.

2.2 Elements of the theory of nuclear reactions

In discussing nuclear reactions it is useful to use the notion of a channel. The entrance channel in a reaction almost always involves two bodies (projectile and target) and, in such cases, is defined by the projectile and target mass and charge numbers, the projectile energy in the laboratory reference system (the target is usually stationary in this frame), and the relative orbital angular momentum. The intrinsic angular momentum (spin) of the participants and possible preferred orientations for these spins (polarized beams or targets) may also be specified, although this latter possibility will not be considered herein. An exit channel usually involves two or more bodies and can be specified in a similar manner. Exit channels may also include specification of the directions of reaction products. In some studies it is necessary to consider multifragment exit channels which require an extension of the two-body formalism.

In this text we will normally use the symbols 'α' and 'β' to refer respectively to entrance and exit channels. However, for two-body channels we will often employ labels to distinguish between projectile (a) and target (A) nuclei in the entrance channel, and a light-emitted particle (b) and heavy residual nucleus (B) in the exit channel. Thus we will speak of a reaction $a + A \rightarrow b + B$ or, if the reaction proceeds via the formation of an intermediate 'compound' nucleus C, of $a + A \rightarrow C \rightarrow b + B$. The advantage of this simple notation is partially offset by the fact that nuclear physicists are in the habit of referring to the mass number, which is an integer equal to the total number of neutrons and protons, using the symbol 'A' and in particular to the projectile and target mass numbers as $A_{\rm p}$ and A_t. This is rarely a source of confusion. However, it is essential to distinguish between the mass number of a nucleus, the nuclear mass and the atomic mass which is the mass of the corresponding neutral atom (conventionally denoted by the symbol 'M' so that, for example, $M_{\rm p}$ and $M_{\rm t}$ represent, respectively, projectile and target masses). We will have more to say concerning the latter distinction in a moment. The need for a convention does not occur with nuclear charge because the charges are integral multiples of the electronic charge. One normally uses the symbol 'Z' to refer to charge numbers. Thus, the positive charge of the nucleus ^{12}C ($Z = 6$) is $6e$ where the basic unit of charge, $e = 1.602177 \times 10^{-19}$ Coulombs. We should also mention that a further expedient which is sometimes used involves labelling exit channel variables with primes.

In the absence of external forces acting on the projectile and target the total momentum, $\boldsymbol{P}$, which is the momentum associated with the motion of the centre of mass, is conserved during the projectile target interaction (see discussion preceding equation (1.2.31)). It is then convenient to describe the system in terms of the coordinates of the centre-of-mass and the relative coordinates of the projectile and target. The motion of the centre-of-mass can be eliminated by making a velocity transformation (see equations (1.2.20)–(1.2.25) which leads to the so-called centre-of-mass system of coordinates (more generally, the centre-of-mass system for any number of particles is defined by the condition that the total momentum, $\boldsymbol{P}$, vanishes). In the present situation, we can specify the system of coordinates so that the z-axis coincides with the incident beam direction. Then $\boldsymbol{P}$ has only one non-zero Cartesian component. Denoting by $v_{\rm p}$ the projectile velocity we first define the velocity, v, in the direction opposed to $v_{\rm p}$, such that the transformed projectile and target momenta are equal and opposite

$$M_{\rm p}(v_{\rm p} - v) = M_t v, \tag{2.2.1}$$

which produces

$$v = \frac{M_{\rm p}}{M_{\rm p} + M_t} v_{\rm p}. \tag{2.2.2}$$

The velocity of centre-of-mass system, $V_{\rm CM}$, is then defined as that velocity which produces the inverse transformation. It therefore has the magnitude v but is directed along the positive z-axis. The kinetic energy associated with the movement of the centre of mass is

$$E_{\rm CM} = \frac{1}{2}(M_{\rm p} + M_t)V_{\rm CM}^2 = \frac{M_{\rm p}}{M_{\rm p} + M_t}\frac{1}{2}M_{\rm p}v_{\rm p}^2 = \frac{M_{\rm p}}{M_{\rm p} + M_t}E_{\rm lab}, \tag{2.2.3}$$

where $E_{\rm lab}$ is the kinetic energy of the projectile in the laboratory system. If we label the energy *in* the centre-of-mass system (sometimes called the relative energy or the available energy) as $E_{\rm cm}$ then the sum $E_{\rm CM} + E_{\rm cm}$ is obviously equal to $E_{\rm lab}$ so that

$$E_{\rm cm} = \frac{M_t}{M_{\rm p} + M_t}E_{\rm lab} = \frac{1}{2}\mu v_{\rm cm}^2, \tag{2.2.4}$$

where the reduced mass

$$\mu = \frac{M_{\rm p}M_t}{M_{\rm p} + M_t} \tag{2.2.5}$$

and the relative velocity in the centre of mass system, $v_{\rm cm}$, is equal to $v_{\rm p}$ (not affected by a Galilean transformation). If the projectile–target interaction at a distance, r, can be characterized by a potential energy $V(r)$ then the kinetic energy in the centre of mass is $E_{\rm cm} - V(r)$.

The (classical) relative angular momentum in the centre of mass is proportional to the impact parameter b which is the distance of closest approach between projectile and target mass centres on a straight line trajectory (see ahead to figure 2.3) . The magnitude of the relative orbital angular momentum is $L = \mu v_{\rm cm} b$ (in terms of equations (1.2.23) and (1.2.24) $\boldsymbol{L} = \boldsymbol{r} \wedge \boldsymbol{p}$) and is quantized in units of $\hbar$. Its expression in terms of the continuous variable b is therefore referred to as a 'semi-classical' approximation. We write

$$\ell = \frac{L}{\hbar} = kb, \tag{2.2.6}$$

($\hbar c = 197.327$ MeV fm) where ℓ is rounded off to the nearest integer and the wavenumber, k, which is the centre-of-mass momentum in units of $\hbar$, is

$$k = \frac{\mu v_{\mathrm{cm}}}{\hbar} = \sqrt{\frac{2\mu E_{\mathrm{cm}}}{\hbar^2}}. \tag{2.2.7}$$

In reaction theory, the fundamental notion is that of a cross-section. We can introduce this concept most simply from the experimental point of view. Consider a beam of $N_{\mathrm{p}}^{\mathrm{in}}$ projectiles per second incident on a foil target and let the number of target particles per unit volume be n_{v} so that the number of target particles per unit area for a 'slice' of the target of infinitesimal thickness $\mathrm{d}z$ is $n_{\mathrm{v}}\,\mathrm{d}z$ (figure 2.1). At some point inside the target, at distance z from the front face ($z = 0$), let the number of incident beam particles per second be $N_{\mathrm{p}}(z)$. The rate of loss of beam particles in $\mathrm{d}z$ due to projectile target interactions, $-\mathrm{d}N_{\mathrm{p}}$, is obviously proportional to the product $N_{\mathrm{p}}n_{\mathrm{v}}\,\mathrm{d}z$. The interaction cross-section, σ, is simply the constant of proportionality. That is,

$$-\mathrm{d}N_{\mathrm{p}} = \sigma N_{\mathrm{p}} n_{\mathrm{v}}\,\mathrm{d}z. \tag{2.2.8}$$

If the full target thickness is t, then integration of (2.2.8) yields

$$N_{\mathrm{p}}^{\mathrm{out}} = N_{\mathrm{p}}^{\mathrm{in}}\,\mathrm{e}^{-\sigma n_{\mathrm{v}} t}, \tag{2.2.9}$$

where $N_{\mathrm{p}}^{\mathrm{out}}$ is the number of beam particles per second which leave the target without having interacted. The interaction rate, W, or number of interactions per second is $N_{\mathrm{p}}^{\mathrm{in}}$–$N_{\mathrm{p}}^{\mathrm{out}}$ which for small target thicknesses reduces to

$$W = \sigma N_{\mathrm{p}}^{\mathrm{in}} n_{\mathrm{v}} t. \tag{2.2.10}$$

Incidentally, fixed foil targets are conveniently specified by their thickness expressed in mg cm^{-2} which we write as t_{m}. In this case the number of target atoms per cm^2 is obtained as

$$n_{\mathrm{v}} t = \frac{N_{\mathrm{A}} t_{\mathrm{m}} \times 10^{-3}}{M_{\mathrm{w}}}, \tag{2.2.11}$$

where M_{w} is the molecular weight (atomic weight for a monatomic substance) and N_{A} is Avogadro's number (6.0221367×10^{23} mol^{-1}) which is the number of molecules (atoms) per mole (a mole is a quantity of a substance whose weight in grams is equal to the molecular (atomic) weight, e.g., 1 mole of carbon contains 12 g).

The interaction cross-section may be subdivided in many ways. Thus, for example, the total interaction cross-section σ_{T} (T for total) can be separated into the sum of cross-sections for elastic scattering and for reactions i.e.,

$$\sigma_{\mathrm{T}} = \sigma_{\mathrm{EL}} + \sigma_{\mathrm{R}}. \tag{2.2.12}$$

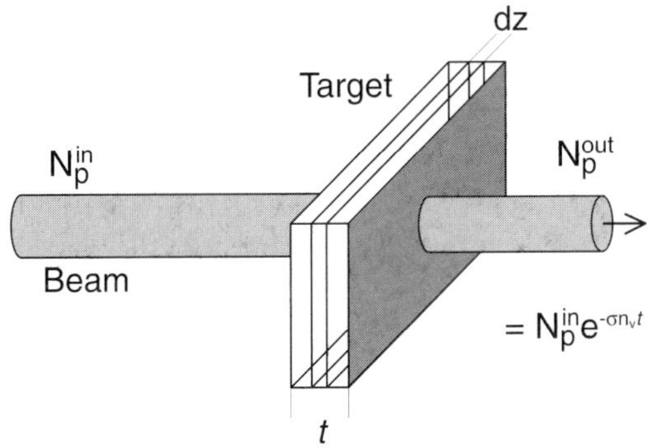

Figure 2.1. Attenuation of a beam of particles passing through a target foil of thickness t. The decrease in intensity is exponential but for thin targets may be considered as linear (i.e., for small x, $e^{-x} \approx 1 - x$).

In elastic scattering the projectile and target nuclei retain their respective identities and are (almost always) both to be found in their lowest energy (ground) states in both entrance and exit channels. Non-elastic channels involve excitation of the reaction partners and possibly transformation of the original nuclei (e.g., by transfer of a nucleon). If we were concerned only with elastic scattering we would use σ_{EL} in (2.2.10).

We may also be interested in more detail, such as the number of particles which are emitted into a solid angle $d\Omega$ at polar angles (θ, ϕ) (see figure 2.2). Then we could write (2.2.10) as

$$W(d\Omega) = N_p^{in} n_v t \frac{d\sigma}{d\Omega} d\Omega. \tag{2.2.13}$$

For obvious reasons, quantities such as $d\sigma/d\Omega$ are referred to as differential cross-sections and the generic term 'cross-section' refers to the interaction products under specified exit channel conditions (type of interaction, solid angle etc).

We can give further meaning to the notion of cross-section which will be useful when we come to the quantum mechanical treatment of scattering. We state that any cross-section may be identified with an interaction rate, per target nucleus, per unit current density of the incoming beam. The beam particles, with velocity v_p and spatial density ρ_p, can be represented by a current density $\rho_p v_p$ which is the number of particles crossing unit area per second. If the beam is uniformly distributed over an area S then $N_p^{in} = S\rho_p v_p$ so that the incident current density is N_p^{in}/S. The rate of interaction corresponding to a cross-section σ is, from equation (2.2.10), $N_p^{in} n_v \sigma t = [\rho_p v_p][n_v S t]\sigma$. The two square bracketed terms represent, respectively, the incident beam current density and the number of target nuclei in the volume swept out by the beam, so indeed we can define σ

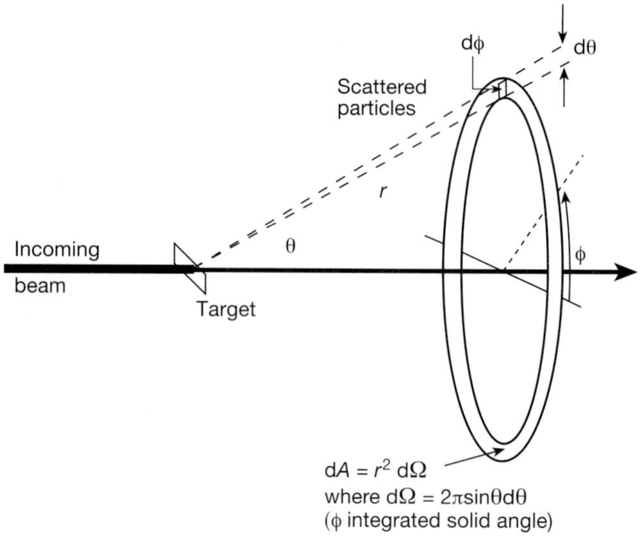

Figure 2.2. Differential cross-section. The figure defines both the double differential cross-section (proportional to the flux of particles produced in the exit channel solid angle, $d\Omega = \sin\theta \, d\theta \, d\phi$ defined by the intervals $d\theta$ and $d\phi$) and the ϕ integrated differential cross-section.

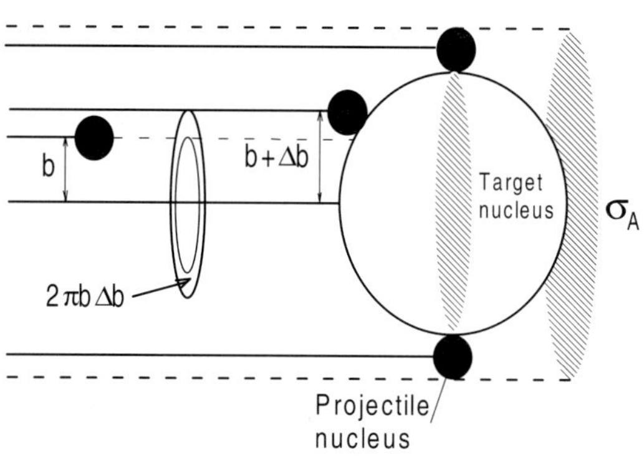

Figure 2.3. Cross-section viewed as a disc obtained by integrating over impact parameter. The area between the two circles on the left of the figure represents the cross-section for impact parameters in the interval $(b, b+\Delta b)$. In most cases interest focusses on a particular kind of reaction whose probability, $P(b)$ can be expressed as a function of b. In this case a 'semi-classical' expression for the differential cross-section is $\frac{d\sigma}{d\Omega} = 2\pi b P(b) \frac{db}{d\Omega}$.

as an observed interaction rate per unit incoming current density and per target nucleus.

A word of warning is appropriate here. Some works confuse the words 'flux' and 'current density'. The flux of a current is the current density integrated over a specified surface. Thus for a constant current density $\boldsymbol{j}$ flowing through a (vector) surface element $\mathbf{d}\boldsymbol{S}$ the flux of the current through $\mathbf{d}\boldsymbol{S}$ is the scalar product $\boldsymbol{j} \cdot \mathbf{d}\boldsymbol{S}$. For a unit surface area perpendicular to the current the flux is numerically equal to the current density. Nevertheless, it is better not to confuse the two quantities.

Note that the usefulness of the notion of cross-section is entirely due to the particular beam target geometry (incident particles move on parallel trajectories). Thus, for example, for a 'beam' produced by an isotropic radioactive source inside a box formed of the target material, the calculation of the interaction rate would, in principle, involve 'interaction solid angles' rather than cross-sections.

Another comment which will be useful later on concerns multiple reaction products. It is not uncommon that a single interaction produces several outgoing particles. In this case it is natural to retain the notion of cross-section in terms of the projectile target interaction rate and simply to use the multiplicity (number) of outgoing particles per interaction to rescale the observed inclusive cross-section.

In more tangible physical terms the cross-section represents a plane surface area corresponding to the interaction of projectile and target (figure 2.3). To see this we represent the reaction rate as $N_{\mathrm{p}}^{\mathrm{in}} n_{\mathrm{V}} \sigma t = N_{\mathrm{p}}^{\mathrm{in}} (n_{\mathrm{V}} S t)\ \sigma / S$ in which the second factor is again the number of target nuclei in the path of the beam and the last factor represents the fractional area 'obscured' by one projectile–target interaction. A typical order of magnitude for a cross-section is thus given by the square of the nuclear size, i.e., 100 fm^2 $= 10^{-28}$ m^2 which is referred to as a barn. Differential cross-sections are usually expressed in barns or millibarns per steradian (b sr^{-1} or mb sr^{-1}).

It is useful to classify the type of interaction (reaction mechanism) according to the impact parameter, b. For large values of b the interaction is due to the Coulomb force between the projectile and target. This long range repulsive force produces essentially elastic scattering. As the impact parameter is reduced, the possibility of absorption from the elastic channel due to short range nuclear forces becomes increasingly important. If we admit that elastic scattering is suppressed and that absorption occurs for impact parameters less than some sharp cut off value b_{A} ('A' for 'absorption') then the total absorption cross section is simply

$$\sigma_{\mathrm{A}} = \int_0^{b_{\mathrm{A}}} 2\pi b \, \mathrm{d}b = \pi b_{\mathrm{A}}^2. \tag{2.2.14}$$

We can obtain the equivalent expression in terms of the quantized angular momentum by associating successive contributions to the absorption cross-section with successive ℓ values (recall from (2.2.6) that ℓ is the angular momentum measured in units of $\hbar$). Thus we define the cross-section for a given value of ℓ as the area of the surface between circles with radii equal to impact parameters $b = \ell / k$ and

$b = (\ell + 1)/k$ which is equal to

$$\frac{\pi}{k^2}[(\ell + 1)^2 - \ell^2] = \frac{(2\ell + 1)\pi}{k^2}.$$

The sum over angular momentum which corresponds to (2.2.14) is then

$$\sigma_{\mathrm{A}} = \frac{\pi}{k^2}\sum_{0}^{\ell_{\mathrm{A}}}(2\ell + 1) = \frac{\pi}{k^2}(\ell_{\mathrm{A}} + 1)^2. \qquad (2.2.15)$$

For large ℓ_{A} we find, with our convention, that $\ell_A = kb_{\mathrm{A}} - 1 \approx kb_{\mathrm{A}}$.

Although the sharp cut-off model is quite useful in practice, the quantal theory of scattering (see following section) in the presence of a short range projectile–target interaction takes into account the fact that, for some values of ℓ, there is a finite probability for both elastic scattering and for absorption from the elastic channel. One can then write, instead of equation (2.2.15),

$$\sigma_{\mathrm{A}} = \frac{\pi}{k^2}\sum_{0}^{\infty}(2\ell + 1)T_\ell, \qquad (2.2.16)$$

where the transmission coefficient T_ℓ can be identified with the probability that a given partial wave (ℓ-wave) is absorbed. A difficulty which occurs at low bombarding energies (less than ~10 MeV for nucleons), and which will be discussed more fully in chapter 3, arises from the fact that some portion of the absorption cross-section may lead to re-emission of the projectile nucleus in the elastic channel. At higher energies this so-called 'compound elastic' process can be neglected so that σ_{A} can be set equal to the reaction cross-section, σ_{R}. We could then replace the subscript 'A' with the subscript 'R' in equations (2.2.14)–(2.2.16). However, even in this case, the formulation expressed by (2.2.16) is still inadequate to deal with many situations of practical interest for which the simple division into elastic and reaction channels is not very useful. For example, if we wish to consider only reactions in which the projectile and target fuse into one composite (compound) nucleus we must again modify (2.2.16) and identify the fusion cross-section as

$$\sigma_F = \frac{\pi}{k^2}\sum_{0}^{\infty}(2\ell + 1)T_\ell F_\ell, \qquad (2.2.17)$$

where the partial fusion probability F_ℓ is the probability that, for a given value of ℓ, absorption from the elastic channel leads to fusion of the projectile and target nuclei.

If fusion does take place, the energy *in* the centre-of-mass system appears as excitation energy in the fused (compound) system which moves in the laboratory frame with the kinetic energy *of* the centre of mass (equation (2.2.3)). The total excitation energy in the compound system is given by

$$E^* = E_{\mathrm{cm}} + Q_{\mathrm{fus}}, \qquad (2.2.18)$$

where the fusion Q-value is defined as the energy released when the projectile and target fuse together

$$Q_{\text{fus}} = [(M_{\text{p}} + M_t) - M_C]c^2. \tag{2.2.19}$$

It would be more correct to refer to Q_{fus} as the ground state Q-value for fusion since it is possible to conceive of reactions in which one of the reactants is excited above the ground state. Throughout this work, however, we will treat excitation energies explicitly so that, unless otherwise stated, the symbol 'Q' should be taken to refer to the Q-value corresponding to a process involving ground state nuclei.

If, as in most cases of practical interest, Q_{fus} is a positive quantity, one might expect a fusion reaction, such as that implied by equation (2.2.19), to proceed spontaneously (i.e., for $E_{\text{cm}} = 0$). This is not so. In practice, the repulsive Coulomb force prevents fusion until the relative velocity is sufficient to bring the projectile and target nuclei into contact. Given that nuclei are roughly spherical with radii 1.2 $A^{1/3}$ fm this happens for small impact parameter (head-on or central) collisions when

$$E_{\text{cm}} = \frac{Z_{\text{p}} Z_t \, \text{e}^2}{1.2(A_{\text{p}}^{1/3} + A_t^{1/3})}, \tag{2.2.20}$$

which, for example, in the case of ^{20}Ne + ^{12}C collisions, evaluates to 14.4 MeV (the square of the electric charge, $e^2 = 1.4399764$ MeV fm). Of course, equation (2.2.20) represents only a first approximation. Fusion probabilities depend on the detailed dynamics of the interaction (including, for example, effects of deformation) between projectile and target nuclei (see chapter 3).

The excited compound nucleus decays by emitting one or more particles or gamma-rays. For heavy compound nuclei fission may also be a competitive mode of decay. Relative probabilities for various decay processes are strongly dependent on the associated Q-values. Decay Q-values are almost always negative quantities if the fused system is a stable nucleus (exceptions occur when particle emission is strongly hindered by a Coulomb barrier so that, despite a positive Q-value for emission, the parent nucleus appears to be stable). If the compound nucleus, C, emits a particle b to leave a residue B,

$$Q(C \to B + b) = [M_C - (M_B + M_b)]c^2. \tag{2.2.21}$$

One might think it natural to discuss such quantities in terms of *nuclear* masses. However the convention currently in use makes use of *atomic* masses in equations such as (2.2.19) and (2.2.21). This choice makes little difference for Q-values because the mass of the electrons can be considered to vanish from such equations (there remains a very small correction due to changes in the binding energy of the electrons).

Atomic masses are usually defined in terms of mass excesses, Δ. If the mass excess is expressed in MeV the mass-energy of a neutral atom with neutron and proton numbers N, Z (mass number $A = N + Z$) is given, in terms of the mass excess $\Delta(N, Z)$ (or $\Delta(A, Z)$) by

$$Mc^2 = A \times uc^2 + \Delta(N, Z) = 931.494A + \Delta(N, Z). \qquad (2.2.22)$$

The exact value of the 'atomic mass unit' $u = 931.49432$ MeV c^{-2} is a matter of convention. Its value has been fixed by declaring the mass excess of the neutral ^{12}C atom to be zero. Mass excesses are tabulated in several works [4, 5] and are also available on the Web [6]. In terms of mass excesses (2.2.19) becomes

$$Q_{\text{fus}} = \Delta(N_{\text{p}}, Z_{\text{p}}) + \Delta(N_t, Z_t) - \Delta(N_C, Z_C). \qquad (2.2.23)$$

For example, the fusion of ^{12}C and ^{20}Ne has a Q-value of

$$Q_{\text{fus}}(^{12}\text{C} + {}^{20}\text{Ne} \to {}^{32}\text{S}) = 0.0 + (-7.042) - (-26.016) = 18.974 \text{ MeV}, \qquad (2.2.24)$$

and the emission of an alpha particle from ^{32}S is associated with a Q-value

$$Q(^{32}\text{S} \to {}^{28}\text{Si} + {}^{4}\text{He}) = -26.016 - (-21.493 + 2.425) = -6.948 \text{ MeV}. \qquad (2.2.25)$$

If all possible exit channels for decay of a compound nucleus are considered (including not only two body decays but also multifragment break-up), the distribution of Q-values is, for nuclei with mass numbers >40, very closely Gaussian (figure 2.4). Explaining this observation is an interesting problem which we leave as a temptation for the reader.

The fact that statistical decay is strongly influenced by Q-values underlines the importance of nuclear or atomic masses in such processes and thereby the utility of a model for estimating masses or mass excesses. In fact, many authors prefer to work with binding energies rather than mass excesses. In principle, the nuclear binding energy is the Q-value for formation of a nucleus (N, Z) from N neutrons and Z protons. However, in practice it is customary to define the binding energy using the mass excesses which, we remind the reader, refer to *atomic* masses. For a nucleus with neutron number N and proton number Z, the nuclear binding energy is given as

$$\begin{aligned} \mathcal{B}(N, Z) &= Z \times \Delta(0, 1) + N \times \Delta(1, 0) - \Delta(N, Z) \\ &= Z \times 7.289 \text{ MeV} + N \times 8.071 \text{ MeV} - \Delta(N, Z), \end{aligned} \qquad (2.2.26)$$

where, of course, in the second line of (2.2.26), $\Delta(N, Z)$ is expressed in MeV. Note that just as $\Delta(N, Z)$ is the mass excess of the neutral atom with N neutrons and Z protons so $\Delta(0, 1)$ is the mass excess of the hydrogen atom (and not that of the proton). The neutron, of course, has no associated atom so that $\Delta(1, 0)$ is the mass excess of a free neutron. Expressed in this way, the binding energy can be thought of as the Q-value for fusion of N neutrons and Z hydrogen atoms to form the neutral atom with nucleus (N, Z). The use of (2.2.26) to estimate the *nuclear* binding energy thus implies the neglect of the change of binding energy of the

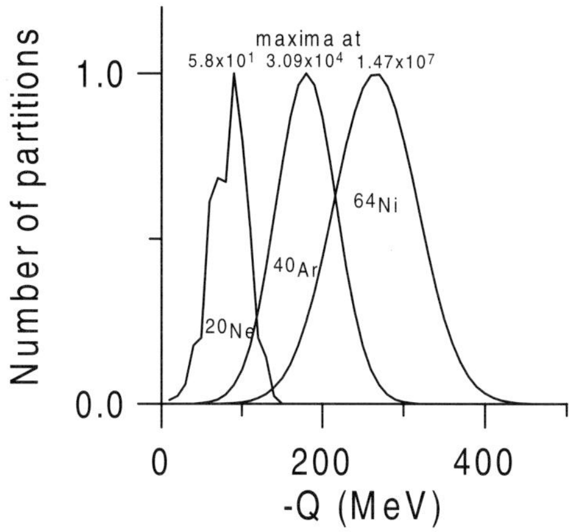

Figure 2.4. The distribution of Q-values (number of partitions per MeV of Q-value) for charge and mass partitions of ^{20}Ne, ^{40}Ar and ^{64}Ni. The curves are shown divided by the maximum values which are given as labels at the maxima of the distributions. For ^{40}Ar and ^{64}Ni the distributions are very close to Gaussian forms. The calculations were performed by D Heuer at the ISN in Grenoble.

Z electrons in going from isolated hydrogen atoms to the nucleus (N, Z). Given that the electron binding energy is approximately $0.0157Z^{7/3}$ keV (and therefore less than 1 MeV even for the heaviest nuclei) this neglect is normally justified in most applications. The nuclear binding energy $\mathcal{B}(N, Z)$ is a more intuitively understandable quantity than the mass excess because it directly expresses the decrease in mass (energy) due to the interactions between nucleons.

For many purposes it is possible to view the nucleus as a charged spherical liquid drop [7]. In such a model, one would expect the nuclear binding energy to be expressed as a bulk or volume term due to the attractive forces between nucleons, a surface correction due to the fact that surface nucleons are surrounded (on average) by a smaller number of nucleons, and a second negative contribution due to repulsive Coulomb forces between protons. Thus

$$\mathcal{B}(N, Z) = a_V A - a_S A^{2/3} - a_C \frac{Z^2}{A^{1/3}}, \qquad (2.2.27)$$

where again $A = N + Z$ and successive terms represent respectively the volume binding energy (proportional to the mass number), the surface correction, and the Coulomb repulsion (the Coulomb energy of a charged sphere).

It turns out to be necessary to correct equation (2.2.27) for two effects which are outside the scope of the liquid drop model. The first concerns nuclei with excess protons or neutrons (referred to as asymmetric nuclei). The effect of asymmetry can be understood within the framework of the so-called independent particle model of the nucleus in which each nucleon is considered to move in a potential well which represents the attractive forces of all other nucleons. If we measure energies from the top of the well then, roughly speaking, the negative energy eigenstates of the Hamiltonian (the bound states) can be considered to occur at the same energies for neutrons and protons with spin projections either $+1/2$ or $-1/2$. We denote the corresponding energies as energy levels and assume them to be evenly spaced (with spacing d), at least over a few levels. By construction, not more than two neutrons or two protons (each pair with oppositely directed spins) may occupy any level (figure 2.5). In the ground state of a nucleus with the same even number of neutrons and protons, all single particle levels up to the Fermi level are occupied (figure 2.5). Let this level be the jth level (at energy $j \times d$) so that two neutrons (or protons) in this level contribute $2dj$ to the summed energy. Changing two neutrons into protons (or protons into neutrons) leads to an energy increase of $2d(j+1)$ and a simultaneous diminution of $2dj$ (figure 2.6). Transformation of the next pair leads to a change of $2d(j+2) - 2d(j-1)$ and, generally, the change in energy produced by transforming n pairs of neutrons into protons is $2d\left[\sum_{i=1}^{n}(j+i) - \sum_{i=1}^{n}(j-(i-1))\right] = 2dn^2$. Given that transforming one pair increases the neutron–proton asymmetry by four we have $n = (N-Z)/4$ so that the energy change is found to be $d(N-Z)^2/8$. It thus appears that nuclei with the same number of neutrons and protons are more stable (have more binding energy) then asymmetric systems. Because (see below) d is expected to vary as $1/A$, the shift in binding energy is proportional to $(N-Z)^2/A$.

The second effect which must be included in the calculation of the binding energy arises because pairs of neutrons (or protons) with spin projections $+1/2$ and $-1/2$ may occupy single particle states with the same energy. The wavefunctions for these nucleons overlap strongly and thus increase the binding energy so that nuclei with even numbers of neutrons and/or protons would be expected to be more stable.

The simple liquid drop binding energy is thus modified by asymmetry and pairing corrections. The corrected formula reads

$$\mathcal{B}(N,Z) = a_{\mathrm{V}}A - a_{S}A^{2/3} - a_{\mathrm{C}}\frac{Z^2}{A^{1/3}} - a_{\mathrm{A}}\frac{(N-Z)^2}{A} + \delta. \tag{2.2.28}$$

The pairing energy, δ, is $+(-)12A^{-1/2}$ MeV for nuclei with even (odd) numbers of neutrons and protons and 0 MeV for even–odd nuclei. Values of the other constants are given in the caption to figure 2.7. This figure compares the predicted binding energies of the most abundant nuclei with the measured values as a function of charge number Z. The typical accuracy of predicted binding energies shown in this figure is of the order of 0.1 MeV per unit mass. The basic parameterization has been much improved and it is now possible to predict binding energies to within about 0.5 MeV [9] even for nuclei lying far from the bottom of the valley of stability (defined on a N–Z plot as the locus of most stable isotopes).

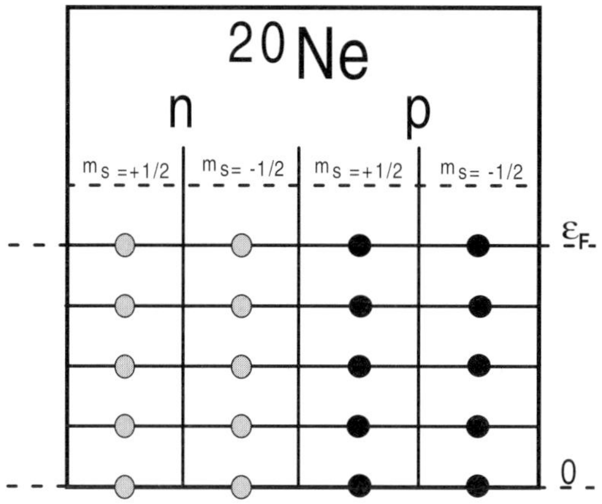

Figure 2.5. Occupation number image of the ^{20}Ne ground state. In this figure neutrons and protons as well as spin up ($m_S = 1/2$) and spin down ($m_S = -1/2$) states are indicated separately. Single particle states are occupied up to the Fermi energy, ε_F. It is useful for some purposes to reduce the single particle level spacing by a factor of four (multiply the single particle density by four) and consider only *nucleons* without specifying the spin and isospin projections.

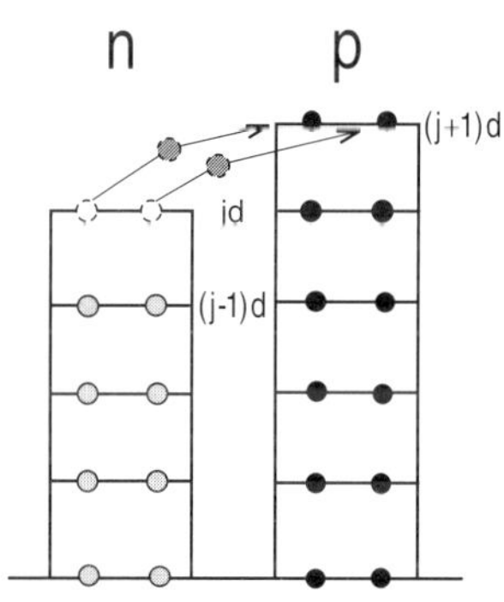

Figure 2.6. Producing an isospin asymmetry by changing two neutrons into protons. The change of two neutrons, rather than one, is convenient because transforming just one neutron involves also a loss of pairing energy.

The single particle level density g (the reciprocal of the spacing) is also the crucial quantity for estimation of microstate densities in excited nuclei. It may be most simply estimated using the Fermi gas model which results from the assumption that nucleons move independently in a box whose volume is taken

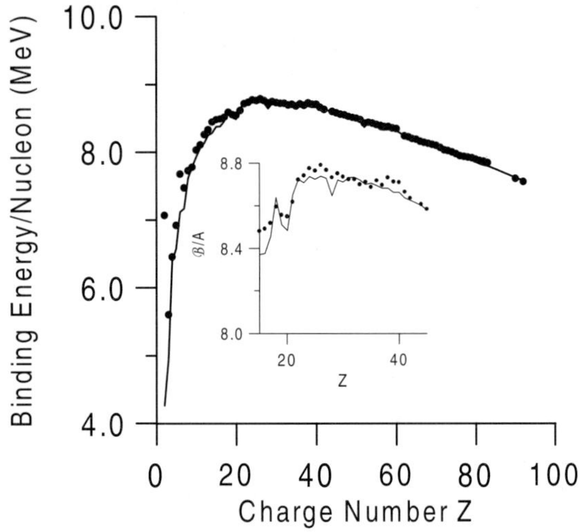

Figure 2.7. Liquid drop prediction for binding energies. The points are measured values for most abundant isotopes as a function of the charge number. The line is the liquid drop prediction using only volume, surface, Coulomb and asymmetry terms. The values of these parameters, $a_V = 15.56$ MeV, $a_S = 17.23$ MeV, $a_C = 0.697$ MeV, and $a_A = 23.29$ MeV, were taken from Williams [8]. The inset shows more detail in the region of the maximum.

from experiment. Consider a single proton (or neutron) in a nucleus of mass number A (volume $V = \frac{4}{3}\pi r_0^3 A$). The average number of energy levels, $d\Omega$, for momenta in the range p to $p + dp$ is (see equation (1.2.15) and succeeding paragraph)

$$d\Omega = \frac{4\pi r_0^3 A}{3} \frac{4\pi p^2 \, dp}{h^3}. \tag{2.2.29}$$

Each energy level may be occupied by two protons with spins 'up' and 'down' (see section 2.6). The lowest energy state of the set of protons occurs when all levels up to a maximum defined by Z are occupied by two protons. Thus we define the Fermi momentum, p_F, by setting twice the integral of (2.2.29) equal to the number of protons, i.e.,

$$Z = 2\frac{4\pi r_0^3 A}{3} \frac{4\pi p_F^3}{3h^3} \approx \frac{A}{2}. \tag{2.2.30}$$

The Fermi energy corresponding to p_F is

$$\varepsilon_F = p_F^2/2m_N = [\hbar^2(9\pi/8)^{2/3}]/(2m_N r_0^2), \tag{2.2.31}$$

where m_N is the nucleon mass. With $r_0 = 1.2$ fm this evaluates to 34.5 MeV. The density of energy levels is obtained directly from (2.2.29) so that the level density at the Fermi momentum (energy) is (with $\mathrm{d}p/\mathrm{d}E = m_N/p$)

$$\left.\frac{\mathrm{d}\Omega}{\mathrm{d}E}\right|_{E=\varepsilon_F} = \frac{4\pi r_0^3 A}{3}\frac{4\pi p^2}{h^3}\left.\frac{\mathrm{d}p}{\mathrm{d}E}\right|_{p=p_F} = \frac{4\pi r_0^3 A}{3}\frac{4\pi m_N p_F}{h^3}. \tag{2.2.32}$$

Since p_F is constant, the level spacing d goes as $1/A$ (as stated above in connection with the asymmetry correction).

It is often useful to define the *average single particle* level density in such a way that each level can be occupied by no more than *one* proton (similarly for neutrons). This is achieved by including a factor of two directly in (2.2.29) and (2.2.32). One can then also form a single nucleon density (one *nucleon* per level) by adding neutron and proton densities. If the proton and neutron Fermi energies are identical, this amounts to including a further factor of two in the single particle density. In both cases the levels are supposed to be evenly spaced (at least over a small energy region).

An obvious defect of the Fermi gas model is that nucleons are assumed to be confined by an essentially infinite potential 'wall'. An extension of this model leads to the independent particle model which, as mentioned above, considers any nucleon in a nucleus to move in a finite potential well resulting from the combined attractive forces of all other nucleons. The well depth is approximately 50 MeV. Each nucleon occupies one of the discrete (bound) states of negative energy with respect to the top of the well and the ground state of the whole nucleus is constructed, as for a Fermi gas, by uniformly filling bound states, in accordance with the Pauli principle, beginning with the state of lowest energy. The highest occupied bound state is, by definition, the Fermi energy, ε_F. The model is quite successful in reproducing the spatial aspects of nucleon motion inside the nucleus. Unfortunately, it fails to reproduce the binding energies and, in particular, the observation which can be made using the semi-empirical liquid drop model that the binding energy of the last nucleon (the energy gap between the Fermi energy and the top of the well) is approximately equal to the average binding energy per nucleon for the whole nucleus (at least for mass numbers $A > 20$). This defect can only be repaired by a detailed consideration of the interaction between pairs of nucleons (see Siemens and Jensen [10]).

If energies are measured from the top of the well the nucleon chemical potential which, following the definition in 1.6, is the increase in internal energy when a nucleon is added to the nucleus, is simply the negative of the binding energy of the nucleon at the Fermi energy and is thus approximately equal to -8 MeV. Futhermore the internal energy of the whole nucleus is the negative of the binding energy and is thus close to $-8A$ MeV for $A > 20$.

Before leaving this section it will be as well to introduce the reader to the notion of transformation of cross-sections. In many cases (complete fusion followed by decay is a good example) we dispose of a model which predicts a differential

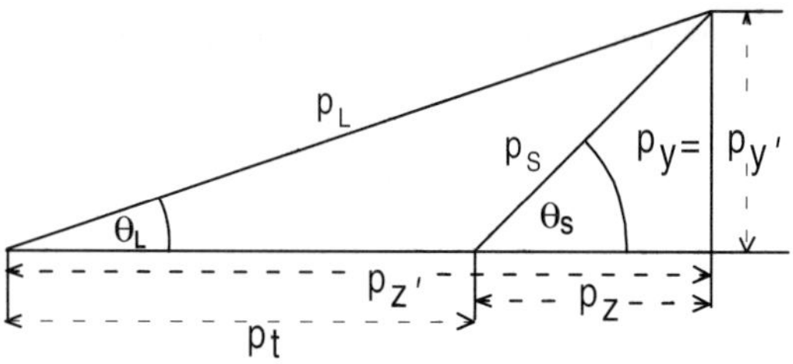

Figure 2.8. Transformation from a cylindrically symmetric source (p_S, θ_S) to a frame fixed in the laboratory (p_L, θ_L). The momentum p_S is resolved into horizontal and vertical components (p_y, p_z). Only p_z is modified by the transformation (see equation (2.2.36)).

cross-section in the centre-of-mass system which is to be compared with a cross-section measurement made in the laboratory frame of reference. To compare the two, it is necessary to transform the model cross-section to the 'lab' system. This direction is usually to be preferred because, in contrast to the model prediction, the data may be incomplete and the inverse transform somewhat ambiguous.

In some studies it may be possible to set up a Monte Carlo version of the model. This technique, which was briefly mentioned in chapter 1, will be more fully explained in chapter 4. For our present needs we can just say that such a *simulation* model generates individual events which, on a computer, appear in much the same fashion as real experimental events. In this case velocity transformations may be imposed on an event by event basis. This technique is most useful when dealing with cross-sections representing correlations between two or more particles. Thus, it is possible to make an analytic transformation of the centre-of-mass cross-section corresponding to the simultaneous emission of, say, three particles

$$\frac{\mathrm{d}^6\sigma}{\mathrm{d}E_1\,\mathrm{d}E_2\,\mathrm{d}E_3\,\mathrm{d}\Omega_1\,\mathrm{d}\Omega_2\,\mathrm{d}\Omega_3}$$

but the resulting expression would be somewhat complicated and would provide little physical insight. The corresponding computer code would, therefore, offer no advantage over the much simpler Monte Carlo method.

This being said, there are less complex situations in which it is worthwhile to carry out the transformation explicitly. We will illustrate this statement with a simple, though much used, example which should help the reader to tackle more complex cases. Our example involves only non-relativistic velocities. The cross-section in a system related to the laboratory system by a (Galilean) velocity transformation is said to characterize a source. A simple source in, say, the centre-of-mass system might be axially symmetric around the beam axis (which we take as the z-axis in a Cartesian reference frame) and would thus be described by a polar angle dependence. Let us take the energy dependence to be of the Maxwellian form and the separable polar angle dependence to be such that $\mathrm{d}\sigma/\mathrm{d}\Omega$ is constant (this corresponds to isotropic emission) so that our source (in its proper frame) is defined by the equation

$$\sigma(E_S, \Omega_S) = \frac{\mathrm{d}^2\sigma}{\mathrm{d}\Omega_S\,\mathrm{d}E_S} = CE_S^{1/2}\,\mathrm{e}^{-\beta E_S}, \tag{2.2.33}$$

where C and β are constants. Noting that $\mathrm{d}\Omega_S = 2\pi \sin\theta_S\,\mathrm{d}\theta_S$ and that for a particle of mass m, $\mathrm{d}E_S = p_S\,\mathrm{d}p_S/m$, (2.2.33) gives, with $\alpha = \beta/2m$

$$\mathrm{d}^2\sigma = \frac{\pi C\sqrt{2}}{m^{3/2}}\,\mathrm{e}^{-\alpha p_S^2} \sin\theta_S\, p_S^2\,\mathrm{d}p_S\,\mathrm{d}\theta_S, \tag{2.2.34}$$

which is an equation in a plane which we take to be the yz plane (figure 2.8). In this plane $p_S^2 = p_y^2 + p_z^2$, $p_S\,\mathrm{d}p_S\,\mathrm{d}\theta_S = \mathrm{d}p_y\,\mathrm{d}p_z$ and $\sin\theta_S = p_y/(p_y^2 + p_z^2)^{1/2}$ so that (2.2.34) can be transformed to

$$\mathrm{d}^2\sigma = \frac{\pi C\sqrt{2}}{m^{3/2}}\,\mathrm{e}^{-\alpha(p_y^2+p_z^2)}\,p_y\,\mathrm{d}p_y\,\mathrm{d}p_z. \tag{2.2.35}$$

In a positive velocity transformation along the z-axis all particles are boosted by the velocity v_t (which is simply V_{CM} if we are transforming from the centre of mass to the laboratory system). In this boosted frame, momenta are denoted with primes. We obviously have $p_{y'} = p_y$ and $p_{z'} = p_z + mv_t$ so that $\mathrm{d}p_y\,\mathrm{d}p_z = \mathrm{d}p_{y'}\,\mathrm{d}p_{z'}$. Putting $p_t = mv_t$ and substituting in (2.2.35) we obtain

$$\mathrm{d}^2\sigma = \frac{\pi C\sqrt{2}}{m^{3/2}}\,\mathrm{e}^{-\alpha[p_{y'}^2+(p_{z'}-p_t)^2]}\,p_{y'}\,\mathrm{d}p_{y'}\,\mathrm{d}p_{z'}. \tag{2.2.36}$$

Now going back to polar coordinates (with the subscript L for 'Lab') in the new frame $\mathrm{d}p_{y'}\,\mathrm{d}p_{z'} = p_{\mathrm{L}}\,\mathrm{d}p_{\mathrm{L}}\,\mathrm{d}\theta_{\mathrm{L}}$, $p_{\mathrm{L}} = (p_{y'}^2 + p_{z'}^2)^{1/2}$ etc, we obtain

$$\mathrm{d}^2\sigma = \frac{\pi C\sqrt{2}}{m^{3/2}}\,\mathrm{e}^{-\alpha[p_{\mathrm{L}}^2\sin^2\theta_{\mathrm{L}}+(p_{\mathrm{L}}\cos\theta_{\mathrm{L}}-p_t)^2]}\,p_{\mathrm{L}}^2\sin\theta_{\mathrm{L}}\,\mathrm{d}p_{\mathrm{L}}\,\mathrm{d}\theta_{\mathrm{L}} \tag{2.2.37}$$

or

$$\frac{\mathrm{d}^2\sigma}{\mathrm{d}E_{\mathrm{L}}\,\mathrm{d}\Omega_{\mathrm{L}}} = CE_{\mathrm{L}}^{1/2}\,\mathrm{e}^{-\beta(E_{\mathrm{L}}-2\sqrt{E_{\mathrm{L}}E_t}\cos\theta_{\mathrm{L}}+E_t)}, \tag{2.2.38}$$

where the 'translation energy' $E_t = p_t^2/2m$. If E_t is large so that all particles are pushed towards small angles then it is advantageous to write (2.2.38) using $E_t = E_t(\sin^2\theta_{\mathrm{L}} + \cos^2\theta_{\mathrm{L}})$. We then obtain

$$\frac{\mathrm{d}^2\sigma}{\mathrm{d}E_{\mathrm{L}}\,\mathrm{d}\Omega_{\mathrm{L}}} = CE_{\mathrm{L}}^{1/2}\,\mathrm{e}^{-\beta E_t\sin^2\theta_{\mathrm{L}}}\,\mathrm{e}^{-\beta(\sqrt{E_t}\cos\theta_{\mathrm{L}}-\sqrt{E_{\mathrm{L}}})^2}, \tag{2.2.39}$$

so that for small angles ($\sin\theta_{\mathrm{L}} \approx \theta_{\mathrm{L}}$, $\cos\theta_{\mathrm{L}} \approx 1$) the angular distribution is approximately Gaussian

$$\frac{\mathrm{d}\sigma}{\mathrm{d}\Omega_{\mathrm{L}}} \propto \mathrm{e}^{-\beta E_t\theta_{\mathrm{L}}^2}, \tag{2.2.40}$$

and depends only on the ratio of the translation energy to the source temperature. Equation (2.2.39) also shows that, at small angles, apart from the slowly changing $E_{\rm L}^{1/2}$ factor, the energy spectrum is a single peaked structure centred around the energy $E_{\rm L} = E_t$. This latter feature is rather sensitive to the source energy spectrum. Thus, if the source is characterized not only by a temperature but also by a minimum (barrier) energy so that the spectrum is of the form

$$\frac{\mathrm{d}^2\sigma}{\mathrm{d}\Omega_S\,\mathrm{d}E_S} = C(E_S - B)^{1/2}\,\mathrm{e}^{-\beta(E_S - B)}, \tag{2.2.41}$$

for $E_S > B$ and zero otherwise, then the energy spectrum at small angles presents a double humped structure. As a straightforward exercise, the reader might like to repeat the derivation for this case using a source angular distribution which is either isotropic as in (2.2.41), or of the form $\mathrm{d}\sigma/\mathrm{d}\Omega_S \sim 1/\sin\theta_S$.

2.3 Quantum mechanical description of scattering from a potential

The simplest example of what is usually referred to as a collision in reaction theory concerns non-relativistic elastic scattering of a spinless projectile by a real or complex potential which represents the projectile–target interaction. If the potential is real then we have the analogue of a conservative force problem in which all projectile target interactions result in elastic scattering. The introduction of an imaginary part in the potential provides a way of treating a loss (or absorption) of elastic flux and thence permits a global treatment of reactions. The so-called Optical Model of elastic scattering which is based on a complex potential will be discussed in the next chapter.

We mentioned previously the simplification of the two-body problem which is achieved by introducing the coordinates of the centre of mass. Since the potential energy is due to the interaction between projectile and target nuclei we may suppose it to be only a function of the relative distance, r. It then follows that the physics of the problem is not influenced by the motion of the centre of mass which, as discussed in the preceding section, is that of a free particle (on which no forces act) with mass M, and momentum, $\boldsymbol{P}$ (see section 1.2 for notation). The time independent Schrödinger equation for the relative motion is simply that for a particle of mass μ and energy $E_{\rm cm}$ moving in a potential $V(r)$.

$$\nabla^2\Psi(\boldsymbol{r}) = \frac{2\mu}{\hbar^2}(V(r) - E_{\rm cm})\Psi(\boldsymbol{r}). \tag{2.3.1}$$

Our first task will be to examine solutions of (2.3.1) for real potentials in order to obtain a 'theoretical' cross-section corresponding to the 'experimental' cross-section introduced in the previous section.

When the projectile is far from the target, the potential energy is assumed to vanish and the solutions of (2.3.1) are referred to as asymptotic solutions. For example, the beam, which is taken as incident along the z-axis (the origin $z = 0$

can be taken to be located at the front face of the target under consideration), can be represented by a plane wavefunction with momentum $p = \hbar k$,

$$\Psi_{\text{in}} = \mathrm{e}^{ikz}. \tag{2.3.2}$$

The choice of normalization, i.e., unity, in equation (2.3.2) is convenient and does not affect the analysis because, as we shall shortly see, we will be interested in *ratios* of squares of incident and scattered waves.

The corresponding quantum mechanical probability current density, j, is given as

$$\boldsymbol{j} = \frac{\hbar}{2\mu\mathrm{i}}(\Psi_{\text{in}}^* \nabla \Psi_{\text{in}} - \Psi_{\text{in}} \nabla \Psi_{\text{in}}^*), \tag{2.3.3}$$

which, with $\Psi_{\text{in}} = \mathrm{e}^{ikz}$, yields a single Cartesian component $j_z = \hbar k/\mu = p/\mu = v_{\text{p}}$, the velocity of the incident projectile nuclei. Note, however, that the square of the wavefunction represents a probability density (dimension L^{-3}) so that j_z has, as expected for a current density, the dimension $L^{-2}T^{-1}$. (Unfortunately, in some works, the quantum mechanical current density is simply referred to as the current.) We could have obtained the same result by observing that the choice of normalization for the incident plane wave implies a beam probability density of $\rho_{\text{p}} = |\Psi_{\text{in}}|^2 = 1$ per unit volume so that the probability current density is $\rho_{\text{p}} v_{\text{p}} = v_{\text{p}}$.

In the presence of the potential, the asymptotic ($r \to \infty$) solution of the Schrödinger equation consists of a superposition of a plane wave and an outgoing spherical wave which represents elastically scattered particles. Given the axial symmetry around the beam (z)-axis we write

$$\Psi(\boldsymbol{r})_{r\to\infty} = \mathrm{e}^{ikz} + f(\theta)\frac{\mathrm{e}^{ikr}}{r}. \tag{2.3.4}$$

The quantity $f(\theta)$ is referred to as the elastic scattering amplitude and is obviously related to the scattering cross-section. The radial current density (except at $\theta = 0°$) is

$$\begin{aligned} j_{\text{r}} &= \frac{\hbar}{2\mu\mathrm{i}}\left[f^*(\theta)\frac{\mathrm{e}^{-ikr}}{r}f(\theta)\frac{\mathrm{d}}{\mathrm{d}r}\left[\frac{\mathrm{e}^{ikr}}{r}\right] - f(\theta)\frac{\mathrm{e}^{ikr}}{r}f^*(\theta)\frac{\mathrm{d}}{\mathrm{d}r}\left[\frac{\mathrm{e}^{-ikr}}{r}\right]\right] \\ &= |f(\theta)|^2\frac{\hbar k}{\mu r^2} = |f(\theta)|^2\frac{v_{\text{p}}}{r^2} \end{aligned} \tag{2.3.5}$$

(which can be obtained from (2.3.4) by noting that the corresponding scattered particle probability density is $|f(\theta)|^2/r^2$). We thus find that the probability current passing through an area $\mathrm{d}A$ (see again figure 2.2), which is also properly referred to as the probability flux through $\mathrm{d}A$, bounded by two circles at (r, θ) and $(r, \theta + \mathrm{d}\theta)$ is

$$j_{\text{r}}\,\mathrm{d}A = j_{\text{r}}\,2\pi r \sin\theta\, r\mathrm{d}\theta = |f(\theta)|^2\frac{v_{\text{p}}}{r^2}r^2\,\mathrm{d}\Omega. \tag{2.3.6}$$

It now follows that the cross-section for scattering into the solid angle $\mathrm{d}\Omega = 2\pi \sin\theta \, \mathrm{d}\theta$ is, dividing (2.3.6) by the current density corresponding to the incident beam,

$$\begin{aligned} \mathrm{d}\sigma &= \frac{\text{flux of scattered particle current through } \mathrm{d}A}{\text{current density of incident particles}} \\ &= \frac{\text{flux of scattered particles into } \mathrm{d}\Omega}{v_\mathrm{p}} \\ &= \frac{\mathrm{d}\sigma}{\mathrm{d}\Omega}\,\mathrm{d}\Omega = |f(\theta)|^2\,\mathrm{d}\Omega. \end{aligned} \tag{2.3.7}$$

Thus, finally, the differential cross-section is simply

$$\frac{\mathrm{d}\sigma}{\mathrm{d}\Omega} = |f(\theta)|^2. \tag{2.3.8}$$

The next task is obviously to calculate the scattering amplitude, $f(\theta)$, for any given potential. The analysis of this problem will furnish the essential background for understanding the transmission coefficients introduced in an *ad hoc* fashion in the previous section. This latter topic is really part of the history of statistical decay and its presentation will therefore be left to the next chapter. We can, however, deal immediately with the formal expression of the scattering amplitude which, we remind the reader, is carried out for a purely real potential.

The link between the scattering amplitude and the potential, $V(r)$, is achieved using the so-called partial wave expansion of $f(\theta)$ in terms of the (complete set of) Legendre polynomials. It is convenient to write this expansion inserting an extra factor $(2\ell + 1)$, i.e.,

$$f(\theta) = \sum_{\ell=0}^{\infty} (2\ell + 1) f_\ell P_\ell(\cos\theta). \tag{2.3.9}$$

The Legendre polynomials can be used quite generally to expand any continuous function of θ. They form a complete set and obey the orthogonality condition

$$\int_{-1}^{+1} P_\ell(x) P_{\ell'}(x)\,\mathrm{d}x = \frac{2\delta_{\ell,\ell'}}{2\ell + 1}. \tag{2.3.10}$$

They also obey recursion relations of which the most important for the present purposes are

$$(x^2 - 1)\frac{\mathrm{d}P_\ell(x)}{\mathrm{d}x} = x\ell P_\ell(x) - \ell P_{\ell-1}(x)$$

and (2.3.11)

$$x\frac{\mathrm{d}P_\ell(x)}{\mathrm{d}x} - \frac{\mathrm{d}P_{\ell-1}(x)}{\mathrm{d}x} = \ell P_\ell(x).$$

The expansion coefficients, f_ℓ, in equation (2.3.9) cannot be simply expressed in terms of the potential. However, progress is possible by solving the Schrödinger equation (2.3.1) for a wavefunction which is, itself, expanded in Legendre polynomials. This expansion could be expressed simply as $\Psi(r,\theta) = \sum_{\ell=0}^{\infty} a_\ell(r) P_\ell(\cos\theta)$ but is more conveniently written

$$\Psi(r,\theta) = \sum_{\ell=0}^{\infty} (2\ell+1)\mathrm{i}^\ell \frac{u_\ell(r)}{kr} P_\ell(\cos\theta). \tag{2.3.12}$$

The advantage of separating radial and angular coordinates can be seen by writing the Laplacian operator in spherical coordinates

$$\nabla^2 = \frac{1}{r^2}\frac{\partial}{\partial r} r^2 \frac{\partial}{\partial r} + \frac{1}{r^2}\left[\frac{1}{\sin\theta}\frac{\partial}{\partial\theta}\left(\sin\theta\frac{\partial}{\partial\theta}\right) + \frac{1}{\sin^2\theta}\frac{\partial^2}{\partial\phi^2}\right]. \tag{2.3.13}$$

The action of the first term on the wavefunction $u_\ell(r)/r$ is quickly found to be

$$\begin{aligned}\left[\frac{1}{r^2}\frac{\mathrm{d}}{\mathrm{d}r} r^2 \frac{\mathrm{d}}{\mathrm{d}r}\right]\frac{u_\ell(r)}{r} &= \left[\frac{1}{r^2}\frac{\mathrm{d}}{\mathrm{d}r} r^2\right]\left[\frac{r\,\mathrm{d}u_\ell(r)/\,\mathrm{d}r - u_\ell(r)}{r^2}\right] \\ &= \frac{1}{r^2}\left[r\frac{\mathrm{d}^2 u_\ell(r)}{\mathrm{d}r^2}\right] = \frac{1}{r}\frac{\mathrm{d}^2 u_\ell(r)}{\mathrm{d}r^2},\end{aligned} \tag{2.3.14}$$

and, dropping the ϕ dependent term (because of the cylindrical symmetry) the action of the θ dependent operator on any Legendre polynomial is, with $x = \cos\theta$,

$$\left[\frac{1}{\sin\theta}\frac{\mathrm{d}}{\mathrm{d}\theta}\left(\sin\theta\frac{\mathrm{d}}{\mathrm{d}\theta}\right)\right] P_\ell(\cos\theta) = \left[-\frac{\mathrm{d}}{\mathrm{d}x}(x^2-1)\frac{\mathrm{d}}{\mathrm{d}x}\right] P_\ell(x), \tag{2.3.15}$$

which, by using the recursion relations, (2.3.11), becomes

$$\begin{aligned} -\frac{\mathrm{d}}{\mathrm{d}x}(x^2-1)\frac{\mathrm{d}}{\mathrm{d}x}P_\ell(x) &= \frac{\mathrm{d}}{\mathrm{d}x}[\ell P_{\ell-1}(x) - x\ell P_\ell(x)] \\ &= -\ell P_\ell(x) - \ell\left[x\frac{\mathrm{d}P_\ell(x)}{\mathrm{d}x} - \frac{\mathrm{d}P_{\ell-1}(x)}{\mathrm{d}x}\right] = -\ell(\ell+1)P_\ell(x).\end{aligned} \tag{2.3.16}$$

Using (2.3.14) and (2.3.16), the Schrödinger equation

$$\left[\nabla^2 + \frac{2\mu}{\hbar^2}(E_{\mathrm{cm}} - V(r))\right]\Psi(r,\theta) = 0, \tag{2.3.17}$$

can be written

$$\sum_{\ell=0}^{\infty}(2\ell+1)\mathrm{i}^\ell \frac{P_\ell(\cos\theta)}{kr}\left[\frac{\mathrm{d}^2}{\mathrm{d}r^2} - \frac{\ell(\ell+1)}{r^2} + \frac{2\mu}{\hbar^2}[E_{\mathrm{cm}} - V(r)]\right]\mu_\ell(r) = 0. \tag{2.3.18}$$

Making use of the orthogonality property of the Legendre polynomials, (2.3.10), (i.e. multiplying by $\mathrm{P}_{\ell'}$ and integrating over $\cos\theta$) we immediately deduce the equation for each of the radial wavefunctions, $u_\ell(r)$,

$$\left[\frac{\mathrm{d}^2}{\mathrm{d}r^2} - \frac{\ell(\ell+1)}{r^2} + \frac{2\mu}{\hbar^2}(E_{\mathrm{cm}} - V(r))\right]u_\ell(r) = 0. \tag{2.3.19}$$

For any value of the angular momentum ℓ the solution of this equation for large r obviously depends on the potential so that comparison of this solution with our asymptotic (cross-section defining) form, (2.3.4), provides a link between the cross-section and the potential. The asymptotic form of $u_\ell(r)$ is easily obtained because, at large r, both $V(r)$ and the centrifugal term $\ell(\ell+1)/r^2$ can be dropped. Thus, recalling the definition of the wavenumber k, we seek the solution of the asymptotic form of (2.3.19)

$$\left[\frac{\mathrm{d}^2}{\mathrm{d}r^2} + k^2\right]u_\ell(r) = 0. \tag{2.3.20}$$

The most general form of the solution of this equation is

$$u_\ell(r) = C_\ell \sin(kr - \ell\pi/2 + \delta_\ell). \tag{2.3.21}$$

(It will soon be apparent why the constant $-\ell\pi/2$ was not absorbed into the 'phase shift' δ_ℓ.) We might guess, at this stage, that the influence of the potential will be manifested in the phase shift δ_ℓ, e.g., for purely elastic scattering (with no absorption) we would expect $|C_\ell|^2 = 1$.

By making a partial wave expansion of the asymptotic wave $\mathrm{e}^{\mathrm{i}kz} + f(\theta)\,\mathrm{e}^{\mathrm{i}kr}/r$ we can now establish the relation between the phase shift, δ_ℓ, and the scattering amplitude, $f(\theta)$. It is only in this restricted sense that we can relate the cross-section to the potential. The expansion coefficients of the plane wave are simply related to spherical Bessel functions [11]

$$\mathrm{e}^{\mathrm{i}kz} = \mathrm{e}^{\mathrm{i}kr\cos\theta} = \sum_{\ell=0}^{\infty}(2\ell+1)\mathrm{i}^\ell j_\ell(kr)P_\ell(\cos\theta). \tag{2.3.22}$$

Furthermore, for large r the Bessel function $j_\ell(kr)$ can be written (multiplying by i^ℓ)

$$\mathrm{i}^\ell j_\ell(kr)_{r\to\infty} \approx \mathrm{i}^\ell \frac{\sin(kr - \ell\pi/2)}{kr} = \frac{\mathrm{e}^{\mathrm{i}kr} - (-1)^\ell\,\mathrm{e}^{-\mathrm{i}kr}}{2\mathrm{i}kr}, \tag{2.3.23}$$

(now you can see why the $\ell\pi/2$ term was included explicitly in the asymptotic form of $u_\ell(r)$) so that, with our partial wave expansion of $f(\theta)$ (2.3.9), the original asymptotic wave can be written as

$$\mathrm{e}^{\mathrm{i}kz} + f(\theta)\frac{\mathrm{e}^{\mathrm{i}kr}}{r} = \frac{1}{2\mathrm{i}kr}\sum_{\ell=0}^{\infty}(2\ell+1)[\mathrm{e}^{\mathrm{i}kr} - (-1)^\ell\,\mathrm{e}^{-\mathrm{i}kr} + 2\mathrm{i}kf_\ell\,\mathrm{e}^{\mathrm{i}kr}]P_\ell(\cos\theta). \tag{2.3.24}$$

This expression can now be compared with the asymptotic form expressed using the phase shifts, δ_ℓ. Developing the sine in (2.3.21) as in (2.3.23) we get

$$u_\ell(r) \to \frac{C_\ell}{2\mathrm{i}}[\mathrm{i}^{-\ell}\,\mathrm{e}^{\mathrm{i}\delta_\ell}\,\mathrm{e}^{\mathrm{i}kr} - \mathrm{i}^\ell\,\mathrm{e}^{-\mathrm{i}\delta_\ell}\,\mathrm{e}^{-\mathrm{i}kr}], \tag{2.3.25}$$

so that the asymptotic form of the full wavefunction is

$$\Psi(r,\theta) \to \frac{1}{2ikr}\sum_{\ell=0}^{\infty}(2\ell+1)\mathrm{i}^{\ell}C_\ell[\mathrm{i}^{-\ell}\,\mathrm{e}^{\mathrm{i}\delta_\ell}\,\mathrm{e}^{ikr} - \mathrm{i}^{\ell}\,\mathrm{e}^{-\mathrm{i}\delta_\ell}\,\mathrm{e}^{-ikr}]P_\ell(\cos\theta)$$
$$= \frac{1}{2ikr}\sum_{\ell=0}^{\infty}(2\ell+1)C_\ell[\,\mathrm{e}^{\mathrm{i}\delta_\ell}\,\mathrm{e}^{ikr} - (-1)^{\ell}\,\mathrm{e}^{-\mathrm{i}\delta_\ell}\,\mathrm{e}^{-ikr}]P_\ell(\cos\theta). \tag{2.3.26}$$

By comparing coefficients of $\mathrm{e}^{+/-ikr}$ in (2.3.24) and (2.3.26) we find that

$$C_\ell = \mathrm{e}^{\mathrm{i}\delta_\ell}, \tag{2.3.27}$$

and that

$$C_\ell\,\mathrm{e}^{\mathrm{i}\delta_\ell} = 1 + 2\mathrm{i}kf_\ell. \tag{2.3.28}$$

Thus, by substituting (2.3.27) in (2.3.28), we finally can express the partial scattering amplitude as

$$f_\ell = \frac{1}{2\mathrm{i}k}(\,\mathrm{e}^{2\mathrm{i}\delta_\ell} - 1), \tag{2.3.29}$$

which was our declared objective. It is usual to define the partial scattering function $S_\ell = \mathrm{e}^{2\mathrm{i}\delta_\ell}$ and thus to write

$$f(\theta) = \frac{1}{2\mathrm{i}k}\sum_{\ell=0}^{\infty}(2\ell+1)(S_\ell - 1)P_\ell(\cos\theta). \tag{2.3.30}$$

We can now use equation (2.3.30) to obtain the so-called shape elastic cross-section, σ_{SE}, which is the integral of the elastic scattering cross-section. We shall have more to say concerning this quantity when, in the next chapter, we discuss the optical model which is concerned with scattering from a complex potential. However, the formal expression is not subject to the choice of a real or complex potential.

$$\sigma_{\mathrm{SE}} = 2\pi\int_{-1}^{+1}|f(\theta)|^2\,\mathrm{d}\cos\theta$$
$$= \frac{1}{4k^2}\sum_{\ell,\ell'=0}^{\infty}(2\ell+1)(2\ell'+1)[(1-S_\ell)(1-S_{\ell'})^*]$$
$$\times\, 2\pi\int_{-1}^{+1}P_\ell(\cos\theta)P_{\ell'}(\cos\theta)\,\mathrm{d}\cos\theta, \tag{2.3.31}$$

which by the orthogonality property (2.3.10) gives

$$\sigma_{\mathrm{SE}} = \frac{\pi}{k^2}\sum_{\ell=0}^{\infty}(2\ell+1)|1-S_\ell|^2. \tag{2.3.32}$$

Equation (2.3.30) and the consequent relation (2.3.32) are the principal results obtained from the partial wave decomposition. In most cases the scattering amplitudes for each partial wave ℓ can only be obtained from a particular potential by numerical integration of the radial Schrödinger equation starting from $r = 0$. The asymptotic wavefunctions are 'matched' (smoothly joined) to these solutions in order to determine the phase shifts. In some particular cases, such as point Coulomb repulsion, an analytical expression of the scattering amplitude (or equivalently of the Coulomb phase shifts) can be found.

In practice, the (complex) nuclear potential, which characterizes a particular entrance channel, and which of course is the physically interesting quantity, is difficult to calculate directly and is usually introduced in a parametrized form. The parameters are then adjusted in sophisticated computer search procedures in order to obtain the final potential. At each step in the search procedure the phase shifts corresponding to the current potential and, thereby, the differential elastic scattering cross-section are calculated and compared with experimental data. The search procedure halts when no further improvement to the quality of fit is obtained (see section 3.4 for examples).

2.4 Decay rates and widths

The decay of a nucleus by emission of radiation is a stochastic process characterized by variables which are subject to discrete changes in continuous time. Consider a one-step decay process such that, at time $t = 0$, we have $N_s(0)$ source nuclei, each of which is susceptible to decay. Any given nucleus decays independently of all others. Furthermore, the probability for decay of any particular nucleus does not vary with time if the nucleus has not decayed and is identically zero if the nucleus has decayed. Let us define the number of decays which occur in the time interval $[0, t]$ to be $n(t)$ and let the probability density function (PDF) for n be $p_n(t)$. If Δt is a small time interval we can write the first order Taylor expansion

$$p_n(t + \Delta t) = p_n(t) + \Delta t \frac{\mathrm{d}p_n(t)}{\mathrm{d}t}. \tag{2.4.1}$$

We now establish the differential equation which governs the time evolution of $p_n(t)$ for *any* continuous process with discrete state space [12]. We can make the general statement that if n decays occur over the interval $[0, t + \Delta t]$, then all, some or none of them must have occurred in the interval $[0, t]$ and the complement in $[t, t + \Delta t]$. Thus if $N(t, t + \Delta t)$ symbolizes the number of decays which occur in the interval $[t, t + \Delta t]$,

$$p_n(t + \Delta t) = \sum_{k=0}^{n} p_{n-k}(t)\,\mathrm{Prob}[N(t, t + \Delta t) = k | N(0, t) = n - k], \tag{2.4.2}$$

where $\mathrm{Prob}[a|b]$ denotes the conditional probability that a occurs, given b. From

(2.4.1) and (2.4.2) we obtain the differential equation

$$\frac{\mathrm{d}p_n(t)}{\mathrm{d}t} = \lim_{\Delta t \to 0} \frac{1}{\Delta t} \left[\sum_{k=0}^{n} p_{n-k}(t) \, \mathrm{Prob}[N(t, t+\Delta t) = k | N(0, t) = n-k] - p_n(t) \right]. \tag{2.4.3}$$

The key quantities in this equation are obviously the conditional probabilities $\mathrm{Prob}[N(t, t + \Delta t) = k|N(0, t) = n - k)]$ which should be understood as being taken in the limit $\Delta t \to 0$. In general, these quantities may require a detailed examination of the history of the process. Radioactivity is one of the simplest processes because, given the independence of decay events,

$$\begin{aligned} \mathrm{Prob}[N(t, t + \Delta t) &= k|N(0, t) = n - k)] = \mathrm{Prob}[N(t, t + \Delta t) = k], \\ \mathrm{Prob}[N(t, t + \Delta t) &= k] = 0 (k > 1). \end{aligned} \tag{2.4.4}$$

The second equation corresponds to the observation that, because decays are uncorrelated, one can always find a Δt which is small enough so that there can be, at most, one decay in the interval $(t, t + \Delta t)$. Equation (2.4.4) defines the Poisson process. Let us write the probabilities for $k = 0, 1$ in terms of the decay rate $\xi(t)$ (which may depend explicitly on time) as

$$\begin{aligned} \lim_{\Delta t \to 0} \mathrm{Prob}[N(t, t + \Delta t) = 0] &= 1 - \xi(t)\Delta t, \\ \lim_{\Delta t \to 0} \mathrm{Prob}[N(t, t + \Delta t) = 1] &= \xi(t)\Delta t. \end{aligned} \tag{2.4.5}$$

The quantity $\xi(t)$ can be identified with the number of decays taking place per unit time. It is obviously proportional to the number of source nuclei N_s. The constant of proportionality, which is usually represented by the symbol λ, can be thought of as a 'decay probability per unit time' or 'specific decay rate' (we shall use both designations interchangeably) and the reciprocal of this quantity as the mean lifetime (τ) of any single nucleus. Thus, in a small time interval Δt

$$\Delta N_s = -\xi(t)\Delta t = -N_s(t)\frac{\Delta t}{\tau}, \tag{2.4.6}$$

which, in the limit $\Delta t \to 0$, yields the familiar radioactive decay law

$$N_s(t) = N_s(0)\,\mathrm{e}^{-t/\tau} = N_s(0)\,\mathrm{e}^{-\lambda t}. \tag{2.4.7}$$

In ordinary radioactive decay from a fixed (unreplenished) source the Poisson process is time dependent because the number of nuclei which make up the source diminishes exponentially with time. In this case one can show that $p_n(t)$ is actually given by the binomial law. However, in cases where the number of source nuclei is maintained constant (e.g., by production of source nuclei using a beam of particles incident on a target) a true Poisson distribution is expected (do not confuse the Poisson process with the Poisson distribution). Thus, putting

$\mathrm{d}\xi / \mathrm{d}t = 0$ and substituting (2.4.5) in (2.4.3), we obtain

$$\begin{aligned} \frac{\mathrm{d}p_n(t)}{\mathrm{d}t} &= \lim_{\Delta t \to 0} \frac{1}{\Delta t}[p_n(1 - \xi\,\Delta t) + p_{n-1}\xi\,\Delta t - p_n] \\ &= -\,\xi(p_n - p_{n-1}), \end{aligned} \tag{2.4.8}$$

the solution of which is the Poisson distribution

$$p_n(t) = \frac{(\xi t)^n\,\mathrm{e}^{-\xi t}}{n!}, \tag{2.4.9}$$

which is characterized by a mean value (and variance) of ξt. Obviously, in this case, (2.4.7) does not apply because the observed number of decays is, on average, simply proportional to the time interval corresponding to the measurement.

It is also useful to define a quantity referred to as the total decay width via the relation

$$\Gamma = \frac{\hbar}{\tau}, \tag{2.4.10}$$

where $\hbar = h/2\pi$. Note that the width is defined to conform with the uncertainty principle, i.e., $\Gamma\tau = \hbar$. If a nucleus can decay by several competing channels (e.g., emission of a neutron, a proton or an alpha particle) then the total decay width is the sum of the partial widths for each channel. Thus

$$\Gamma = \Sigma_i \Gamma_i. \tag{2.4.11}$$

The branching ratio (probability) associated with a particular decay channel is then given by

$$P_i = \frac{\Gamma_i}{\Gamma}. \tag{2.4.12}$$

The decay width, Γ, has a more physical significance than so far implied. This may be seen using a qualitative argument. The time dependence of a non-decaying (stationary) state can be represented by a forever oscillating wavefunction of the form $\Psi(t) = \mathrm{e}^{-\mathrm{i}\omega_{\mathrm{R}} t}$ where the real constant frequency ω_{R} is obtained using the Schrödinger equation, i.e.,

$$\mathrm{i}\hbar\frac{\mathrm{d}\Psi(t)}{\mathrm{d}t} = H\Psi(t) = E\Psi(t), \tag{2.4.13}$$

so that $\omega_{\mathrm{R}} = E/\hbar$. Instead of a single fixed frequency let us consider a wavefunction composed of a superposition of waves of different frequencies.

$$\Psi(t) = \int_{-\infty}^{\infty} A(\omega)\,\mathrm{e}^{-\mathrm{i}\omega t}\,\mathrm{d}\omega. \tag{2.4.14}$$

The Schrödinger equation applied to such a wavefunction clearly implies a range of energies rather than the fixed energy E. Such a 'wavepacket', which can be constructed to be localized in space at a given moment in time (say $t = 0$), will 'spread' with time, i.e., will not remain localized, and this behaviour can be taken to represent a decaying state for which we write the wavefunction as

$$\Psi_{\rm D}(t) = {\rm e}^{-{\rm i}\omega_{\rm R}t}\,{\rm e}^{-t/2\tau}. \tag{2.4.15}$$

The square of this wavefunction 'decays' with the time constant τ, i.e. $|\Psi_{\rm D}(t)|^2 = {\rm e}^{-t/\tau}$. Now equation (2.4.14) represents the Fourier transform of the frequency spectrum, $A(\omega)$. This spectrum may therefore be recovered by carrying out the inverse transform. However, by equating (2.4.14) and (2.4.15) we can perform the inverse transform on (2.4.15), which we recall, is taken to exist for all $t > 0$. Thus, using the fact that $\int_0^\infty {\rm e}^{-ax}\,{\rm d}x = 1/a$, we find

$$A(\omega) = \frac{1}{2\pi}\int_0^\infty {\rm e}^{-{\rm i}\omega_{\rm R}t}\,{\rm e}^{-t/2\tau}\,{\rm e}^{{\rm i}\omega t}\,{\rm d}t = \frac{-1}{2\pi[{\rm i}(\omega-\omega_{\rm R}) - 1/2\tau]}. \tag{2.4.16}$$

Equation (2.4.16) reveals the 'energy profile' of the decaying state. Thus, setting $E = \hbar\omega$, we obtain the probability amplitude to find a particular energy in the wavepacket

$$A(E) = \frac{\hbar}{2\pi}\left[\frac{\rm i}{(E - E_{\rm R}) + {\rm i}\hbar/2\tau}\right], \tag{2.4.17}$$

and thence, with $\Gamma = \hbar/\tau$, the energy profile (proportional to the square of A) of the decaying state

$$P(E) \sim \frac{1}{(E - E_{\rm R})^2 + \Gamma^2/4}. \tag{2.4.18}$$

Equation (2.4.18) (which is referred to as the Breit–Wigner equation [13]) describes a structure which is peaked at $E_{\rm R}$ with a characteristic width, Γ. It thus appears that the width, introduced in an *ad hoc* manner by (2.4.10) corresponds to the energy width of a wavepacket which represents a decaying state and can thus be assimilated to the width of the state itself.

2.5 Level and state densities in atomic nuclei

Not surprisingly, the calculation of microstate densities and, in particular, of microstate densities corresponding to internal excitation of nuclei is an essential element of statistical calculations of nuclear decay. The two approaches considered in this section are based on the occupation number formalism introduced in section 1.6. Both of them assume a constant single particle level density and thus fall under the general heading of 'equidistant spacing models (ESM)'.

The first approach, which we may refer to as the combinatorial method, provides an exact solution for equidistant spacing of single particle levels. In its simplest form it expresses the level density (see below) as a sum over contributions from states with different numbers of particles excited to single particle levels above the Fermi energy. Each contribution may be calculated separately and the method thus provides densities for a fixed number of excited nucleons.

For this reason it may readily be applied to studies of the temporal evolution of an excited nucleus towards equilibrium. The second approach is based on the grand canonical ensemble and involves the (double) Laplace transform of the density, the latter being expressed as a function of energy and nucleon number. An analytical expression for the the density is ‘recovered’ using a saddle point approximation to the inverse transform. The resulting expression has been, and continues to be, of paramount importance in the investigation of nuclear decay.

Before beginning our study we need to emphasize a point made in chapter 1 by distinguishing between level and state densities. The theoretical developments to be exposed are mostly concerned with establishing the number of states uniquely as a function of excitation energy, E^* (total energy minus the ground state energy). The corresponding density is usually designated by the symbol $\rho(E^*)$ and referred to as the level density. We retain this convention and adopt a second rule in which the term *state density*, will be taken to refer to a degree of definition which may be less than complete but goes beyond the simple specification of the energy. For example, it is often useful to consider the contribution to $\rho(E^*)$ corresponding to states with a given spin projection M, but with total angular momentum J and parity unspecified, usually referred to as $W(E^*, M)$.

The most useful partial specification involves the state density $\rho(E^*, J)$. As will be shown at the end of this section, this quantity can be obtained with little difficulty from the density $W(E^*, M)$ which, in turn, can be deduced starting from the level density $\rho(E^*)$. That is why the analysis begins by considering $\rho(E^*)$. For this discussion, the widths (lifetimes) discussed in the previous section will be ignored so that each state or level will be considered to be characterized by a precise energy. In other words, the calculated density will be that corresponding to all states irrespective of their lifetimes.

The calculation of the level density or of the number of levels constrained only by a fixed total energy E (the distinction between E and E^* will soon be evident) starting from the spectrum of single particle (s.p.) energy levels may be framed in terms of occupation numbers. Any distribution of nucleons over their respective s.p. levels is characterized by an energy which is the sum over s.p. energies

$$E(\boldsymbol{n}) = \sum_k n_k \varepsilon_k, \tag{2.5.1}$$

where the set of occupation numbers $\boldsymbol{n} = \{n_k\}$ defines the configuration. Obviously the total number of nucleons is equal to the sum of occupation numbers,

$$N(\boldsymbol{n}) = \sum_k n_k. \tag{2.5.2}$$

In addition the single particle level density is specified in such a way that no more than one nucleon can be present at any given s.p. energy level (i.e., $n_k = 0$ or 1).

States with energies near the ground state of a given nucleus can usually be understood as excitations of single nucleons to unoccupied single particle energy levels in the nuclear potential well. For higher energies several single particles may be excited simultaneously so that, as the excitation energy increases, there are many distinct ways of exciting the nucleus to any given small energy region. For this 'combinatorial' reason, level densities increase very rapidly with excitation energy.

We could define a level density by selecting all configurations $\boldsymbol{n}$ for which (2.5.1) yields an energy between say E and $E + \delta E$. That choice is made in the derivation based on the the Laplace transform of the level density to be discussed later. Alternatively, we could state that the problem to be solved is simply to find the number of N-nucleon configurations, $\Omega_N(E)$, which have exactly the energy E. We refer here to the number of configurations rather than a density, and an exact energy E because the energy levels constructed from any given single particle spectrum would be expected to be highly degenerate ($\Omega_N(E)$ would be expected to be large for some set of precise energies and otherwise zero). More realistically, residual interactions would be expected to introduce small shifts in the various states ($\boldsymbol{n}$) and thus to smooth the final energy level spectrum. However these interactions would not change the *number* of states. We may also expect that the *average* level density would not be much affected by the detailed structure of the single particle spectrum although a particular structure in this spectrum may induce local fluctuations. These considerations have lead to calculations of nuclear level densities based on single particle levels with an (average) *constant* spacing (ESM). If the spacing of single particle levels is d, then $\Omega_N(E)$ is non-zero only for integral values of $K_E = E/d$. It is then natural to define a level density as $\Omega_N(E)/d$. We have already met with the equidistant spacing approximation in connection with the asymmetry term in the binding energy formula (see discussion preceding equation (2.2.32)).

The single particle level density may be obtained from the single nucleon Fermi gas model. We have seen earlier that this model may be used to estimate the single particle level density in nuclei although of course the absence of spherical symmetry does not allow the single particle states to be differentiated in orbital angular momentum. Fortunately, this is not important in the present context. Thus, in practice, a single particle density is obtained by adding the densities corresponding to states for a single proton and a single neutron (the two projections of the isospin quantum number) and including the degeneracy associated with the two orientations of the nucleon spin. Recalling (2.2.31) and including a factor of four to account for spin and isospin degeneracy, the single particle level density at ε_F is $g_0 = 4Am_N r_0^2/[\hbar^2(3\pi^2)^{1/3}] = 3A/2\varepsilon_F$ which, if we take ε_F to be 34.5 MeV, yields $g_0 = A/23$ (MeV^{-1}) where A is the mass number. It would, of course, be possible to distinguish neutrons and protons, in which case the final level density would be obtained as a convolution of neutron and proton densities.

We begin by obtaining the number of states at any given energy level for a set of N nucleons. The total energy of any state is specified by the integer K_E. Thus, naively, one might guess that the number of states at energy E is simply the number of ways that N nucleons may be distributed over s.p. levels in such a way that the summed energy is $E = K_E d$. Obviously this guess would be wrong because, in the process of distributing nucleons over the s.p. levels, zero, one, or more than one nucleon may occupy any given level. Thus the number of energy levels is overestimated because the Pauli principle is not respected. As an example, we may consider the situation where K_E is partitioned into just one part of size K_E which corresponds to the configuration where the occupation number $n_{K_E} = 1$ and $n_0 = N - 1$, all other occupation numbers being zero.

If we insist on strict application of the Pauli principle, the energy of the lowest level (the ground state) of N fermions can be written in terms of the s.p. level spacing, d, as

$$E_0 = d \sum_{k=0}^{N-1} k = d \frac{N(N-1)}{2}. \tag{2.5.3}$$

The ground state for 20 nucleons with $E = 190d$ is depicted in figure 2.9(a). The Fermi energy ε_F is defined as the energy which lies just above the last occupied level in this state.

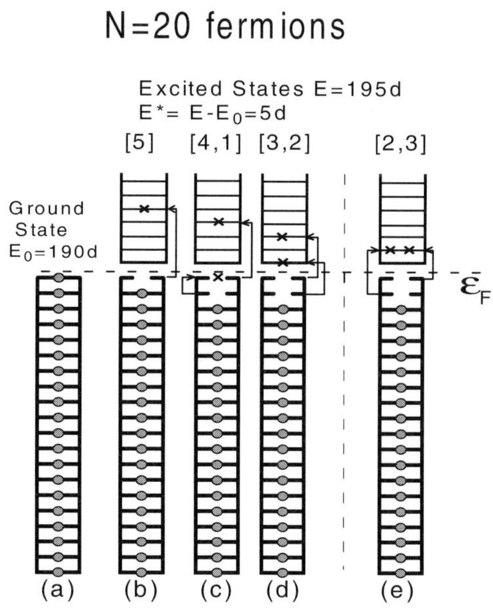

Figure 2.9. Excitation of $5d$ MeV above the ground state in a nucleus composed of 20 nucleons. The single particle spacing refers to nucleons irrespective of spin and isospin projections. There is therefore only one nucleon per s.p. level. (a) Ground state. (b) Excitation $5d$ produced by exciting one nucleon. (c) and (d) Two nucleon excitations. (e) A state with excitation $5d$ which does not obey the Pauli principle.

The key to building states which are consistent with the Pauli principle is to begin with the ground state. Let us consider a specific example in which the excitation energy $E^* = E - E_0$ is $5d$ (figure 2.9). We can label states by the number of jumps (each of energy d) made by each contributing nucleon and refer to states by these configuration labels. Possible states are obtained from the ground state by

(1) Moving the highest energy nucleon in the ground state to the state lying at $5d$ above the Fermi energy, ε_F (figure 2.9(*b*)). The configuration label in this case is simply $\{5\}$.
(2) Moving the highest energy nucleon to the state at $4d$ above ε_F and the second highest to the state left unoccupied by the first move (figure 2.9(*c*)). Configuration $\{4, 1\}$.
(3) Moving the highest energy nucleon to the state lying at $3d$ above ε_F and the second highest to the state immediately above the state left unoccupied by the first move (figure 2.9(*d*)): $\{3, 2\}$.
(4) Moving the highest energy nucleon to the state lying at $3d$ above ε_F, the second highest to the state left unoccupied by the first move and the third highest to the state left unoccupied by the second move: $\{3, 1, 1\}$.
(5) Moving the highest energy nucleon to the state lying at $2d$ above ε_F, the second highest to the state just above the state left unoccupied by the first move and the third highest to the state left unoccupied by the second move: $\{2, 2, 1\}$.
(6) Moving the highest energy nucleon to the state lying at $2d$ above ε_F, the second highest to the state left unoccupied by the first move, the third highest to the state left unoccupied by the second move and the fourth highest to the state left unoccupied by the third move: $\{2, 1, 1, 1\}$.
(7) Successively moving five nucleons up $1d$: $\{1, 1, 1, 1, 1\}$.

There are three important points to notice. The first is that by arranging moves in order of size we avoid violation of the Pauli principle. Thus consider the configuration $\{2, 2, 1\}$. The second particle falls just below the first. Now consider the configuration $\{3, 2\}$ and imagine the alternative $\{2, 3\}$. In this case (figure 2.9(*e*)) the first particle would lie at $2d$ above the Fermi energy but the second particle would need to be placed at the *same energy* thus violating the Pauli principle.

The second point, which is of considerable consequence, is that the set of configurations is just the set of partitions of the number five into integer summands arranged in order of decreasing size. The number of such configurations is identical with the number of partitions of five into integer summands *without regard to order*. This latter quantity is well known in combinatorial analysis.

The third important point is that any particular excitation energy, $E^* = Kd$, may be built, in accordance with the Pauli principle, using different numbers of nucleons. In all cases the number of excited nucleons cannot exceed the total number N. It follows that there cannot be more than N parts in any configuration.

Consequently we may generalize the above example by stating that the number of states at an excitation energy Kd (i.e., the energy above the ground state), constructed in a nucleus with N nucleons in accordance with the Pauli principle, is the number of partitions of K into not more than N parts without regard to order. We denote this quantity by the symbol $\mathbb{P}_N(K)$ and write

$$\Omega_N(E^* = Kd) = \mathbb{P}_N(K). \tag{2.5.4}$$

We now show how $\mathbb{P}_N(K)$ may be obtained using a recursion relation. For convenience and for future reference, we will refer to a part as a *fragment*. The quantity $\mathbb{P}_N(K)$ may be written as a sum $\mathbb{P}_N(K) = \sum_{n=1}^{N} P(n, K)$ where $P(n, K)$ is the number of ways of partitioning K into exactly n fragments without regard to order. We can separate $P(n, K)$ into partitions with at least one fragment of size 1 and partitions which contain no such fragment. The number of partitions of the first kind is obviously $P(n-1, K-1)$. The second set of partitions is a little more subtle. Consider the set of partitions of the integer $(K - n)$ into n fragments. The set contains partitions including one or more fragments of size 1. It follows that if we attach a fragment of size 1 to each of the n fragments we can be sure that the new set includes all partitions of an integer $(K - n) + n = K$ with no fragment of size 1. This addition process obviously does not change the *number* of partitions. It follows that the number of partitions of the integer K into n fragments all of size >1 is simply $P(n, K - n)$. Thus, we can write

$$P(n, K) = P(n - 1, K - 1) + P(n, K - n). \tag{2.5.5}$$

Equation (2.5.5) can be used as the basis for an iteration. Replacing the last term on the right-hand side of this equation we find, successively,

$$P(n, K) = P(n - 1, K - 1) + P(n - 1, K - 1 - n) + P(n, K - 2n)$$
$$P(n, K) = P(n - 1, K - 1) + P(n - 1, K - 1 - n) + P(n - 1, K - 1 - 2n) + P(n, K - 3n)$$
$$\vdots$$
$$P(n, K) = \sum_{k=0}^{k_{\max}} P(n - 1, K - 1 - kn) + P(n, K - (k + 1)n). \tag{2.5.6}$$

The end point of the iteration can be obtained by noting that a partition of an integer into $n - 1$ parts implies that the integer is greater than or equal to $n - 1$ which, in the sum in equation (2.5.6), implies that $K - 1 - kn \geq n - 1$ or

$$k_{\max} = \mathrm{INT}\left(\frac{K}{n}\right) - 1, \tag{2.5.7}$$

where INT(x) is the highest integer $\leq x$. This condition also ensures that the last term in equation (2.5.6) is zero. This is guaranteed if $K - (k_{\max} + 1)n < n$.

Using (2.5.7), we find that this condition is equivalent to $K - \mathrm{INT}\left(\frac{K}{n}\right) < 1$, and is thus always satisfied because the left-hand side is a fraction (or 0 if K is divisible by n). It follows that the residual term in square brackets in the equation is always 0 if k_{max} is given by (2.5.7). We therefore find

$$P(n, K) = \sum_{k=0}^{\mathrm{INT}(K/n)-1} P(n-1, K-kn-1). \tag{2.5.8}$$

With (2.5.8), all terms in the sum $\mathbb{P}_N(K) = \sum_{n=1}^{N} P(n, K)$ can be constructed and the ESM density follows immediately. Note that, when $K < N$, $\mathbb{P}_N(K)$ is simply $P(K) = \sum_{n=1}^{K} P(n, K)$, the total number of partitions of the integer K (because K cannot be partitioned into more than K parts). This was the case in our example with $K = 5d$.

We now consider the level density for a fixed number of particles and holes. A particle is defined as any occupied s.p. level lying above the Fermi energy and a hole as any unoccupied s.p. level below ε_F. Let us consider a particular excitation energy partitioned such that n_{p} particles represent an energy $\kappa_{\mathrm{p}}d$ and n_{h} holes represent an energy, $\kappa_{\mathrm{h}}d$. Following Bohning [14] we apply the technique used for the full set of fermions to the set of particles. We construct excited states involving n_{p} particles starting from the lowest energy n_{p} particle state. This, of course, is the state with n_{p} particles in the first n_{p} single particle levels above ε_F. The number of energy quanta (number of energy units each equal to d) to be distributed is therefore

$$k_{\mathrm{p}} = \kappa_{\mathrm{p}} - \frac{n_{\mathrm{p}}(n_{\mathrm{p}}+1)}{2}, \tag{2.5.9}$$

where the plus sign (cf equation (2.5.3)) occurs because the lowest particle level has 1 rather than 0 quanta. The number of such states is again the number of ways that k_{p} can be partitioned into *no more* than n_{p} parts since an arbitrary number of particles may be promoted above the $n_{\mathrm{p}}{}^{\mathrm{th}}$ s.p. level above the Fermi surface in order to construct the energy $k_{\mathrm{p}}d$. That is

$$\Omega_{n_{\mathrm{p}}}(\kappa_{\mathrm{p}}) - \mathbb{P}_{n_{\mathrm{p}}}(k_{\mathrm{p}}). \tag{2.5.10}$$

If we turn the diagram which represents energies below the Fermi surface upside down (figure 2.10) we see that it is equivalent to the particle diagram except that hole energies start at 0 rather than at d. In other words, if the excitation energy allocated to holes is κ_{h} and

$$k_{\mathrm{h}} = \kappa_{\mathrm{h}} - \frac{n_{\mathrm{h}}(n_{\mathrm{h}}-1)}{2}, \tag{2.5.11}$$

(note the $n_{\mathrm{h}} - 1$) then the number of energy levels constructed from holes is

$$\Omega_{n_{\mathrm{h}}}(\kappa_{\mathrm{h}}) = \mathbb{P}_{n_{\mathrm{h}}}(k_{\mathrm{h}}). \tag{2.5.12}$$

The total energy (measured from the lowest single particle level) can be written by defining a particular particle contribution as $\eta_{\mathrm{p}}d = \varepsilon_{\mathrm{p}} - \varepsilon_F$ and a particular hole contribution as $\eta_{\mathrm{h}}d = \varepsilon_F - \varepsilon_{\mathrm{h}}$. Then denoting the sum over occupied particle states by $\sum_{\mathrm{p}}$ (with a similar notation for hole states)

$$E = \mathrm{d}\frac{N(N-1)}{2} + \mathrm{d}\left(\sum_{\mathrm{p}} \eta_{\mathrm{p}} + \sum_{\mathrm{h}} \eta_{\mathrm{h}}\right), \tag{2.5.13}$$

so that the energy *above* the ground state is simply the sum over occupied particle and hole states,

$$E^* = \mathrm{d}\left(\sum_{\mathrm{p}} \eta_{\mathrm{p}} + \sum_{\mathrm{h}} \eta_{\mathrm{h}}\right). \tag{2.5.14}$$

Now consider a level specified as being constructed using exactly n_{p} particles and n_{h} holes (this is a little formal because in our example $n_{\mathrm{p}} = n_{\mathrm{h}}$) and let the energy above the ground state be specified (in units of d) as

$$\kappa = \kappa_{\mathrm{p}} + \kappa_{\mathrm{h}}, \tag{2.5.15}$$

so that $\kappa_{\mathrm{p}} = \sum_{\mathrm{p}} \eta_{\mathrm{p}}$ and $\kappa_{\mathrm{h}} = \sum_{\mathrm{h}} \eta_{\mathrm{h}}$. Then the total number of levels for n_{p} particles and n_{h} holes (which obey the Pauli principle) is the convolution sum

$$\Omega_{n_{\mathrm{p}},n_{\mathrm{h}}}(\kappa) = \sum_{k_{\mathrm{p}}=0}^{k} \mathbb{P}_{n_{\mathrm{p}}}(k_{\mathrm{p}})\mathbb{P}_{n_{\mathrm{h}}}(k - k_{\mathrm{p}}), \tag{2.5.16}$$

where

$$k = k_{\mathrm{p}} + k_{\mathrm{h}} = \kappa - \frac{n_{\mathrm{p}}(n_{\mathrm{p}}+1)}{2} - \frac{n_{\mathrm{h}}(n_{\mathrm{h}}-1)}{2}.$$

Equation (2.5.16), complemented by (2.5.4), represents the convolution expression for the particle-hole level density. These equations are used in models of pre-equilibrium nucleon emission which will be studied in chapter 3. The disadvantage of constructing the density in this way is, of course, that it can be done only by direct numerical calculation. In some circumstances it may therefore be useful to consider approximations which lead to densities expressed as (more or less) simple functions. In this context, the usefulness of the exact expressions lies in the fact that approximations can be tested and hopefully validated. It is worth noting that a highly accurate approximation for $P(K)$ was published in 1918 by Hardy and Ramanujan [15].

The particle–hole densities do not usually appear in statistical theories of nuclear decay (pre-equilibrium emission is an exception) mainly because it is usually not necessary to single out contributions for fixed numbers of particles and holes. Thus, the density is given directly as $\mathbb{P}_N(K)/d$ or $P(K)/d$.

We now turn our attention to the obtention of an analytic approximation for the density $P(K)/d$. As before, the states are to be constructed by distributing fermions over single particle levels. The occupation numbers obey equations (2.5.1) and (2.5.2) and the Pauli principle is strictly obeyed in that each n_k

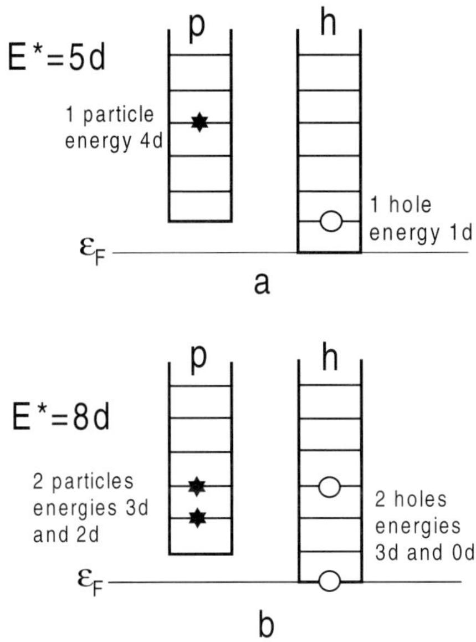

Figure 2.10. Examples of particle hole configurations for a total excitation of (*a*) $5d$(1p, 1h) and (*b*) $8d$ (2p, 2h).

can take on only the values 0, 1. The approximation is based on the grand partition function which can be written, following Ericson [3] as

$$\mathbb{Z}(\alpha, \beta) = \sum_{n} \mathrm{e}^{\alpha N(\boldsymbol{n}) - \beta E(\boldsymbol{n})}, \tag{2.5.17}$$

where, as before, $\boldsymbol{n} = \{n_k\}$ designates any configuration of fermions distributed over the single particle levels. The usefulness of this approach lies in the fact that $\mathbb{Z}(\beta, \alpha)$ is the (double) Laplace transform of the required level density (see section 1.5). To see how the calculation is set up we can begin by studying the transform with respect to the energy variable alone.

Let us suppose the true level density (expressed as a function of E rather than E^*) is $\rho(E)$. Then, although there is no heat bath, we can conceive of the corresponding canonical partition function as the Laplace transform

$$Z(\beta) = \int_0^{\infty} \rho(E)\,\mathrm{e}^{-\beta E}\,\mathrm{d}E \tag{2.5.18}$$

and we recover the density via the inverse transform

$$\rho(E) = \frac{1}{2\pi\mathrm{i}} \int_{c-\mathrm{i}\infty}^{c+\mathrm{i}\infty} Z(\beta)\,\mathrm{e}^{\beta E}\,\mathrm{d}\beta = \frac{1}{2\pi\mathrm{i}} \int_{c-\mathrm{i}\infty}^{c+\mathrm{i}\infty} \mathrm{e}^{\ln[Z]+\beta E}\,\mathrm{d}\beta. \tag{2.5.19}$$

One may construct a level density by approximating the exponent in the integral on the right-hand side of (2.5.19). To this end, let us develop $f(\beta) = \ln[Z] + \beta E$ about the maximum value which occurs at, say, β_0. Given that β_0 defines the maximum, the first derivative of $f(\beta)$ is zero so that the Taylor expansion of $f(\beta)$ can be written

$$f(\beta) = f(\beta_0) + \frac{(\beta - \beta_0)^2}{2!} \left.\frac{\mathrm{d}^2 f}{\mathrm{d}\beta^2}\right|_{\beta=\beta_0}. \tag{2.5.20}$$

Then, making the substitution $\beta = c + \mathrm{i}y$, and noting that the second derivative $\mathrm{d}^2 f/\mathrm{d}\beta^2 = -\mathrm{d}^2 f/\mathrm{d}y^2$, the integral (2.5.19) is simply that of a Gaussian distribution and thus evaluates to

$$\rho_{\mathrm{C}}(E) = \mathrm{e}^{f(\beta_0)} \frac{1}{2\pi} \int_{-\infty}^{\infty} \mathrm{e}^{\frac{f_{yy}(y-y_0)^2}{2}} \,\mathrm{d}y = \frac{\mathrm{e}^{f(\beta_0)}}{\sqrt{-2\pi f_{yy}}}, \tag{2.5.21}$$

where $f_{yy} \equiv \mathrm{d}^2 f/\mathrm{d}y^2 = -\mathrm{d}^2 f/\mathrm{d}\beta^2$ evaluated at β_0. Given that $f = \ln[Z] + \beta E$ we find $-f_{yy} = \mathrm{d}^2 \ln[Z(\beta)]/\mathrm{d}\beta^2$ (at $\beta = \beta_0$) so that (2.5.21) can be written (the C stands for canonical)

$$\rho_{\mathrm{C}}(E) = \frac{\mathrm{e}^{\ln[Z(\beta_0)] + \beta_0 E}}{\sqrt{2\pi \left.\frac{\mathrm{d}^2 \ln[Z(\beta)]}{\mathrm{d}\beta^2}\right|_{\beta=\beta_0}}}. \tag{2.5.22}$$

Provided we have a model for $Z(\beta)$ (not easily constructed in the canonical approximation scheme) the density can thus be obtained.

The corresponding expression for the grand canonical partition function (2.5.17) is derived in the same way. However, the algebra is a little more fastidious. Omitting the first derivative terms, the second order Taylor expansion around a maximum for a function of two variables is

$$\begin{aligned} f(\alpha, \beta) = f(\alpha_0, \beta_0) &+ \frac{(\alpha - \alpha_0)^2}{2!} \left.\frac{\partial^2 f}{\partial \alpha^2}\right|_{\alpha=\alpha_0} + \frac{(\beta - \beta_0)^2}{2!} \left.\frac{\partial^2 f}{\partial \beta^2}\right|_{\beta=\beta_0} \\ &+ \frac{(\alpha - \alpha_0)(\beta - \beta_0)}{2!} \left.\frac{\partial^2 f}{\partial \alpha \partial \beta}\right|_{\alpha=\alpha_0, \beta=\beta_0} \\ &+ \frac{(\beta - \beta_0)(\alpha - \alpha_0)}{2!} \left.\frac{\partial^2 f}{\partial \beta \partial \alpha}\right|_{\alpha=\alpha_0, \beta=\beta_0}. \end{aligned} \tag{2.5.23}$$

Then the equation corresponding to (2.5.21) which we obtain with $\alpha = c + \mathrm{i}x$ and $\beta = c' + \mathrm{i}y$, and further simplify by putting $X = x - x_0$, $Y = y - y_0$ is

$$\rho_{\mathrm{GC}}(N, E) = \frac{\mathrm{e}^{f(\alpha_0, \beta_0)}}{4\pi^2} \int_{-\infty}^{\infty} \mathrm{e}^{\frac{f_{xx} X^2}{2}} \left[\int_{-\infty}^{\infty} \mathrm{e}^{\frac{f_{yy} Y^2 + (f_{xy} + f_{yx}) XY}{2}} \,\mathrm{d}Y \right] \mathrm{d}X, \tag{2.5.24}$$

(GC for grand canonical) which, by performing the integration in the square brackets (by completing the square) and assuming continuous mixed derivatives so that $f_{xy} = f_{yx}$, gives

$$\begin{aligned}\rho_{\mathrm{GC}}(N,E) &= \frac{\mathrm{e}^{f(\alpha_0,\beta_0)}}{4\pi^2}\sqrt{\frac{-2\pi}{f_{yy}}}\int_{-\infty}^{\infty}\mathrm{e}^{\frac{X^2(f_{xx}f_{yy}-f_{xy}^2)}{2f_{yy}}}\,\mathrm{d}X \\ &= \frac{\mathrm{e}^{\ln[\mathbb{Z}(\alpha_0,\beta_0)]-\alpha_0 N+\beta_0 E}}{2\pi\sqrt{f_{xx}f_{yy}-f_{xy}^2}} \\ &= \frac{\mathrm{e}^{\ln[\mathbb{Z}(\alpha_0,\beta_0)]-\alpha_0 N+\beta_0 E}}{2\pi\sqrt{\frac{\partial^2\ln[\mathbb{Z}]}{\partial\alpha^2}\frac{\partial^2\ln[\mathbb{Z}]}{\partial\beta^2}-\left(\frac{\partial^2\ln[\mathbb{Z}]}{\partial\alpha\partial\beta}\right)^2\Big|_{\alpha_0,\beta_0}}},\end{aligned} \tag{2.5.25}$$

in which we have indicated that the derivatives are to be evaluated at the point (α_0, β_0).

To proceed beyond the formal approximation represented by (2.5.25) we need to model the partition function. With the conditions (2.5.1) and (2.5.2) we can write

$$\mathbb{Z}(\alpha,\beta) = \sum_{\{n_k\}} \mathrm{e}^{\alpha\sum n_k - \beta\sum n_k\varepsilon_k}, \tag{2.5.26}$$

where the sum is over all configurations for which $n_k = 0, 1$. Furthermore, in (2.5.26), the occupation numbers are not further constrained because N and E are not fixed (this is the essential advantage of the grand canonical distribution and the reason why we cannot fully develop the canonical approximation). We have already encountered the Fermi–Dirac grand partition function in section 1.6. We write equation (1.6.20) (dropping the FD subscript to conform with the current notation) as

$$\mathbb{Z}(\alpha,\beta) = \prod_k (1 + \mathrm{e}^{\alpha-\beta\varepsilon_k}). \tag{2.5.27}$$

In (2.5.27) the '1' on the right-hand side comes from the $n_k = 0$ terms and the exponential factor from the contribution with $n_k = 1$. We now seek the logarithm of $\mathbb{Z}(\alpha, \beta)$. This is, of course,

$$\ln[\mathbb{Z}(\alpha,\beta)] = \sum_k \ln[1 + \mathrm{e}^{\alpha-\beta\varepsilon_k}]. \tag{2.5.28}$$

This is not quite what we need. We wish to calculate the numerator in (2.5.25) and thus to obtain $\ln[\mathbb{Z}(\alpha,\beta)] + \alpha N - \beta E$. To this end, we write the sum (2.5.28) as an integral by introducing the exact single particle level density (cf section 1.3)

$$g(\varepsilon) = \sum_k \delta(\varepsilon - \varepsilon_k). \tag{2.5.29}$$

Note that $g(\varepsilon)$ is non-zero only for positive ε. Now the sum in (2.5.28) can be expressed as an integral

$$\ln[\mathbb{Z}(\alpha,\beta)] = \int_0^\infty g(\varepsilon)\ln[1+\mathrm{e}^{\alpha-\beta\varepsilon}]\,\mathrm{d}\varepsilon. \tag{2.5.30}$$

Given that each single particle level can be occupied by only one fermion, the ground state energy is

$$E_0 = \int_0^{\varepsilon_F} \varepsilon g(\varepsilon)\,\mathrm{d}\varepsilon \tag{2.5.31}$$

and the number of fermions is

$$N = \int_0^{\varepsilon_F} g(\varepsilon)\,\mathrm{d}\varepsilon. \tag{2.5.32}$$

In order to make use of these identities we divide the domain of integration in (2.5.30) into two parts. Thus, setting $\mu = \alpha/\beta$

$$\ln[\mathbb{Z}(\alpha,\beta)] = \int_0^\mu g(\varepsilon)\ln[1+\mathrm{e}^{\beta(\mu-\varepsilon)}]\,\mathrm{d}\varepsilon + \int_\mu^\infty g(\varepsilon)\ln[1+\mathrm{e}^{\beta(\mu-\varepsilon)}]\,\mathrm{d}\varepsilon, \tag{2.5.33}$$

we transform the second integral with the substitution $\varepsilon' = \varepsilon - \mu$ and, at the same time add an integral in which $g(\varepsilon)$ appears only with a negative argument (i.e., we add 0). We then obtain

$$\begin{aligned}\ln[\mathbb{Z}(\alpha,\beta)] = {} & \int_0^\mu g(\varepsilon)\ln[1+\mathrm{e}^{\beta(\mu-\varepsilon)}]\,\mathrm{d}\varepsilon + \int_0^\infty g(\mu+\varepsilon')\ln[1+\mathrm{e}^{-\beta\varepsilon'}]\,\mathrm{d}\varepsilon' \\ & + \int_\mu^\infty g(\mu-\varepsilon')\ln[1+\mathrm{e}^{-\beta\varepsilon'}]\,\mathrm{d}\varepsilon',\end{aligned} \tag{2.5.34}$$

where the last term is identically 0. We now add and subtract a non-zero term which has the same integrand as the last term of (2.5.34) integrated from 0 to μ. We thus arrive at

$$\begin{aligned}\ln[\mathbb{Z}(\alpha,\beta)] = {} & \int_0^\mu g(\varepsilon)\ln[1+\mathrm{e}^{\beta(\mu-\varepsilon)}]\,\mathrm{d}\varepsilon \\ & + \int_0^\infty [g(\mu+\varepsilon')+g(\mu-\varepsilon')]\ln[1+\mathrm{e}^{-\beta\varepsilon'}]\,\mathrm{d}\varepsilon' \\ & - \int_0^\mu g(\mu-\varepsilon')\ln[1+\mathrm{e}^{-\beta\varepsilon'}]\,\mathrm{d}\varepsilon'.\end{aligned} \tag{2.5.35}$$

Now changing the variable of integration in the last term ($\varepsilon = \mu - \varepsilon'$) and grouping the first and last terms we obtain

$$\begin{aligned}\ln[\mathbb{Z}(\alpha,\beta)] = {} & \int_0^\infty [g(\mu+\varepsilon')+g(\mu-\varepsilon')]\ln[1+\mathrm{e}^{-\beta\varepsilon'}]\,\mathrm{d}\varepsilon' \\ & + \int_0^\mu g(\varepsilon)\ln[1+\mathrm{e}^{\beta(\mu-\varepsilon)}]\,\mathrm{d}\varepsilon - \int_0^\mu g(\varepsilon)\ln[1+\mathrm{e}^{-\beta(\mu-\varepsilon)}]\,\mathrm{d}\varepsilon,\end{aligned} \tag{2.5.36}$$

which, using the identity $x = \ln[1 + e^{x}] - \ln[1 + e^{-x}]$ to combine the last two terms, becomes

$$\ln[\mathbb{Z}(\alpha, \beta)] = \int_0^\infty [g(\mu+\varepsilon')+g(\mu-\varepsilon') \ln[1+e^{-\beta\varepsilon'}]\, d\varepsilon' + \beta \int_0^\mu g(\varepsilon)(\mu-\varepsilon)\, d\varepsilon. \tag{2.5.37}$$

The part of the last integral on the right corresponding to the limits $(0, \varepsilon_F)$ is, according to equations (2.5.31) and (2.5.32), simply equal to $\alpha N - \beta E_0$. We thus 'extract' these terms and write

$$\begin{aligned}\ln[\mathbb{Z}(\alpha, \beta)] = & -\beta E_0 + \alpha N \\ & + \int_0^\infty [g(\mu+\varepsilon) + g(\mu-\varepsilon)] \ln[1 + e^{-\beta\varepsilon}]\, d\varepsilon \\ & + \beta \int_{\varepsilon_F}^{\mu} g(\varepsilon)(\mu - \varepsilon)\, d\varepsilon. \end{aligned} \tag{2.5.38}$$

Equation (2.5.38) is an exact expression of the grand canonical partition function. Specific cases depend on the form of $g(\varepsilon)$. For equidistant spacing, d, between single particle levels $g(\varepsilon)$ is obviously a constant which we represent as g_0. In this case the integrals in (2.5.38) can be easily evaluated. We obtain (using the fact that $\int_0^\infty \ln[1 + e^{-x}]\, dx = \pi^2/12$),

$$\ln[\mathbb{Z}(\alpha, \beta)] = -\beta E_0 + \alpha N + \frac{\pi^2 g_0}{6\beta} + \frac{\beta g_0}{2}(\mu - \varepsilon_F)^2. \tag{2.5.39}$$

The point (α_0, β_0) at which the exponent of the numerator of equation (2.5.25) maximizes is now obtained by setting its derivatives, with respect to α and β, equal to zero. The derivatives of $\ln[\mathbb{Z}(\alpha, \beta)]$ are, of course, easily calculated from (2.5.39). Adding βE to both sides of this equation and setting $\frac{\partial}{\partial\alpha}(\ln[\mathbb{Z}(\alpha, \beta)] - \alpha N + \beta E) = 0$ yields

$$\mu_0 - \frac{\alpha_0}{\beta_0} = \varepsilon_F. \tag{2.5.40}$$

The corresponding equation for β determines β_0 in terms of the excitation energy above the ground state (again denoted as E^*) i.e.,

$$E^* = E - E_0 = \frac{\pi^2 g_0}{6\beta_0^2}. \tag{2.5.41}$$

The denominator in (2.5.25) can be straightforwardly evaluated using (2.5.39). Not forgetting that $\mu = \alpha/\beta$, we find

$$\partial^2 \ln[\mathbb{Z}]/\, d\alpha^2 = g_0/\beta, \partial^2 \ln[\mathbb{Z}]/\, d\beta^2 = \left(\frac{\pi^2 g_0}{3\beta^3} - \frac{g_0\alpha^2}{\beta^3}\right)$$

and

$$\partial^2 \ln[\mathbb{Z}]/\,\mathrm{d}\alpha\,\mathrm{d}\beta = -\frac{g_0\alpha}{\beta^2}$$

so that, at the point (α_0, β_0)

$$2\pi\sqrt{\frac{\partial^2 \ln[\mathbb{Z}]}{\partial\alpha^2}\frac{\partial^2 \ln[\mathbb{Z}]}{\partial\beta^2} - \left(\frac{\partial^2 \ln[\mathbb{Z}]}{\partial\alpha\,\mathrm{d}\beta}\right)^2\Bigg|_{\alpha_0,\beta_0}} = \frac{2\pi^2 g_0}{\sqrt{3\beta_0^4}}. \tag{2.5.42}$$

Finally, defining the so-called level density parameter

$$a = \frac{g_0\pi^2}{6}, \tag{2.5.43}$$

we obtain the level density expressed in terms of the excitation energy above the ground state as

$$\rho_{\mathrm{GC}}(E^*) = \frac{\mathrm{e}^{\beta_0 E^* + \sqrt{aE^*}}}{2\pi^2 g_0/\sqrt{3}\beta_0^2} = \frac{\mathrm{e}^{2\sqrt{aE^*}}}{\sqrt{48}E^*}. \tag{2.5.44}$$

The above theory is based on the notion that the excitation energy E^* is the sum of several single particle excitations and thus breaks down at low excitation energies (of the order of g_0^{-1}). The leading order behaviour is obviously embodied in the exponential factor which dominates more and more as E^* is increased. In figure 2.11 we show the logarithm of (2.5.44) calculated with and without the E^* factor in the denominator. In many studies, approximations of this order are used to avoid the unphysical singularity at $E^* = 0$ and to take advantage of the analytical simplicity of the exponential form. Also shown in this figure is the density calculated as the convolution of separate neutron and proton densities. As can be seen the difference in the two calculations is not significant.

Interestingly enough, the statistical mechanical result (2.5.44) is exactly the asymptotic form of the Hardy–Ramanujan approximation (see [15] or chapter 24 in reference [10]).

A further interesting point concerns the notion of temperature. We can define a 'nuclear' temperature as the inverse of β_0, i.e.

$$t_{\mathrm{nuc}} = \sqrt{\frac{E^*}{a}}. \tag{2.5.45}$$

This value differs from the microcanonical temperature which was defined by (1.3.14). Indeed with (2.5.44), the microcanonical temperature is

$$\frac{1}{T} = \frac{\mathrm{d}\ln[\rho_{\mathrm{GC}}(E^*)]}{\mathrm{d}E^*} = \frac{-1}{E^*} + \sqrt{\frac{a}{E^*}}. \tag{2.5.46}$$

In principle, as discussed in the combinatorial approach, the use of (2.5.44) is limited to excitation energies which involve only single particle levels around

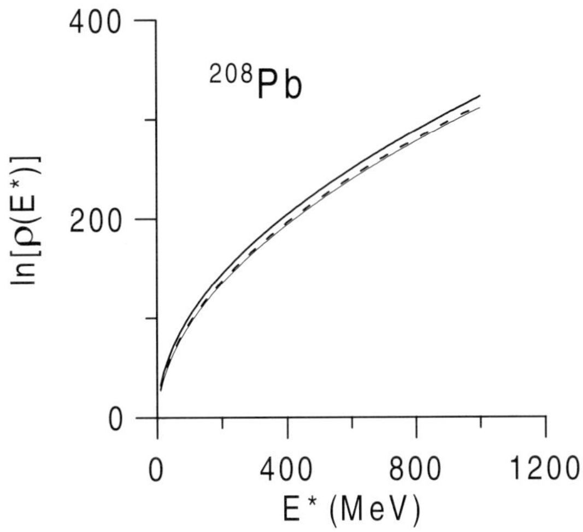

Figure 2.11. Level densities in ^{208}Pb. The heavy continuous line represents the approximation $\ln[\rho] = 2\sqrt{aE^*}$. The dashed curve was obtained using equation (2.5.44) and the light continuous trace (which lies very close to the dashed curve) represents the density obtained by calculating the proton and neutron densities separately and performing the appropriate convolution.

the Fermi energy (the depth of the particle 'reservoir' below the Fermi energy is finite). For high excitation energies one may also question the use of a constant g. Indeed the value of this parameter has been the object of much discussion. Several authors (e.g. Toke and Swiatecki [16]) have shown that corrections to the 'bulk' nuclear matter estimation of the single particle spacing have a large effect. In particular, a correction due to the contribution of the nuclear surface increases values of the level density parameter, 'a', and brings the theory more into line with the experimental values which can be globally represented by $a \approx A/8$ (see figure 2.12). Other corrections are due to shell effects and to pairing [18,19]. These corrections again mainly affect the s.p. level spacing at the Fermi energy and, as shown in figure 2.12, may be of considerable importance especially near closed shells.

So far, in this section, we have not considered the angular momentum dependence of the nuclear state density. This development is suprisingly simple and depends on the fact that, quite generally, for states with angular momentum quantum number J

$$\rho(E^*, J) = W(E^*, M = J) - W(E^*, M = J + 1), \qquad (2.5.47)$$

where $W(E^*, M = J)$ is the density of states for which the quantum number corresponding to the projection of the angular momentum on an arbitrarily chosen

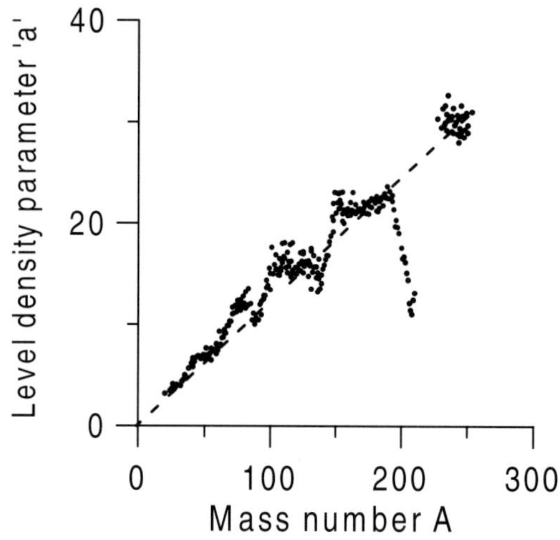

Figure 2.12. Level density parameters obtained by counting neutron resonances and fitting the formula $\ln[\rho] = 2\sqrt{aE^*}$. The drawing was constructed using a table provided by Iljinov *et al* [17]. Strong deviations from the overall trend ($a = A/8$) are mainly due to shell effects.

axis is M. The states which possess a projection $M = J$ are all those states whose angular momentum quantum number is at least J (see section 2.6). Equation (2.5.47) follows directly.

Our problem is thus to calculate $W(E^*, M)$. The projection M is the sum of contributions from each occupied s.p. state and can therefore be taken to be the sum of N random variables each of which may take on any one of the set of $2j+1$ values of m_j corresponding to the total angular momentum of the s.p. state. We have no need to discuss the details of the distribution of angular momentum in the s.p. states. It is sufficient to remark that, for each s.p. state, the mean value of $m_j = 0$ and that the probability distribution for values of m_j would be expected to be symmetric about the mean. If N is not too small the central limit theorem then tells as that the resulting distribution of M values would be expected to be approximately Gaussian with mean $M = \Sigma\langle m\rangle = 0$ and dispersion $\sigma^2 = N\langle m^2\rangle$ where the means are taken over all projections and over all contributing single particle states. We thus find (we drop the subscript 'GC' from hereon)

$$W(E^*, M = J) = \rho(E^*)\frac{1}{\sqrt{2\pi\sigma^2}}\,\mathrm{e}^{\frac{-J^2}{2\sigma^2}}. \tag{2.5.48}$$

It now follows from (2.5.47) that

$$\rho(E^*, J) = \frac{\rho(E^*)}{\sqrt{2\pi\sigma^2}}\left[\mathrm{e}^{\frac{-J^2}{2\sigma^2}} - \mathrm{e}^{\frac{-(J+1)^2}{2\sigma^2}}\right]$$

$$= \frac{\rho(E^*)}{\sqrt{2\pi\sigma^2}} \mathrm{e}^{\frac{-(J+1/2)^2}{2\sigma^2}} \left[\mathrm{e}^{\frac{(J+1/4)}{2\sigma^2}} - \mathrm{e}^{\frac{-(J+3/4)}{2\sigma^2}} \right]. \tag{2.5.49}$$

Approximating $(J+1/2)^2$ as $J(J+1)$ and expanding the exponentials in the square brackets we get

$$\rho(E^*, J) = \frac{\rho(E^*)(2J+1)}{2\sigma^3\sqrt{2\pi}} \mathrm{e}^{\frac{-J(J+1)}{2\sigma^2}}. \tag{2.5.50}$$

An estimation of σ^2 is obtained using the canonical approximation discussed in chapter 1. Classically, the angular momentum, $\boldsymbol{J}$, of a rotating object is linked to the rotational energy by

$$E_{\mathrm{ROT}} = \frac{|\boldsymbol{J}|^2}{2I}, \tag{2.5.51}$$

where I is the moment of inertia. If we assume the density $\rho(E^*)$ behaves as a heat reservoir characterized by the temperature T then the probability to find a given rotational energy would go as $\mathrm{e}^{-E_{\mathrm{ROT}}/T}$. Replacing the square of the angular momentum with the quantum mechanical equivalent, the density would be

$$\rho(E^*, J) = \rho(E^*, J=0)\,\mathrm{e}^{-J(J+1)\hbar^2/2IT}. \tag{2.5.52}$$

Comparison with (2.5.50) leads to the conclusion that

$$\rho(E^*, J=0) = \frac{\rho(E^*)}{2\sqrt{2\pi}\sigma^3}, \tag{2.5.53}$$

with the spin cut-off parameter

$$\sigma^2 = N\langle m^2 \rangle = \frac{IT}{\hbar^2}. \tag{2.5.54}$$

Notice that equation (2.5.52) can be written only as an approximation to $\rho(E^* - E_{\mathrm{ROT}})$. For some angular momentum J_{Y}, which is obviously given by

$$\frac{J_{\mathrm{Y}}(J_{\mathrm{Y}}+1)\hbar^2}{2I} = E^*, \tag{2.5.55}$$

the rotational energy is equal to the excitation energy E^* and the density, despite the appearance of (2.5.52), will fall to zero. A plot of J_Y versus E^* is referred to as the Yrast line (figure 2.13). There are no states with $J > J_Y$, i.e., 'beyond' the Yrast line.

It is usual to express the quantity $I/\hbar^2$ in MeV^{-1}. Thus if we consider the moment of inertia of a uniform density sphere of mass number, A, and radius, R fm, about an axis passing through its centre, we would write

$$\frac{I}{\hbar^2} = \frac{2}{5}\frac{AR^2 m_0 c^2}{(\hbar c)^2} = 9.569 \times 10^{-3} AR^2 \mathrm{MeV}^{-1}. \tag{2.5.56}$$

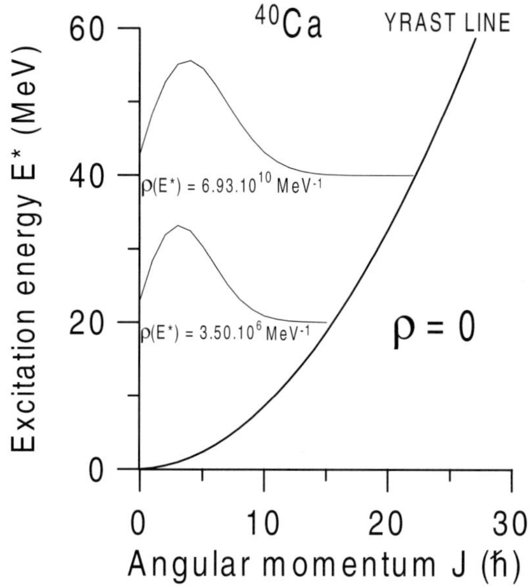

Figure 2.13. Yrast line in ^{40}Ca. The variation of the density of states as a function of angular momentum is indicated at excitation energies $E^* = 20$ MeV and $E^* = 40$ MeV. There are no states to the right of the Yrast line.

To obtain (2.5.56), we assumed the mass to be A atomic mass units ($m_0c^2 = u = 931.494$ MeV) and used the constant $\hbar c = 197.327$ MeV fm.

We now briefly discuss the separation of the state density into positive and negative parity states. An analysis of this problem which is similar in spirit to that used for angular momentum leads to the conclusion that it is generally safe to assume that level densities for positive and negative parities are identical except at low excitation energies where individual levels may be resolved. It thus appears that, apart from decays to resolved final states, decay probabilities are not differentiated by this quantum number.

The importance of the microstate densities presented above will quickly become apparent in the next chapter where we discuss some of the 'historical' applications of methods of statistical physics to the decay of excited nuclei. These studies, and indeed modern studies, rely heavily on numerical estimates of numbers of microstates or analytic forms for the corresponding densities. This dependence may turn out to constitute something of a stumbling block simply because theoretical calculations of densities at high excitation energies cannot be directly verified by experiment. For this reason the subject is far from closed and work continues to improve the quality of predictions. Among recent works we may cite contributions by Mustafa, Blann, Ignatyuk and Grimes [20] and by Hilaire and coworkers [21].

2.6 Angular momentum in quantum mechanics

In quantum mechanics the description of the state of motion of a physical system is made in terms of the wavefunction $\psi(t)$ which obeys the Schrödinger equation

$$i\hbar\frac{d\psi(t)}{dt} = H\psi(t). \tag{2.6.1}$$

The measurement of an observable is represented by the action of an operator on the wavefunction. There is a special class of operators ($\hat{o}$) for which

$$i\hbar\frac{d(\hat{o}\psi(t))}{dt} = H(\hat{o}\psi(t)), \tag{2.6.2}$$

i.e., $\hat{o}\psi(t)$ represents a state of motion of the system in the same sense as $\psi(t)$. If $\hat{o}$ is not explicitly time dependent and the inverse of $\hat{o}$ is defined then (2.6.2) can be multiplied by $\hat{o}^{-1}$ to give

$$i\hbar(\hat{o}^{-1}\hat{o})\frac{d\psi(t)}{dt} = i\hbar\frac{d\psi(t)}{dt} = \hat{o}^{-1}H(\hat{o}\psi(t)), \tag{2.6.3}$$

which with (2.6.1) implies

$$H\psi(t) = \hat{o}^{-1}H(\hat{o}\psi(t)), \tag{2.6.4}$$

or again

$$(H\hat{o} - \hat{o}H)\psi(t) = 0. \tag{2.6.5}$$

Operators which commute with the Hamiltonian (a commutator is usually represented by square brackets), i.e., $[H\hat{o}] = (H\hat{o} - \hat{o}H)$, are said to generate symmetry transformations [22]. Angular momentum operators are in fact associated with rotational symmetries. However, there is another interpretation. The Schrödinger equation can be written for small Δt as

$$\psi(t + \Delta t) - \psi(t) - \frac{i\Delta t H\psi(t)}{\hbar}. \tag{2.6.6}$$

It follows that any function of the Hamiltonian (which obviously commutes with H) or any other observable whose operator commutes with the Hamiltonian is a constant of the motion (conserved quantity) for the system considered. Symmetry transformations therefore correspond to conserved quantities. There are normally only a few such quantities. For example, a closed system in classical mechanics is usually characterized by the total energy, momentum and angular momentum. In quantum mechanics the set of all such constants is referred to as the set of quantum numbers. We will label this set using the single symbol α and write the wavefunction simply as ψ_α or in the bra–ket notation as the basis state $|\alpha\rangle$. If all quantum numbers are numerically specified there is only one corresponding state. Furthermore, the set is complete in the sense that any state of the system can be written as a linear superposition of states each characterized by particular quantum numbers. Such basis states (or the corresponding wavefunctions) are linearly independent. They may be normalized so that the integral of $\psi_\alpha^* \psi_\alpha$ is unity. An orthonormal basis is defined by

$$\langle \alpha \mid \alpha' \rangle = \int \psi_\alpha^* \psi_{\alpha'} \, \mathrm{d}\tau = \delta_{\alpha,\alpha'}, \tag{2.6.7}$$

where the integral is over all the space in which the wavefunction is defined ($\psi_\alpha = \psi_\alpha(\tau)$). In some cases it is possible to find an alternative set of basis vectors each of which can be written as a linear combination of vectors of the original basis. Then any member of the set of vectors $|\beta\rangle$ can be expanded as

$$|\beta\rangle = \sum_\alpha |\alpha\rangle \langle \alpha | \beta \rangle. \tag{2.6.8}$$

Note that (2.6.8) implies that the closure operator, $\Sigma_\alpha |\alpha\rangle\langle\alpha| = 1$.

We will now establish the set of quantum numbers needed to specify basis states for angular momentum. We therefore pull these numbers out of α and represent them explicitly. For single particles with no intrinsic spin, or for single particles with spin but no orbital angular momentum, only two such numbers are necessary, the square of the angular momentum and its projection on an arbitrarily chosen axis (usually referred to as the z-axis). We can provide a plausibility argument in support of this statement by establishing the correspondence between a classical and quantum mechanical description of a system with orbital angular momentum (spin angular momentum has no simple counterpart in classical mechanics). The classical angular momentum of a point-like particle $\boldsymbol{L} = \boldsymbol{r} \wedge \boldsymbol{p}$ is written in component form

$$\begin{aligned} L_x &= y p_z - z p_y, \\ L_y &= z p_x - x p_z, \\ L_z &= x p_y - y p_x. \end{aligned} \tag{2.6.9}$$

The operator form of these equations can be constructed by noting that the z-component position and momentum operators are just z and $-\mathrm{i}\hbar\frac{\partial}{\partial z}$ with similar expressions for the other components. The action of the momentum operator can be understood by considering a plane wave such as $\mathrm{e}^{\mathrm{i}kz}$. This wavefunction represents a particle with momentum $p_z = \hbar k$. We find, as expected, that the plane wave is an eigenfunction of the momentum operator with eigenvalue p_z

$$-\mathrm{i}\hbar \frac{\partial \, \mathrm{e}^{\mathrm{i}kz}}{\partial z} = \hbar k \, \mathrm{e}^{\mathrm{i}kz}. \tag{2.6.10}$$

We can write down the operator expressions for the components of the orbital angular momentum. Thus, for example, the operator $\mathbb{L}_z$ is

$$\mathbb{L}_z = -\mathrm{i}\hbar \left(x \frac{\partial}{\partial y} - y \frac{\partial}{\partial x} \right), \tag{2.6.11}$$

and the commutator $[\mathbb{L}_x, \mathbb{L}_y]$ is quickly found to be

$$
\begin{aligned}
[\mathbb{L}_x, \mathbb{L}_y] = (-\mathrm{i}\hbar)^2 \bigg[\left(y\frac{\partial}{\partial z} - z\frac{\partial}{\partial y} \right) \left(z\frac{\partial}{\partial x} - x\frac{\partial}{\partial z} \right) \\
- \left(z\frac{\partial}{\partial x} - x\frac{\partial}{\partial z} \right) \left(y\frac{\partial}{\partial z} - z\frac{\partial}{\partial y} \right) \bigg] \\
= -(-\mathrm{i}\hbar)^2 \left(x\frac{\partial}{\partial y} - y\frac{\partial}{\partial x} \right) = \mathrm{i}\hbar\mathbb{L}_z,
\end{aligned} \tag{2.6.12}
$$

with similar expressions for $[\mathbb{L}_y, \mathbb{L}_z]$ and $[\mathbb{L}_z, \mathbb{L}_x]$ so that in a complete set of commuting observables only one projection of the angular momentum can be included (conventionally, the z-projection). On the other hand, the square of the total angular momentum commutes with all three components of the angular momentum. To show this we write

$$
[\mathbb{L}^2, \mathbb{L}_z] = [\mathbb{L}_x^2, \mathbb{L}_z] + [\mathbb{L}_y^2, \mathbb{L}_z] + [\mathbb{L}_z^2, \mathbb{L}_z]. \tag{2.6.13}
$$

The last commutator is obviously 0. The first term can be simplified by using the commutation relation $(\mathbb{L}_z\mathbb{L}_x - \mathbb{L}_x\mathbb{L}_z) = \mathrm{i}\hbar\mathbb{L}_y$. Thus

$$
\begin{aligned}
[\mathbb{L}_x^2, \mathbb{L}z] &= \mathbb{L}_x(\mathbb{L}_x\mathbb{L}_z) - (\mathbb{L}_z\mathbb{L}_x)\mathbb{L}_x \\
&= \mathbb{L}_x(\mathbb{L}_z\mathbb{L}_x - \mathrm{i}\hbar\mathbb{L}_y) - (\mathbb{L}_z\mathbb{L}_x)\mathbb{L}_x \\
&= -[\mathbb{L}_z\mathbb{L}_x]\mathbb{L}_x - \mathrm{i}\hbar\mathbb{L}_x\mathbb{L}_y \\
&= -\mathrm{i}\hbar(\mathbb{L}_y\mathbb{L}_x + \mathbb{L}_x\mathbb{L}_y).
\end{aligned} \tag{2.6.14}
$$

Using a similar development, the second term is easily shown to be the same as the first except for a change in sign so that the first and second terms cancel. We thus arrive at the result

$$
[\mathbb{L}^2, \mathbb{L}_z] = 0, \tag{2.6.15}
$$

so that the square of the angular momentum can be included in the set of commuting observables. The eigenfunctions of the operators $\mathbb{L}^2$ and $\mathbb{L}_z$ are the so-called spherical harmonics. The spherical harmonic with angular momentum quantum number, ℓ, and z-projection quantum number, m, is [23]

$$
Y_{\ell m}(\theta, \phi) = (-1)^m \sqrt{\frac{(2\ell+1)(\ell-m)!}{4\pi(\ell+m)!}} P_\ell^m(\cos\theta)\,\mathrm{e}^{\mathrm{i}m\phi}, \tag{2.6.16}
$$

where the associated Legendre functions $P_\ell^m(x)$ can be obtained from the recursion relation

$$
(2\ell+1)xP_\ell^m(x) = (\ell-m+1)P_{\ell+1}^m(x) + (\ell+m)P_{\ell-1}^m(x), \tag{2.6.17}
$$

with

$$
P_\ell^{m+1}(x) = \frac{1}{\sqrt{x^2-1}}[(\ell-m)xP_\ell^m(x) - (\ell+m)P_{\ell-1}^m(x)], \tag{2.6.18}
$$

and $P_\ell^0(x)$ identified with the Legendre polynomial $P_\ell(x)$. The eigenvalues of $\mathbb{L}^2(\ell(\ell+1)\hbar^2)$ and $\mathbb{L}_z(m\hbar)$ can be obtained by expressing these operators in spherical coordinates (see, for example, equation (2.3.16)). However, as shown below they may be obtained more simply from the commutation relations.

The results (2.6.12) and (2.6.15) apply also to wavefunctions which represent particles with intrinsic angular momentum (spin). In this case the angular momentum quantum number is usually denoted by the letter j (or J). We thus have the commutation relations $[\mathbb{J}^2, \mathbb{J}_z] = 0$ and $[\mathbb{J}_x, \mathbb{J}_y] = i\hbar\mathbb{J}_z$ etc. We may also define the well-named raising and lowering operators $\mathbb{J}_+ = \mathbb{J}_x + i\mathbb{J}_y$ and $\mathbb{J}_- = \mathbb{J}_x - i\mathbb{J}_y$. It is easy to show that these operators satisfy the commutation relations

$$\begin{aligned}[][\mathbb{J}_+, \mathbb{J}_z] &= -\hbar\mathbb{J}_+, \\ [\mathbb{J}_-, \mathbb{J}_z] &= \hbar\mathbb{J}_-,\end{aligned} \tag{2.6.19}$$

and that

$$\mathbb{J}^2 = \mathbb{J}_-\mathbb{J}_+ + \mathbb{J}_z^2 + \hbar\mathbb{J}_z = \mathbb{J}_+\mathbb{J}_- + \mathbb{J}_z^2 - \hbar\mathbb{J}_z. \tag{2.6.20}$$

Assuming J is finite we now look for the set of eigenvalues of $\mathbb{J}^2$ and $\mathbb{J}_z$. Leaving aside the specification of the eigenvalue of $\mathbb{J}^2$ (and indeed, all other quantum numbers) for the moment, let us simply label eigenstates of $\mathbb{J}_z$ using the ket $|m\rangle$ so that $\mathbb{J}_z|m\rangle = m\hbar|m\rangle$. Now let $j\hbar$ be the largest eigenvalue of $\mathbb{J}_z$ and let the corresponding state be $|j\rangle$. Then

$$\mathbb{J}_z\mathbb{J}_-|j\rangle = \mathbb{J}_-\mathbb{J}_z|j\rangle - \hbar\mathbb{J}_-|j\rangle = (j-1)\hbar\mathbb{J}_-|j\rangle. \tag{2.6.21}$$

It follows that $\mathbb{J}_-|j\rangle$ is an eigenvector of $\mathbb{J}_z$ with eigenvalue $m\hbar = (j-1)\hbar$. Obviously, the game can be replayed so that successive applications of $\mathbb{J}_-$ generate a succession of eigenvectors of $\mathbb{J}_z$ with eigenvalues $j\hbar$, $(j-1)\hbar$, $(j-2)\hbar$ etc. Furthermore, since j is the largest value of the set of eigenvalues of $\mathbb{J}_z$ the operation $\mathbb{J}_+|j\rangle$ leads to the null vector. It then follows from (2.6.20) that

$$\mathbb{J}^2|j\rangle = j(j+1)\hbar^2|j\rangle, \tag{2.6.22}$$

so that the state with eigenvalue $m = j$ is also an eigenvector of $\mathbb{J}^2$ with eigenvalue $j(j+1)\hbar^2$. However since $\mathbb{J}^2$ and $\mathbb{J}_-$ commute, $|j-1\rangle$ and more generally $|j-p\rangle$ (with p a positive integer) is also an eigenvector of $\mathbb{J}^2$ with the same eigenvalue $j(j+1)\hbar^2$ ($\sqrt{j(j+1)\hbar^2}$ can be thought of as the magnitude of the total angular momentum). The end of the lowering sequence occurs for some state with $m = j - k$ and is symbolized by the operation $\mathbb{J}_-|j-k\rangle = |\text{nul}\rangle$. Then, given that

$$\mathbb{J}^2|j-k\rangle = (\mathbb{J}_+\mathbb{J}_- + \mathbb{J}_z^2 - \hbar\mathbb{J}_z)|j-k\rangle, \tag{2.6.23}$$

we find

$$j(j+1)\hbar^2 = (j-k)^2\hbar^2 - (j-k)\hbar^2, \tag{2.6.24}$$

from which we obtain $j = k/2$ which means that j is a positive integral or half integral number. It follows that the eigenvectors of $\mathbb{J}^2$ and $\mathbb{J}_z$ are characterized by a single eigenvalue of $\mathbb{J}^2$, i.e., $j(j+1)\hbar^2$ and a set of integral or half integral multiples of $\hbar$, $m\hbar$ which are possible eigenvalues of $\mathbb{J}_z$. The so-called 'magnetic' quantum numbers m, change in unit steps from the maximum value $m = j$ to a minimum value $m = (j - k) = (j - 2j) = -j$.

We can now label basis wavefunctions with the quantum numbers m (corresponding to the eigenvalue of $\mathbb{J}_z$) and j (corresponding to the maximum eigenvalue of $\mathbb{J}_z$), it being understood that the latter quantity directly provides the eigenvalue of $\mathbb{J}^2$. Then, grouping other eigenvalues in the symbol α, a basis state of angular momentum is represented as $|\alpha, j, m\rangle$. It is as well to note here that physicists often refer to 'a state with angular momentum $j\hbar$' meaning, of course 'the state with eigenvalue of $\mathbb{J}^2$ equal to $j(j+1)\hbar^2$'. This convention as well as that of refering to j as an angular momentum rather than an angular momentum quantum number usually causes no confusion but one should be aware of the distinction.

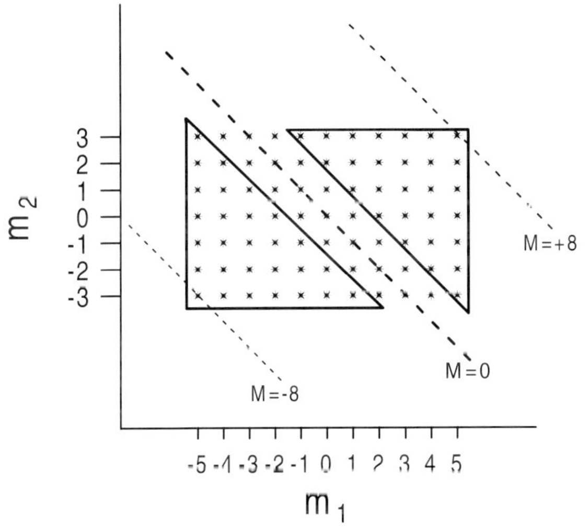

Figure 2.14. Addition of angular momenta $j_1 = 5$, $j_2 = 3$. Each point corresponds to given projections m_1 and m_2 (in units of $\hbar$) of j_1 and j_2. The diagonal lines correspond to fixed values of the projection of the vector sum $j_1 + j_2$, i.e., $M = m_1 + m_2$. Starting from the bottom left-hand corner, there is only one state with $M = -8$, i.e., the state with $J = 8$. Then two states with $M = -7$, (1 with $J = 8$ and one with $J = 7$). Then three states with $M = -6 \ldots$. This behaviour continues until $M = -2$ ($J = 8, 7, 6, 5, 4, 3, 2$). There is no state with $J = 1$ so that the number of states for $M = -1$ is equal to that for $M = -2$, i.e., seven. We get seven states ($J = 8, 7, 6, 5, 4, 3, 2$) until $M = +2$. For $M = 3$ the state $J = 2$ cannot contribute and the number of states falls to six ($J = 8, 7, 6, 5, 4, 3$), then five for $M = 4$, four for $M = 3$ etc.

A problem that often occurs involves wavefunctions which represent states in which two or more angular momenta are added. The most obvious example concerns orbital angular momentum and intrinsic angular momentum (spin) for a single particle (nucleus). Let us therefore consider the set of vectors which represent states obtained by adding two angular momenta. The basis functions are the product of the spaces $|\alpha, j_1, m_1\rangle$ and $|\alpha, j_2, m_2\rangle$.They can be written as $|\alpha, j_1, m_1, j_2, m_2\rangle$. In classical mechanics the possible values of the vector sum $\boldsymbol{J} = \boldsymbol{j}_1 + \boldsymbol{j}_2$ are determined by the lengths and orientations of the vectors $\boldsymbol{j}_1$ and $\boldsymbol{j}_2$. In quantum mechanics, as we have seen, only the z-component of the orientation is specified. However the z-projections are additive because $\mathbb{J}_z = j_{1z} + j_{2z}$. Let us denote the operators corresponding to j_1 and j_2 as $\mathbb{J}_1$ and $\mathbb{J}_2$, respectively, and let the sum J be represented simply by the operator $\mathbb{J}$. It is easy to see that $\mathbb{J}^2$ commutes with $\mathbb{J}_1^2$ and $\mathbb{J}_2^2$ so that the eigenvectors of $\mathbb{J}^2$ are also those of $\mathbb{J}_1^2$ and $\mathbb{J}_2^2$. Possible values of J can be determined by a simple argument presented by Messiah [1, chapter XIII]. The essential step is identical with that used to determine the angular momentum dependence of the density of states in the previous section (see (2.5.47)). The number of states having a given value of J is

$$N(J) = n(M = J) - n(M = J + 1), \tag{2.6.25}$$

where $n(M)$ is the number of states having a given z-projection M. Plotting m_1 versus m_2 (figure 2.14) we see that fixed values of $M = m_1 + m_2$ are situated on diagonal cuts. Over the centre part of this diagram the number of states is $2j_2 + 1$ (we can choose j_2 as the lesser of the two j) and is constant so that the subtraction in (2.6.25) yields 0. In fact, $n(M)$ depends on M only for $|j_1 - j_2| \leq |M| \leq j_1 + j_2$ (the two triangular regions shown in the figure). Within this range, the dependence on M is quickly seen to be

$$n(M) = j_1 + j_2 + 1 - |M|, \tag{2.6.26}$$

so that $N(J)$ is exactly 1 for $|j_1 - j_2| \leq J \leq j_1 + j_2$. We have thus proved that every pair of vectors with given (j_1, m_1, j_2, m_2) represents a state with total angular momentum in the range $J = |j_1 - j_2|$ to $j_1 + j_2$ and z-projection $M = m_1 + m_2$. It follows that an alternative set of basis states is provided by the vectors $|\alpha, j_1, j_2, J, M\rangle$. Using the closure relation we can express these vectors as a linear superposition of vectors of the original basis vectors $|\alpha, j_1, m_1, j_2, m_2\rangle$ (as in (2.6.8))

$$|\alpha, j_1, j_2, J, M\rangle = \sum_{m_1, m_2} |\alpha, j_1, m_1, j_2, m_2\rangle\langle\alpha, j_1, m_1, j_2, m_2|\alpha, j_1, j_2, J, M\rangle. \tag{2.6.27}$$

Because the angular momentum subspaces are independent of α the transformation coefficients in (2.6.27) are also independent of this set of eigenvalues. The coefficients $\langle j_1, m_1, j_2, m_2|j_1, j_2, J, M\rangle$ which are thus purely geometrical quantities are referred to as Clebsch–Gordan or Wigner coefficients.They appear with different notations in the literature. We will use the form $(j_1, m_1, j_2, m_2|J, M)$. These coefficients are obviously zero unless J lies in the range $|j_1 - j_2|$ to $j_1 + j_2$ and $M = m_1 + m_2$. They are the matrix elements of a unitary transformation and obey orthogonality relations

$$\sum_{m_1,m_2} (j_1, m_1, j_2, m_2|J, M)(j_1, m_1, j_2, m_2|J', M') = \delta_{JJ'}\delta_{MM'}, \qquad (2.6.28)$$

and

$$\sum_{J,M} (j_1, m_1, j_2, m_2|J, M)(j_1, m_1', j_2, m_2'|J, M) = \delta m_1 m_1' \delta m_2 m_2'. \qquad (2.6.29)$$

There are some further subtleties concerned with the relative phases of the angular momentum eigenvectors. The essential result is that it is possible to adopt a phase convention under which the Clebsch–Gordan coefficients are real positive numbers. The interested reader is referred to the work of Edmonds [24] or to that of Messiah [1]. The main use of angular momentum occurs in the next two chapters in relation to the presentation and application of the Hauser–Feshbach theory of compound nucleus decay.

This concludes our presentation of basic nuclear physics. In the next chapter we will be concerned with the historical developement of statistical models of nuclear decay.

References

[1] Messiah A 1961 *Quantum Mechanics* (Amsterdam: North-Holland)
[2] Frobrich P and Lipperheide R 1996 *Theory of Nuclear Reactions* (Oxford: Oxford University Press)
[3] Ericson T 1960 *Adv. Phys.* **9** 425
[4] Tuli J K 1995 *Nuclear Wallet Cards* 5th edn (Upton, NY: Brookhaven National Laboratory)
[5] Hodgson P E, Gadioli E and Gadioli Urba E 1997 *Introductory Nuclear Physics* (Oxford: Pergamon)
[6] Current address http://nucleardata.nuclear.lu.se/database/masses
[7] von Weizacker C F 1935 *Z. Phys.* **96** 431; Bethe H A and Bacher R F 1936 *Rev. Mod. Phys.* **8** 82
[8] Williams W S C 1991 *Nuclear and Particle Physics* (Oxford: Clarendon)
[9] Moller P, Nix J R, Myers W D and Swiatecki W J 1995 *Atomic Data and Nuclear Data Tables* **59** 185
[10] Siemens P J and Jensen A S 1987 *Elements of Nuclei* (New York: Addison-Wesley)
[11] Abramowitz M and Stegun I A 1970 *Handbook of Mathematical Functions* 9th edn (New York: Dover) Ch 10
[12] Cox D R and Miller H D 1965 *The Theory of Stochastic Processes* (London: Chapman and Hall)
[13] Breit G and Wigner E 1936 *Phys. Rev.* **49** 519
[14] Bohning M 1970 *Nucl. Phys.* A **152** 529

[15] Hardy G H and Ramanujan S 1918 *Proc. London Math. Soc. 2* **17** 75
[16] Toke J and Swiatecki W J 1981 *Nucl. Phys.* A **372** 141
[17] Iljinov J, De Sanctis E, Guaraldo C, Lucherini V, Muccifora V, Polli E, Reolon A R and Rossi P 1992 *Nucl. Phys.* A **543** 517
[18] Dilg W, Schantl W, Vonach H and Uhl M 1973 *Nucl. Phys.* A **217** 269
[19] Gilbert A and Cameron A G W 1965 *Can. J. Phys.* **43** 1446
[20] Mustafa M G, Blann M, Ignatyuk A and Grimes S M 1992 *Phys. Rev.* C **45** 1078; Weidenmueller H 1964 *Phys. Lett.* **10** 331
[21] Hilaire S, Delaroche J P and Koning A J 1998 *Nucl. Phys.* A **632** 417
[22] Brink D M and Satchler G R 1968 *Angular Momentum* (Oxford: Clarendon)
[23] Brandt S and Dahmen H D 1991 *Quantum Mechanics on the Macintosh* (New York: Springer)
[24] Edmonds A R 1957 *Angular Momentum in Quantum Mechanics* (Princeton: Princeton University Press)

PART 2

SINGLE AND MULTISTEP EVAPORATION

Chapter 3

History of statistical models of nuclear decay processes

A model is nothing else but the overstressing of certain factors which are responsible for the phenomenon under consideration.

Victor F Weisskopf

3.1 Introduction

In this and the following chapter we present elements of the history of statistical models of nuclear decay processes. We mainly concentrate on those subjects which are 'statistical' in the restricted sense that the theoretical treatment involves estimation of microstate densities. Thus some topics will be left aside. Studies pertaining to low excitation energies such as neutron resonances, and fluctuation cross-sections will not be discussed. The interested reader is referred to the works of Vogt [1] and Von Witsch *et al* [2], which provide rather detailed examples of these areas of research.

Any historical account must, of course, have a beginning and an end. One can naturally take the beginning to coincide with Bohr's statement of the independence hypothesis (section 3.2). The 'end' is harder to define especially in the presence of enthusiastic ongoing studies. We have decided somewhat arbitrarily to terminate our account somewhere in the first half of the eighties. This means that, in addition to the Weisskopf and Hauser–Feshbach models (sections 3.3–3.5), we consider as history fusion studies (section 3.6), pre-equilibrium emission (section 3.7) and the statistical model for fission at low excitation energies (section 3.8). Fusion is somehow the odd man out of this collection because the current understanding of this mechanism depends on dynamical rather than statistical considerations. Dynamical models are constantly being developed and improved so that fusion is as much a modern area of research as it is part of the history of nuclear reaction theories. However, inasmuch as the fusion mechanism determines the initial conditions for statistical decay a brief account is included herein.

The Weisskopf theory of compound nucleus decay [3], which seems to be the earliest statistical theory of decay of excited nuclei, can be explained essentially using classical statistical and geometrical arguments. The same is not true of the much used Hauser–Feshbach (HF) theory [4] which relies on the notion of the transmission coefficient. To provide a proper understanding of this quantity and of its use in the HF theory, we review essential aspects of the highly successful optical model of elastic scattering (section 3.4). The presentation of the HF theory itself is limited to the basic approximations and derivations. This is because we devote a significant part of the following chapter to the practical implementation of the theory for single and multistep evaporation. In chapter 4 we will also present examples of experimental investigations.

3.2 The Bohr independence hypothesis

In an address delivered before the Copenhagen Academy and published in the February 1936 issue of *Nature* [5], Niels Bohr presented the ideas which provided the conceptual basis for a statistical theory of nuclear reactions. The central notion is that of the compound nucleus. We can do no better than to quote Bohr directly.

> The phenomena of neutron capture thus force us to assume that a collision between a high speed neutron and a heavy nucleus will in the first place result in the formation of a compound system of remarkable stability. The possible later breaking up of this intermediate system by the ejection of a material particle or its passing with emission of radiation to a stable final state must in fact be considered as separate competing processes which have no immediate connection with the first stage of the encounter.

The reason offered by Bohr in support of this hypothesis is the 'close packing' of nucleons in atomic nuclei. Put slightly differently, the independence hypothesis relies on the short mean free path for the incident neutron and any scattered nucleon which implies multiple scatterings and a rapid sharing of the incident energy among several nucleons. At this stage the specific 'memory' of the entrance channel is lost and the subsequent evolution of the compound system depends only on conserved quantities. Bohr observed that the concentration of excitation energy in one particle which would be required for it to escape from the compound system would require a long time interval (compared with the 'loss of memory' time). He further emphasized the 'free competition between all the different possible processes of liberation of material particles and of radiative transitions which can take place from the semi-stable intermediate state of the compound system'.

Niels Bohr also speculated on the eventual modifications of the reaction scenario with increasing incident energy. He correctly predicted that for compound nuclei excited to several tens of MeV (he actually cited 100 MeV incident neutrons), not one, but several particles and gamma rays could be emitted. Even more remarkably he wrote 'For still more violent impacts, with particles of energies of about a thousand million volts, we must even be prepared for the collision to lead to the explosion of the whole nucleus'.

Given the degree of understanding of nuclear structure then available this set of predictions must be considered as a truly outstanding feat of intuition. Indeed, the ideas presented in Bohr's paper have stimulated experimental and theoretical investigations for over half a century.

We can resume the Bohr hypothesis by representing a reaction involving the compound nucleus as

$$a + A \to C \to B + b. \tag{3.2.1}$$

In (3.2.1) the collision of the projectile, a, on a target nucleus, A, forms the excited compound nucleus, C, which subsequently emits a particle (or a gamma ray), b, leaving a residual nucleus B. The independence hypothesis implies that the cross-section associated with the reaction $A(a, b)B$ can be written

$$\sigma(a+A \to B+b) = \sigma(a+A \to C)P(C \to B+b) = \sigma(a+A \to C)\frac{\Gamma_{C\to B+b}}{\Gamma}, \tag{3.2.2}$$

where $\sigma(a + A \to C)$ is the cross-section for formation of the compound system and $P(C \to B+b)$ is the probability for decay into the particular channel $(B+b)$ which, as explained in section 2.4, may be expressed as the ratio of partial to total widths. The excitation energy of the compound nucleus (assuming, as is usually the case, that a and A are in their respective ground states) is simply given by equation (2.2.18), i.e., $E_{\mathrm{cm}} + Q_{a+A\to C}$. The decay probabilities (or widths) can be written using various degrees of detail. Thus we may be interested simply in the relative probabilities for say, alpha and proton emission. Or, in more detail, we may wish to investigate the energy spectra for any given emitted particle (which is measurable) or even the relative angular momentum between b and B (which is not directly measurable).

Later work in the 1950s established the existence of a new class of nuclear reactions referred to as direct reactions [6,7]. Such 'fast' reactions take place for large impact parameter collisions when the products of the primary interaction of the projectile with the target nucleus have a high probability of escaping. This was one point that Bohr seems not to have considered, i.e., that the energy sharing process is not efficient (even from simple geometrical considerations) at the surface of target nuclei.

The problem of experimental verification of the independence hypothesis seems to have been first tackled by Ghoshal in Berkeley [8]. A given compound nucleus may be formed using different entrance channels. Ghoshal used the reactions p + ^{63}Cu and α + ^{60}Ni to form the compound system ^{64}Zn. If the independence hypothesis holds up, the ratios of cross-sections for different exit channels (n, $2n$, pn ...) should be independent of the entrance channel for a fixed compound nucleus excitation energy E_C^*. The difference in this excitation for the two entrance channels with the *same* beam energy was taken to be a fixed shift in energy due to the difference in reaction Q-values for the $p + {}^{63}Cu$ and $\alpha + {}^{60}Ni$ fusion processes. By displaying cross-sections as a function of the projectile energy and shifting the proton-induced cross-sections down by 7 MeV, Ghoshal concluded that the independence hypothesis was verified. In fact, because of the centre-of-mass energy correction (equation (2.2.18)) it is easy to see that the introduction of a universal shift is incorrect. Using this equation together with (2.2.4) the correct shifts are 4.30 MeV at an alpha particle energy of 10 MeV and 5.72 MeV at 40 MeV.

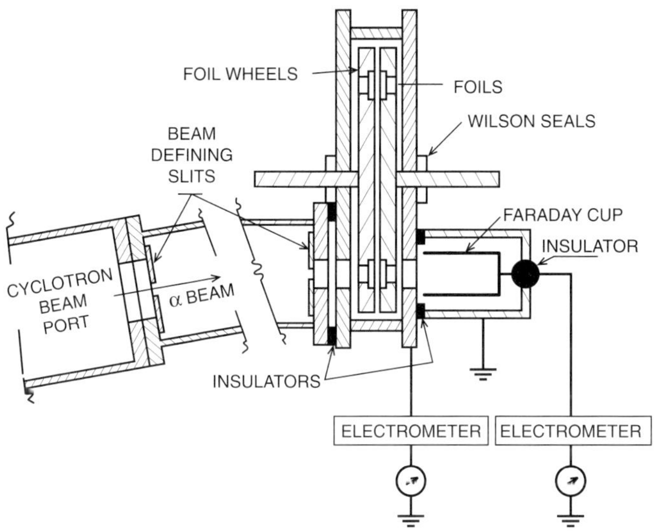

Figure 3.1. Experimental set-up used by Walter John for measurement of evaporation excitation functions [9]. The α particle beam (shown on the left hand side of the figure) impinges on one of several target positions arranged around the circumference of two rotatable foil wheels. Beam energy degrader foils are situated on the left-hand wheel, and the right-hand wheel holds the Pb target foils. The nuclei produced by evaporation were identified and the production rate estimated, by measuring radioactive decays 'off-line' after typically 10 h of bombardment. The time-integrated intensity of the beam is measured in the Faraday cup which is depicted on the lower right-hand side of the figure. (Reprinted from *Phys. Rev.*, **103**, John W, p 704, Copyright 1956, by the American Physical Society.)

Ghoshal's experiment was repeated by John [9] who compared (α, Xn) reactions on lead isotopes with earlier measurements made using the p + Bi entrance channel. A sketch of the apparatus used by John is shown in figure 3.1. With the help of aluminium absorber foils, the α beam energy was varied between

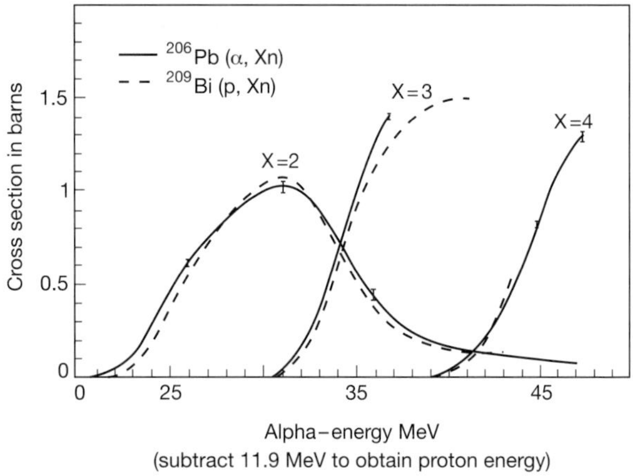

Figure 3.2. Experimental measurements [9] of ^{206}Pb(α, $2n$)^{208}Po, ^{206}Pb(α, $3n$)^{207}Po and ^{206}Pb(α, $4n$)^{206}Po excitation functions compared with those leading to the same isotopes produced in the ^{209}Bi(p, Xn) reaction. (Reprinted from *Phys. Rev.*, **103**, John W, p 704. Copyright 1956, by the American Physical Society.)

48 and 18 MeV. The target and beam energy degraders were mounted on two 'foil wheels'. The various activities were measured off-line after typically 10 h bombardment and used to estimate cross-sections. Results from John's paper are shown in figure 3.2. Good agreement between the excitation functions of (α, Xn) and (p, Xn) channels is obtained using an energy shift of 11.9 MeV which may be compared with the expected shift of 10.4 MeV. Part of this discrepancy (0.5 MeV) can be attributed to the centre-of-mass energy correction, but the remaining 1.0 MeV (1.6 MeV in John's paper due to imprecisions in the then available mass excesses) was not explained. The results obtained thus agreed qualitatively with those of Ghoshal.

The unexplained energy shift (from 1 to 2 MeV) observed in these two experiments is nowadays attributed to differences in angular momentum brought into the respective compound systems by the α and proton projectiles. Indeed, equation (2.2.6) shows that an α particle brings in about twice as much angular momentum as a proton at the same beam energy and impact parameter. The compound nuclei created are therefore not 'identical'. We thus conclude that a *precise* test of the independence hypothesis along the lines envisaged by Ghoshal or John, is, in fact, impossible. On the other hand, the striking similarity of the observed excitation functions together with the plausibility of an angular momentum dependent energy shift leads most observers to consider these results as strong support for Bohr's hypothesis.

Classically, the angular momentum brought into the compound nucleus is a vector which lies in the plane perpendicular to the beam direction. Even if the projectile and target have intrinsic angular momenta (spins) this remains true on average. Loss of memory of the entrance channel (except for conserved quantities) thus implies that angular distributions of emitted particles are symmetric with respect to the plane perpendicular to the beam direction in the centre-of-mass system. This was verified in several experiments. Figure 3.3 which is taken from measurements of Durham and Halbert [10] on the $^{16}O(^{14}N, \alpha)^{26}Al$ reaction is one example. This measurement thus provides evidence for the formation of the compound nucleus ^{30}P.

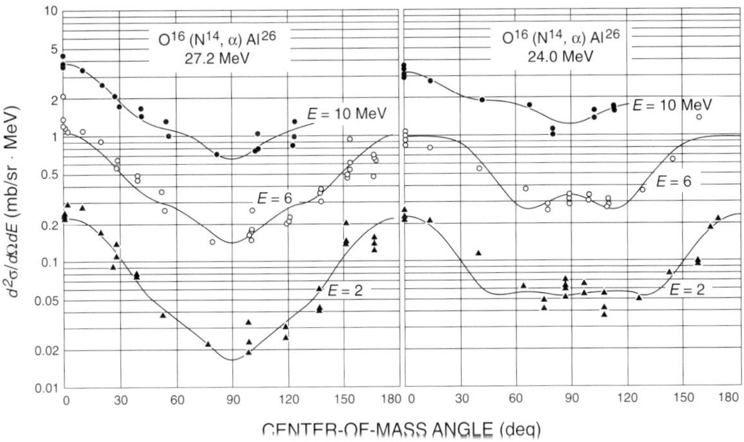

Figure 3.3. Differential cross-sections measured in the centre-of-mass system for the reaction $^{16}O(^{14}N, \alpha)^{26}Al$ at two energies [10]. The labels '$E = 10$ MeV' etc, refer to the excitation energy in the residual nucleus. In each case the cross-sections were averaged over intervals of 1 MeV. The observed forward backward asymmetry is expected for compound nucleus decay as explained in the text. (Reprinted from *Phys. Rev.*, **137**, Durham E F and Halbert M L, p B850. Copyright 1965, by the American Physical Society.)

3.3 The Weisskopf theory of evaporation from the compound nucleus

At the time of Bohr's work, it seems that Frenkel [11] had already realized that it may be possible to compare the emission of particles from an excited compound nucleus with the evaporation of molecules from a fluid. However, the first statistical theory of compound nuclear decay is generally associated with the name of Weisskopf [3]. We now present and comment on this theory which is based on consideration of the 'inverse' reaction $b + B \rightarrow C$. Although the original paper gave few details concerning the physical picture involved, it is not difficult to obtain the result in a rather transparent manner and to relate the derivation to the transition state method which will appear later on in connection with the Bohr–Wheeler theory of neutron emission and fission.

We consider decays of the compound nucleus, C, with excitation energy E_C^* in which a particle, b, is emitted with kinetic energy, ε_b, to leave a residual nucleus, B. For simplicity of presentation, we assume the nucleus B to be sufficiently massive so that it may be considered to be stationary in the b–B centre-of-mass system (i.e., the velocity of the centre of mass ≈ 0). The energy, ε_b, should therefore be thought of as the kinetic energy of b *relative* to B. After a time interval, Δt, the particle b is assumed to lie within a fixed volume V centred on B. We take this volume to be a sphere although there is no obligation to do so. Decays from an ensemble of similarly prepared compound nuclei which take place in a time interval Δt will populate the volume V in a rather particular manner (see figure 3.4(a)). Furthermore, there is a one-to-one correspondence between *decaying* compound nucleus microstates, whose number we write as $\Omega_{C\to bB}(\Delta t)$, and the *decayed* states produced in V so that $\Omega_{C\to bB}(\Delta t)$ also represents states in which b and B are to be found somewhere in the volume V. Assuming the defining property of equilibrium, the probability of decay in the time interval Δt from any state of the compound nucleus is given as the number of compound nuclear states which decay in Δt divided by the total number. Defining $W_{C\to bB}$ as the decay probability per unit time (or specific decay rate) we can thus write

$$\Delta t W_{C\to bB} = \frac{\Omega_{C\to bB}(\Delta t)}{\Omega_C(E_C^*)} = \frac{\rho_{C\to bB}(\Delta t)}{\rho_C(E_C^*)}. \tag{3.3.1}$$

For a fixed (unreplenished) number of source nuclei, the quantity W corresponds to the radioactive decay constant, λ (section 2.4). In writing equation (3.3.1) we have assumed that the number of microstates in the compound nucleus, $\Omega_C(E_C^*)$, together with the subset of these states which decays to $b + B$ in Δt, can be accurately represented by continuous microstate densities. We now *invent* a new set of microstates for the pair b, B with density ρ_{bB} constrained by the volume V such that $\rho_{C\to bB}(\Delta t)$ forms a subset of ρ_{bB}. This condition can be fulfilled by choosing a suitably small value of Δt (or a large volume V). Multiplying equation (3.3.1) by unity (ρ_{bB}/ρ_{bB}) and dividing by Δt we obtain

$$W_{C\to bB} = \frac{\Gamma_{C\to bB}}{\hbar} = \frac{1}{\Delta t \rho_C(E_C^*)} \frac{\rho_{C\to bB}(\Delta t)}{\rho_{bB}} \rho_{bB}, \tag{3.3.2}$$

in which, for future reference, we have defined the decay width $\Gamma_{C\to bB}$. We can identify the fraction $\rho_{C\to bB}(\Delta t)/\rho_{bB}$ as the probability that states belonging to ρ_{bB} result from decay of the compound nucleus. We can also *reverse the arrow of time* and identify this same quantity as the fraction of all states belonging to ρ_{bB} which lead to *formation* of the compound nucleus in time Δt. In the classical picture, this simply amounts to reversing the directions of trajectories (running the 'film' backwards).

Following Weisskopf, we consider a particular contribution to the decay probability per unit time for which the density of states ρ_{bB} corresponds to the particle b constrained to lie in V with a randomly oriented velocity restricted

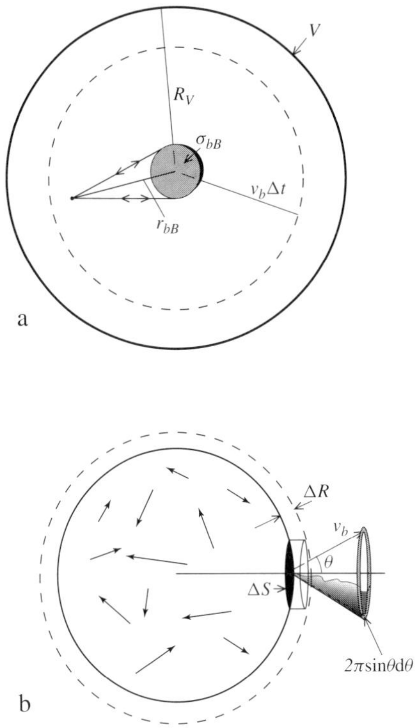

Figure 3.4. Evaporation. (*a*) The Weisskopf picture. Only those trajectories contained in the solid angle subtended by the cross-section σ_{bB} represent phase space states resulting from the decay of the compound nucleus. It does not matter whether the corresponding trajectories are considered to be time reversed although this picture was employed by Weisskopf. The remaining symbols are explained in the text. (*b*) Evaporation considered as a (small) flux of non-interacting particles through a surface element of a container held at constant temperature, T. The solid angle associated with evaporation at an angle, θ, is represented as a partially shaded conical shell.

to the interval $(v_b, v_b + \mathrm{d}v_b)$. Of all possible microstates corresponding to this specification, we now seek the fraction of microstates which will certainly lead to formation of a compound nucleus in a time Δt. In order for the particle b, at distance r_{bB} from the centre of B, to intercept the section, σ_{bB} (the cross-section for the inverse reaction $b + B \to C$), its direction must be restricted to a fraction

$$p(r_{bB}) = \frac{\sigma_{bB}}{4\pi r_{bB}^2}, \tag{3.3.3}$$

of the total solid angle (4π). The average value, $\langle p \rangle$, is the required fraction of microstates. In the interval Δt, only those microstates for which the separation, r_{bB}, is less than $v_b\Delta t$ can possibly lead to formation of compound nuclei. Thus, if $r_{bB} > v_b \Delta t$, $p(r_{bB}) = 0$. In calculating $\langle p \rangle$ we may further take the volume, V, to be much greater than the volume occupied by the nucleus B so that little error is introduced by assuming equation (3.3.3) to hold down to $r_{bB} = 0$. The average of $p(r_{bB})$ over the volume V (with radius R_V) is then given by

$$\begin{aligned} \langle p \rangle &= \frac{1}{V} \int_{r_{bB}=0}^{R_V} p(r_{bB}) 4\pi r_{bB}^2 \, \mathrm{d}r_{bB} \\ &= \frac{1}{V} \int_{r_{bB}=0}^{v_b \Delta t} \frac{\sigma_{bB}}{4\pi r_{bB}^2} 4\pi r_{bB}^2 \, \mathrm{d}r_{bB} = \frac{\sigma_{bB} v_b \Delta t}{V}, \end{aligned} \tag{3.3.4}$$

which we now identify with the average transition probability in time Δt from states of the system $(b + B)$ to the time-reversed decaying states of the compound nucleus, C or inversely, from states of C to states of $(b + B)$

$$\frac{\rho_{C \to bB}(\Delta t)}{\rho_{bB}} = \frac{\rho_{bB \to C}(\Delta t)}{\rho_{bB}} = \frac{\sigma_{bB} v_b \Delta t}{V}. \tag{3.3.5}$$

Thus, from (3.3.2), the contribution to the specific transition rate for emitted particles with velocity in the range $(v_b, v_b + \mathrm{d}v_b)$ is given by

$$\begin{aligned} \frac{\mathrm{d}W_{C \to bB}}{\mathrm{d}v_b} \mathrm{d}v_b &= \frac{\sigma_{bB} v_b}{V} \frac{1}{\rho_C(E_C^*)} \frac{\mathrm{d}\rho_{bB}}{\mathrm{d}v_b} \mathrm{d}v_b \\ &= \frac{\mathrm{d}W_{C \to bB}}{\mathrm{d}\varepsilon_b} \mathrm{d}\varepsilon_b = \frac{\sigma_{bB}}{V} \left(\frac{2\varepsilon_b}{m_b} \right)^{1/2} \frac{1}{\rho_C(E_C^*)} \frac{\mathrm{d}\rho_{bB}}{\mathrm{d}\varepsilon_b} \mathrm{d}\varepsilon_b, \end{aligned} \tag{3.3.6}$$

in which the second line has been written in terms of the kinetic energy of the particle b, $\varepsilon_b = \frac{1}{2} m_b v_b^2$, rather than the velocity.

The calculation of the density of microstates, ρ_{bB}, to be associated with the configuration $b + B$ is straightforward. We assume the particle b to have been emitted in its ground state and denote the excitation energy of the residual nucleus as E_{B}^*. The density of microstates corresponding to the macrostate $b + B$ is then obtained as the convolution of two densities. The first is the phase space density for the pair (b, B). Given that the (heavy) residue, B, is considered as stationary (with zero energy) and located at the centre of V, the phase space contribution to the density of microstates is given by the one particle phase space integral (corresponding to b) which, for a particle of mass, m_b, and energy, ε_b, is expressed, following equation (1.3.7), as

$$\rho_b(\varepsilon_b) = \frac{g_b V (2\pi m_b)^{3/2} \varepsilon_b^{1/2}}{h^3 \Gamma(3/2)}, \tag{3.3.7}$$

where the factor, g_b, is a quantum mechanical correction which gives the number of possible spin orientations of particle b (i.e., $g_b = 2s_b + 1$). The second density is, of course, the density of microstates in the excited nucleus B which we write as $\rho_{\mathrm{B}}(E_{\mathrm{B}}^*)$.

Let the total energy in the system $b+B$ be $E_{bB} = \varepsilon_b + E^*_{\mathrm{B}} = E^*_C + Q_{C\to b+B}$. The corresponding density of microstates is then the convolution

$$\rho_{bB}(E_{bB}) = \int_0^{E_{bB}} \int_0^{E_{bB}} \rho_b(\varepsilon_b)\rho_{\mathrm{B}}(E^*_{\mathrm{B}})\delta(\varepsilon_b + E^*_{\mathrm{B}} - E_{bB})\,\mathrm{d}\varepsilon_b\,\mathrm{d}E^*_{\mathrm{B}}$$
$$= \int_0^{E_{bB}} \rho_b(\varepsilon_b)\rho_{\mathrm{B}}(E_{bB} - \varepsilon_b)\,\mathrm{d}\varepsilon_b. \qquad (3.3.8)$$

The density of states corresponding to a kinetic energy of b contained in an interval $(\varepsilon_b, \varepsilon_b + \mathrm{d}\varepsilon_b)$ is thus

$$\frac{\mathrm{d}\rho_{bB}}{\mathrm{d}\varepsilon_b}\,\mathrm{d}\varepsilon_b = \rho_b(\varepsilon_b)\rho_{\mathrm{B}}(E_{bB} - \varepsilon_b)\,\mathrm{d}\varepsilon_b = \frac{g_b V (2\pi m_b)^{3/2}\varepsilon_b^{1/2}}{h^3\Gamma(3/2)}\rho_{\mathrm{B}}(E_{bB} - \varepsilon_b)\,\mathrm{d}\varepsilon_b. \qquad (3.3.9)$$

Using (3.3.6) and (3.3.9), the transition probability per unit time for emitted particle kinetic energies in the range $(\varepsilon_b, \varepsilon_b + \mathrm{d}\varepsilon_b)$ is obtained from the corresponding transition probability (3.3.4) for the inverse process as

$$\frac{\mathrm{d}W_{c\to b+B}}{\mathrm{d}\varepsilon_b}\,\mathrm{d}\varepsilon_b = \frac{\sigma_{bB}}{V}\left(\frac{2\varepsilon_b}{m_b}\right)^{1/2} \frac{g_b V (2\pi m_b)^{3/2}\varepsilon_b^{1/2}}{h^3\Gamma(3/2)} \frac{\rho_{\mathrm{B}}(E_{bB} - \varepsilon_b)}{\rho_C(E^*_C)}\,\mathrm{d}\varepsilon_b$$
$$= \sigma_{bB}\frac{8\pi g_b m_b \rho_{\mathrm{B}}(E_{bB} - \varepsilon_b)}{h^3 \rho_C(E_{bB} - Q_{C\to b+B})}\varepsilon_b\,\mathrm{d}\varepsilon_b, \qquad (3.3.10)$$

where we have used $\Gamma(3/2) = \sqrt{\pi}/2$. Equation (3.3.10) represents the Weisskopf theory for compound nuclear decay (this equation is not *literally* the formula given by Weisskopf because he denoted by 'h' the symbol which is nowadays referred to as $\hbar = h/2\pi$). Notice that by defining ε_b as the energy of b *relative* to B the compound nucleus excitation energy in the inverse reaction is correctly given by $E^*_C = E^*_{\mathrm{B}} + Q_{b+B\to C} + \varepsilon_b$.

The choice of the phase space configuration of $b + B$ is of little importance provided that one can calculate consistently both the density of microstates and the average fraction of microstates leading to formation of the compound nucleus in the interval Δt. We shall see later (in section 3.8) that a rather different construction produces precisely the same result.

Weisskopf further remarked that, quite generally, by using the definition of microcanonical entropy, $S(E) = \ln[\rho(E)]$, equation (3.3.10) can be written

$$\frac{\mathrm{d}W_{C\to b+B}}{\mathrm{d}\varepsilon_b}\,\mathrm{d}\varepsilon_b = \frac{\sigma_{bB} 8\pi g_b m_b}{h^3}\,\mathrm{e}^{S_{\mathrm{B}}(E_{bB}-\varepsilon_b) - S_C(E_{bB}-Q_{C\to b+B})}\varepsilon_b\,\mathrm{d}\varepsilon_b. \qquad (3.3.11)$$

Then, *assuming* the nuclei B and C have the same microstate densities in the energy region around E_{bB}, i.e., $S_{\mathrm{B}}(E_{bB}) = S_C(E_{bB})$, and that we can use the first order development $S(E_{bB} + \Delta E) = S(E_{bB}) + \beta\Delta E$ we obtain, as a rather strong approximation

$$\frac{\mathrm{d}W_{C\to b+B}}{\mathrm{d}\varepsilon_b}\,\mathrm{d}\varepsilon_b = \frac{\sigma_{bB} 8\pi g_b m_b}{h^3}\,\mathrm{e}^{\beta Q_{C\to b+B}}\varepsilon_b\,\mathrm{e}^{-\beta\varepsilon_b}\,\mathrm{d}\varepsilon_b. \qquad (3.3.12)$$

Equation (3.3.12) not only predicts an exponential dependence of the specific decay rate on corresponding Q-values but also provides the form of the expected kinetic energy spectrum.

A particular point of interest in the Weisskopf decay theory is that the kinetic energy dependence is of the same form as the spectrum of particles evaporated from an ideal fluid composed of non-interacting, structureless particles held at a fixed temperature T inside a volume V. Given that the particles are assumed to be independent (not to interact), the physical properties of such a system can be investigated by considering the case of a single particle. If the particle mass is m, then, by extending the reasoning given in section 1.2 (see after (1.2.15)), the total number of microstates for which the particle momentum lies in the range $(p, p + \mathrm{d}p)$ is given (to within a constant equal to the number of heat bath microstates at total energy E) by

$$\frac{\mathrm{d}\Omega(p)}{\mathrm{d}p}\,\mathrm{d}p = \frac{V}{h^3}\,\mathrm{e}^{-p^2/2mT}\,4\pi p^2\,\mathrm{d}p, \tag{3.3.13}$$

which by noting that the energy $\varepsilon = p^2/2m$ can immediately be rewritten as

$$\rho(\varepsilon)\,\mathrm{d}\varepsilon = \frac{2\pi V}{h^3}(2m)^{3/2}\varepsilon^{1/2}\,\mathrm{e}^{-\varepsilon/T}\,\mathrm{d}\varepsilon, \tag{3.3.14}$$

so that the energy spectrum measured in the interior of the container goes as $\varepsilon^{1/2}\,\mathrm{e}^{-\varepsilon/T}$. From (3.3.14) we may also obtain (by integrating over ε) that the total number of single particle microstates is

$$\Omega = \frac{V}{h^3}(2\pi mT)^{3/2}. \tag{3.3.15}$$

Suppose now that the particle can escape (evaporate) through a small hole with surface area ΔS in the container wall, and let us define a set of 'transition' microstates such that if a particle with velocity, $v = p/m$, occupies any one of these states it will certainly escape in the time interval Δt. As in the Weisskopf model, the probability that the particle will evaporate in a time Δt is equal to the number of transition microstates divided by the total number. In order to isolate transition microstates we consider a volume element $\Delta R \Delta S$ where ΔR is the thickness of a shell at the surface of the container such that $\Delta R/\Delta t > v$ (this condition can be ensured by suitable choice of Δt). We select microstates in this volume such that v makes an angle θ with the outwardly pointing normal to the surface element ΔS. The states included are thus all those in a solid angle $\mathrm{d}\omega = 2\pi \sin\theta\,\mathrm{d}\theta$ (figure 3.4(b)). The fraction of microstates corresponding to particles in the interval $(\theta, \theta + \mathrm{d}\theta)$ crossing ΔR in the outward direction, and thereby escaping from the container in time Δt is then

$$f(\theta) = \Delta t\frac{v\cos\theta}{\Delta R} \qquad \left[0 \le \theta \le \frac{\pi}{2}\right],$$
$$f(\theta) = 0 \qquad \left[\frac{\pi}{2} < \theta \le \pi\right]. \tag{3.3.16}$$

The average value of this fraction taken over all directions is obtained by integrating over solid angles and dividing by 4π

$$\langle f \rangle = \frac{1}{4\pi}\int_{-1}^{1} 2\pi f(\theta)\,\mathrm{d}\cos\theta = \frac{v\Delta t}{\Delta R}\frac{1}{4\pi}\int_{0}^{1} 2\pi\cos\theta\,\mathrm{d}\cos\theta = \frac{v\Delta t}{4\Delta R}. \tag{3.3.17}$$

Now the number of states contained in $\Delta V = \Delta R \Delta S$ with momenta in the range $(p, p + \mathrm{d}p)$ is

$$\frac{\mathrm{d}\Omega_{\Delta V}(p)}{\mathrm{d}p}\,\mathrm{d}p = \frac{\Delta R \Delta S\,\mathrm{e}^{-\varepsilon/T}\,4\pi p^2\,\mathrm{d}p}{h^3}, \tag{3.3.18}$$

so that the corresponding number of transition microstates which is given by multiplying (3.3.17) by (3.3.18) is

$$\frac{\mathrm{d}\Omega_t(\Delta t)}{\mathrm{d}p}\,\mathrm{d}p = \frac{v\Delta t}{4\Delta R} \times \frac{\Delta R \Delta S\,\mathrm{e}^{-\varepsilon/T}\,4\pi p^2\,\mathrm{d}p}{h^3}. \tag{3.3.19}$$

Noting that $\mathrm{d}\varepsilon = v\,\mathrm{d}p$ and that

$$\frac{\mathrm{d}\Omega_t}{\mathrm{d}p}\,\mathrm{d}p = \frac{\mathrm{d}\Omega_t}{\mathrm{d}\varepsilon}\,\mathrm{d}\varepsilon,$$

(3.3.19) can be written

$$\rho_t(\Delta_t)\,\mathrm{d}\varepsilon = \frac{\mathrm{d}\Omega_t(\Delta t)}{\mathrm{d}\varepsilon}\,\mathrm{d}\varepsilon = \frac{\Delta t \Delta S\,2\pi m\varepsilon\,\mathrm{e}^{-\varepsilon/T}\,\mathrm{d}\varepsilon}{h^3}. \tag{3.3.20}$$

The decay probability per unit time and per unit surface for a single particle with kinetic energy in the range $(\varepsilon, \varepsilon + \mathrm{d}\varepsilon)$ is now obtained by dividing (3.3.20) by the total number of microstates, Ω (3.3.15), and by Δt and ΔS. Thus

$$\begin{aligned}\frac{\mathrm{d}^3 P}{\mathrm{d}S\,\mathrm{d}t\,\mathrm{d}\varepsilon}\,\mathrm{d}\varepsilon &= \frac{\rho_t(\Delta t)}{\Omega \Delta t \Delta S}\,\mathrm{d}\varepsilon \\ &= \frac{2\pi m\varepsilon\,\mathrm{e}^{-\varepsilon/T}\,\mathrm{d}\varepsilon}{\Omega h^3} = \frac{\varepsilon\,\mathrm{e}^{-\varepsilon/T}\,\mathrm{d}\varepsilon}{VT^{3/2}(2\pi m)^{1/2}},\end{aligned} \tag{3.3.21}$$

so that if the volume, V, encloses a large number of independent particles, N, the number which evaporate per unit surface and per unit time is obtained by multiplying (3.3.21) by N and performing the integral over ε. We thus obtain (with the particle density $\rho_{\mathrm{p}} = N/V$)

$$\frac{\mathrm{d}^2 N}{\mathrm{d}S\,\mathrm{d}t} = \frac{\rho_{\mathrm{p}} T^{1/2}}{(2\pi m)^{1/2}}. \tag{3.3.22}$$

Alternatively, we could have reasoned in terms of particles rather than in terms of microstates. The flux through ΔS for particles with velocity v directed such that the angle between v and the outward normal to the surface is θ is simply $\Delta S\ \rho_{\mathrm{p}} v \cos\theta$. It is then sufficient to multiply this flux by the probability that a particle has velocity and direction in the interval $(v, v + \mathrm{d}v; \theta, \theta + \mathrm{d}\theta)$. The appropriate probability distribution which is referred to as Maxwell's distribution (obtained, assuming randomly directed velocities, by normalizing $\mathrm{e}^{-\varepsilon/T}$ and expressing the result in terms of polar angle, θ, and velocity, v) is

$$\frac{\mathrm{d}^2 P_{\mathrm{M}}}{\mathrm{d}\theta\,\mathrm{d}v}\,\mathrm{d}\theta\,\mathrm{d}v = \frac{m^{3/2}}{(2\pi T)^{3/2}}\,\mathrm{e}^{-mv^2/2T}\,2\pi\sin\theta\,\mathrm{d}\theta\,v^2\,\mathrm{d}v, \tag{3.3.23}$$

which by transforming to Cartesian coordinates ($2\pi\sin\theta\,\mathrm{d}\theta\,v^2\,\mathrm{d}v = \mathrm{d}v_x\,\mathrm{d}v_y\,\mathrm{d}v_z$ and $\mathrm{e}^{-mv^2/2T} = \mathrm{e}^{-mv_x^2/2T}\,\mathrm{e}^{-mv_y^2/2T}\,\mathrm{e}^{-mv_z^2/2T}$) is easily shown to be correctly normalized so that

$$\int_0^\infty \mathrm{d}v \int_0^\pi \mathrm{d}\theta\,\frac{\mathrm{d}^2 P_{\mathrm{M}}}{\mathrm{d}v\,\mathrm{d}\theta} = 1.$$

The sum over angles of the flux is obtained by multiplying the flux by (3.3.23) and integrating over (solid) angles for which $\cos\theta$ is positive ($0 \le \theta \le \frac{\pi}{2}$). We obtain

$$\frac{\mathrm{d}^2 N}{\mathrm{d}t\,\mathrm{d}v} = \frac{m^{3/2}}{(2\pi T)^{3/2}}\,2\pi\,\Delta S\ \rho_{\mathrm{p}} v^3\,\mathrm{e}^{-mv^2/2T}\int_0^1 \cos\theta\,\mathrm{d}\cos\theta. \tag{3.3.24}$$

By dividing by ΔS, and substituting $\varepsilon = mv^2/2$ equation (3.2.24) yields $\left(v^3\,\mathrm{d}v = v^2\,\mathrm{d}v^2/2 = 2\varepsilon\,\mathrm{d}\varepsilon/m^2\right)$

$$\frac{\mathrm{d}^3 N}{\mathrm{d}S\,\mathrm{d}t\,\mathrm{d}\varepsilon} = \frac{2\pi\rho_{\mathrm{p}}}{m^2}\,\frac{m^{3/2}\varepsilon\,\mathrm{e}^{-\varepsilon/T}}{(2\pi T)^{3/2}} = \frac{\rho_{\mathrm{p}}}{T^{3/2}}\,\frac{\varepsilon\,\mathrm{e}^{-\varepsilon/T}}{(2\pi m)^{1/2}}, \tag{3.3.25}$$

which, when further divided by N, is identical with (3.3.21). The energy spectrum is therefore identical with the Weisskopf kinetic energy spectrum (equation (3.3.12)). This observation has led to the use of the term 'evaporation' in describing particle emission from the compound nucleus. We also note that the rate of evaporation through a surface element ΔS is obtained by multiplying (3.3.21) by ΔS and that the corresponding rate through a spherical surface is obtained by replacing ΔS by $4\pi R^2$. Then, by associating the cross-section, σ_{bB}, with πR^2, we obtain a result which is similar to that of Weisskopf. This will be more apparent when we discuss the transition state theory for neutron emission and fission in section 3.8.

Weisskopf himself appears to have considered equation (3.3.10) as a qualitative rather than a quantitative implementation of the Bohr hypothesis [12]. Nevertheless, it is still considered as a useful starting point for estimating cross-sections. Figure 3.5(a) shows an example of a neutron energy spectrum which nicely exhibits the expected evaporation form, and in figure 3.5(b) we show the excitation function for the (α, p) reaction on ^{58}Ni together with the Weisskopf model prediction made by Ait-Tahar [13]. For light compound systems one should improve the theory by removing the approximation of treating the residual nucleus B as a stationary heavy mass. Another possible improvement concerns the specification of the inverse cross-section σ_{bB}. In principle, this should be the cross-section for formation of the compound nucleus when an *excited* nucleus B is bombarded with a projectile b. In practice the effect of the excitation energy of the nucleus B is usually ignored and σ_{bB} is taken as the absorption cross-section (2.2.16). Finally, we note that, as formulated, there is no *a priori* inclusion of angular momentum. This is a fundamental drawback because, as we learnt in the previous chapter, the densities of nuclear states are rather strongly angular momentum dependent. This difficulty is perhaps the main reason why the Weisskopf theory was relatively rapidly superseded by the quantum mechanical Hauser–Feshbach theory.

We have now finished with the description of the Weisskopf evaporation theory. Before leaving this topic, however, it is most instructive to consider a development of Weisskopf's model which leads directly to an important result known as the reciprocity theorem. We begin by removing the restriction of treating the residual nucleus B as a stationary heavy mass (the one particle phase space approximation) and, at the same time, work in the centre-of-mass coordinate system in which the position of the centre of mass and the total momentum of the system b, B are fixed. Using the phase space density obtained in section 1.2 (equation (1.2.32)) we thus rewrite equation (3.3.10). In so doing, it is convenient to use the symbol β to refer to the channel $b+B$ so that the centre-of-mass kinetic energy which replaces ε_b in the calculation of the density of phase space states is written ε_β and the reduced mass, which replaces m_b, is written as μ_β (note that v_b in equation (3.3.10) is the relative velocity of the pair (bB) and is thus independent of the choice of reference system so that $v_b = v_\beta$). We also make a change of variable. Thus, using the fact that $E_{bB} = E^*_{\mathrm{B}} + \varepsilon_b (|\,\mathrm{d}\varepsilon_b/\,\mathrm{d}E^*_{\mathrm{B}}| = 1)$ we can replace the differential element $\mathrm{d}\varepsilon_b$ in the original Weisskopf result by $\mathrm{d}E^*_{\mathrm{B}}$. Under these circumstances, for the evaporation of spinless particles ($g_b = 1$), the (improved) Weisskopf result can be written

$$\frac{\mathrm{d}W_{C\to b+B}}{\mathrm{d}E^*_{\mathrm{B}}}\,\mathrm{d}E^*_{\mathrm{B}} = \frac{1}{\hbar}\frac{\mathrm{d}\Gamma_\beta}{\mathrm{d}E^*_{\mathrm{B}}}\,\mathrm{d}E^*_{\mathrm{B}} = \sigma_{bB}\frac{8\pi\varepsilon_\beta\mu_\beta\rho_{\mathrm{B}}(E^*_{\mathrm{D}})}{h^3\rho_C(E^*_C)}\,\mathrm{d}E^*_{\mathrm{B}}$$

$$= \frac{\sigma_{bB}}{2\pi^2\hbar}\frac{2\mu_\beta\varepsilon_\beta}{\hbar^2}\frac{\rho_{\mathrm{B}}(E^*_{\mathrm{B}})}{\rho_C(E^*_C)}\,\mathrm{d}E^*_{\mathrm{B}}. \qquad (3.3.26)$$

We could further compress the notation by writing the cross-section for formation of the compound nucleus from $b + B$ as σ_β. For the present, however, we retain the more explicit σ_{bB}. Now suppose we integrate over a small region of E^*_{B} (over which ε_β can be considered to be constant) which we assume contains only one excited state of the residual nucleus B. We then find (the square of the wave number in channel β is $k^2_\beta = 2\mu_\beta\varepsilon_\beta/\hbar^2$)

$$\Gamma_\beta(k_\beta) = \frac{\sigma_{bB}}{2\pi^2}\frac{k^2_\beta}{\rho_C(E^*_C)}. \qquad (3.3.27)$$

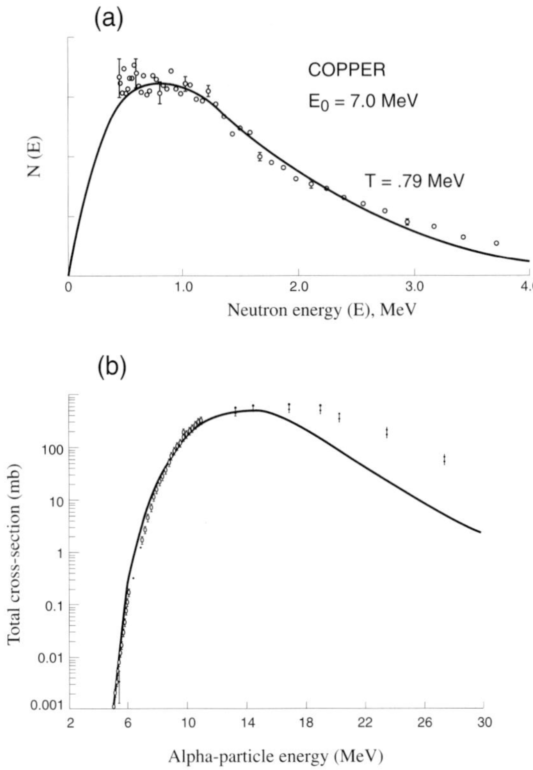

Figure 3.5. (*a*) Energy spectrum of neutrons emitted in the (n, n') reaction on copper for incident neutrons of 7 MeV. The measurement was made at Los Alamos by D B Thomson and reported by V F Weisskopf (1961 *Physics Today* **14** 18) The continuous curve is a prediction using the evaporation model. (*b*) Weisskopf model prediction (continuous line) of the excitation function for the ^{58}Ni (α, p) reaction compared with experimental measurements made by Ait-Tahar [13]. (Reprinted from *Pre-Equilibrium Nuclear Reactions*, Gadioli E and Hodgson P E (1992) by permission of Oxford University Press.)

Since β is an arbitrary channel equation (3.3.27) provides a general relation between the width for decay and the cross-section for formation of the compound nucleus. If we now consider any two channels with projectile energies which lead to the same excitation, E_C^*, in the compound nucleus we find that

$$\frac{\sigma_{bB} k_\beta^2}{\Gamma_\beta(k_\beta)} = 2\pi^2 \rho_C(E_C^*) = \text{const.} = \frac{\sigma_{aA} k_\alpha^2}{\Gamma_\alpha(k_\alpha)}, \tag{3.3.28}$$

which, by dividing on both sides by the total width $\Gamma = \sum_\gamma \Gamma_\gamma$ and noting that $\sigma_{aA}\Gamma_\beta/\Gamma$ is the expression of the Bohr hypothesis for the cross-section $\sigma_{\alpha\beta}$

(explicitly, $\sigma_{a+A\to C\to b+B}$), leads directly to the relation

$$\frac{\sigma_{\alpha\beta}}{k_\beta^2} = \frac{\sigma_{\beta\alpha}}{k_\alpha^2}, \tag{3.3.29}$$

which is known as the reciprocity theorem. The Weisskopf result implies the reciprocity theorem and (to within a constant) vice versa. The theorem is, in fact, of more general validity than implied by the above derivation.

Equation (3.3.29) refers to cases for which the residual nucleus B may be considered to be in a single non-degenerate excited state. A generalization is obtained straightforwardly by considering the integration over the small region of $E_{\rm B}^*$ to include $\Omega_{\rm B}(E_{\rm B}^*)$ states. The analogue of (3.3.27) is obviously

$$\Gamma_\beta(k_\beta) = \frac{\sigma_{bB}}{2\pi^2}\frac{k_\beta^2\Omega_{\rm B}(E_{\rm B}^*)}{\rho_C(E_C^*)}, \tag{3.3.30}$$

and the generalization of the reciprocity theorem is

$$\frac{\sigma_{\alpha\beta}}{k_\beta^2\Omega_{\rm B}(E_{\rm B}^*)} = \frac{\sigma_{\beta\alpha}}{k_\alpha^2\Omega_A(E_A^*)}, \tag{3.3.31}$$

where we can now think of Ω_A (or $\Omega_{\rm B}$) as representing one or more than one state.

Finally, we may examine the consequences of making a partial wave expansion of the cross-section, σ_{bB}, in equation (3.3.27). Writing this cross-section as

$$\sigma_{bB} = \frac{\pi}{k_\beta^2}\sum_\ell (2\ell+1)T_\ell(\varepsilon_\beta)$$

and including a finite number of states in the integral over $E_{\rm B}^*$ we find

$$\Gamma_\beta(E_{\rm B}^*) = \frac{1}{2\pi\rho_C(E_C^*)}\sum_\ell (2\ell+1)T_\ell(\varepsilon_\beta)\Omega_{\rm B}(E_{\rm B}^*). \tag{3.3.32}$$

It may be useful, in certain circumstances, to think of the quantity $T_\ell(\varepsilon_\beta)/2\pi\rho_C(E_C^*)$ as a partial width, $\Gamma_\ell(\varepsilon_\beta)$, in which case

$$\Gamma_\beta(E_{\rm B}^*) = \sum_\ell (2\ell+1)\Gamma_\ell(\varepsilon_\beta)\Omega_{\rm B}(E_{\rm B}^*). \tag{3.3.33}$$

This formulation does not take into account the angular momenta of the participating nuclei. If, however, the transmission coefficients which characterize the decay depend only on the relative orbital angular momentum, ℓ, the definition remains useful. Thus, for example, for a spinless emitted particle, we could extend the Weissskopf theory by writing the decay width from a compound nucleus state of angular momentum J_i to final states of angular momentum J_f as

$$\Gamma_\beta(J_i, J_f, E_{\rm B}^*) = \frac{1}{2\pi\rho_C(E_C^*, J_i)}\sum_{\ell=|J_i-J_f|}^{J_i+J_f}(2\ell+1)T_\ell(\varepsilon_\beta)\Omega_{\rm B}(E_{\rm B}^*, J_f), \tag{3.3.34}$$

and define the partial width for compound nucleus states with given J_i as

$$\Gamma_\ell(J_i, \varepsilon_\beta) = \frac{T_\ell(\varepsilon_\beta)}{2\pi\rho_C(E_C^*, J_i)}. \tag{3.3.35}$$

This extension, though plausible, requires a little more thought. There are some (quantum mechanical) subtleties associated with equations like (3.3.34) which are more properly discussed is the context of the Hauser–Feshbach to be presented in section 3.5. The importance of a proper treatment of the angular momentum dependence of the state density has been emphasized by Thomas [14].

3.4 The Optical Model of elastic scattering

The origins of the optical model date back to the late forties notably with the work of Fernbach, Serber and Taylor [15]. These authors proposed a model of elastic scattering which was constructed by analogy with the scattering of light by a refracting and absorbing sphere. The model was successfully applied to scattering of 90 MeV neutrons [16] which showed more 'transparence' than that predicted by the 'strong absorption' Bohr model.

The quantum mechanical version of the model, developed in the early fifties, involves the solution of the Schrödinger equation for a short range, spherically symmetric, *complex* potential.

$$\left[\nabla^2 + \frac{2\mu}{\hbar^2}(E_{\rm cm} - V(r) - {\rm i}W(r))\right]\Psi(\boldsymbol{r}) = 0. \tag{3.4.1}$$

An example of such studies (using a square well form for the real and imaginary parts of the potential) is to be found in the work of Feshbach, Porter and Weisskopf [17]. Hodgson [18] gives a simple argument to show that the imaginary part of the potential is responsible for loss of elastic flux. The divergence of the quantum mechanical probability current density, $\nabla \cdot \boldsymbol{j}$, is

$$\begin{aligned}\nabla \cdot \boldsymbol{j} &= \frac{\hbar}{2\mu{\rm i}}\nabla \cdot [(\Psi^*\nabla\Psi - \Psi\nabla\Psi^*)] \\ &= \frac{\hbar}{2\mu{\rm i}}(\Psi^*\nabla^2\Psi - \Psi\nabla^2\Psi^*),\end{aligned} \tag{3.4.2}$$

which, with the help of (3.4.1) becomes

$$\nabla \cdot \boldsymbol{j} = \frac{2W(r)}{\hbar}\Psi\Psi^* = \frac{2W(r)}{\hbar}|\Psi(\boldsymbol{r})|^2. \tag{3.4.3}$$

In fluid mechanics, the divergence of a current is positive for a source and negative for a sink. Equation (3.4.3) thus shows that, if W is negative, there is loss of flux (a sink) providing incident particles 'feel' the imaginary potential. Fröbrich and Lipperheide [19] show how this argument may be further developed. They define the cross-section for absorption from the elastic channel as the surface integral of the net probability current density (the quantum mechanical flux) penetrating some closed surface which contains the potential, per unit incident current density. Gauss's theorem (also known as Green's theorem) allows us to write such a surface integral as a volume integral. For an arbitrary flow defined by a current density $\boldsymbol{j}$ and an outwardly directed set of surface elements $\mathrm{d}\boldsymbol{S}$,

$$\int_S \boldsymbol{j} \cdot \mathrm{d}\boldsymbol{S} = \int_V \nabla \cdot \boldsymbol{j}\, \mathrm{d}V, \tag{3.4.4}$$

where V is the volume enclosed by the surface S. Given that the incident current density is simply equal to the projectile–target relative velocity, v_p (section 2.3), we find that the cross-section for absorption is

$$\sigma_\mathrm{A} = -\frac{1}{v_\mathrm{p}} \int_V \nabla \cdot \boldsymbol{j}\, \mathrm{d}V = -\frac{2}{\hbar v_\mathrm{p}} \int_V W(r) |\Psi(\boldsymbol{r})|^2\, \mathrm{d}V, \tag{3.4.5}$$

which of course is a positive quantity for negative W. The minus sign appears because we are considering the net flux *into* the volume V (i.e. a sink) whereas Gauss's theorem, as written above with the elements $\mathrm{d}\boldsymbol{S}$ defined as outwardly pointing areas, refers to flux flowing outward through the surface (source).

We now develop a partial wave expansion for σ_A which will lead, on the one hand, to a definition for the transmission coefficients T_ℓ (equation (2.2.16)) in terms of the imaginary potential, $W(r)$, and on the other hand, to a relation between T_ℓ and the elastic scattering function S_ℓ. It will probably be realized that, in contrast to the case of a purely real potential, the presence of an imaginary potential implies a complex phase shift (the modulus $|S_\ell| < 1$) and that this feature will appear as the essential ingredient in the expression of the absorption cross-section.

Let us begin with the partial wave equivalent of equation (3.4.5). The probability density, $|\Psi(\boldsymbol{r})|^2$, can be directly written using the partial wave expansion of the (cylindrically symmetrical) wavefunction (equation (2.3.12)).

$$\Psi(r, \theta) = \sum_{\ell=0}^{\infty} (2\ell + 1) \mathrm{i}^\ell \frac{u_\ell(r)}{kr} P_\ell(\cos\theta), \tag{3.4.6}$$

where the functions $P_\ell(\cos\theta)$ are Legendre polynomials and $u_\ell(r)$ is referred to as the radial wave function. From (3.4.6) we obtain

$$|\Psi(r, \theta)|^2 = \frac{1}{(kr)^2} \sum_{\ell,\ell'=0}^{\infty} (2\ell + 1)(2\ell' + 1)\, \mathrm{i}^\ell (-\mathrm{i})^{\ell'} \times u_\ell(r)\, u^*_{\ell'}(r)\, P_\ell(\cos\theta)\, P_{\ell'}(\cos\theta). \tag{3.4.7}$$

Given that W is spherically symmetrical, this density is the only angle dependent factor in (3.4.5) ($\mathrm{d}V = 2\pi\,\mathrm{d}\cos\theta\, r^2\,\mathrm{d}r$) so that it can be directly integrated. Using the orthogonality property (2.3.10) of the Legendre polynomials is the integral

$$\int_{-1}^{+1} |\Psi(r,\theta)|^2 2\pi\,\mathrm{d}\cos\theta = \frac{4\pi}{(kr)^2}\sum_{\ell=0}^{\infty}(2\ell+1)|u_\ell(r)|^2. \tag{3.4.8}$$

With (3.4.8) we can make a partial wave expansion of σ_A. Thus by substituting this equation in (3.4.5) we obtain

$$\sigma_A = -\frac{8\pi}{\hbar v_\mathrm{p} k^2}\sum_{\ell=0}^{\infty}(2\ell+1)\int_{r=0}^{\infty} W(r)|u_\ell(r)|^2\,\mathrm{d}r, \tag{3.4.9}$$

which finally allows us to define the transmission coefficient T_ℓ (see (2.2.16)) as

$$T_\ell = -\frac{8}{\hbar v_\mathrm{p}}\int_{r=0}^{\infty} W(r)|u_\ell(r)|^2\,\mathrm{d}r, \tag{3.4.10}$$

in terms of which, repeating equation (2.2.16)

$$\sigma_A = \frac{\pi}{k^2}\sum_{\ell=0}^{\infty}(2\ell+1)T_\ell. \tag{3.4.11}$$

Equation (3.4.10) is useful if one is interested in studying the radial region which contributes to any given transmission coefficient. It obviously requires knowledge of the radial wave function, $u_\ell(r)$. We may remark, however, that the process of calculating the radial wave function by numerical integration yields the phase shift which, as we shall now see, is sufficient to determine T_ℓ.

In section 2.3 we expressed the (shape) elastic scattering cross-section in terms of the scattering function S_ℓ. The absorption cross-section may be similarly expressed (thus avoiding explicit reference to the imaginary potential) by again considering the net flux into a spherical surface located at a large distance r but this time working in terms of partial waves. The radially directed current density *into* a spherical surface (which contains the potential) is

$$\frac{-\hbar}{2\mu\mathrm{i}}\left(\Psi^*\frac{\partial}{\partial r}\Psi - \Psi\frac{\partial}{\partial r}\Psi^*\right) = -\mathrm{Re}\left[\Psi^*\frac{\hbar}{\mu\mathrm{i}}\frac{\partial}{\partial r}\Psi\right] = \frac{\hbar}{\mu}\mathrm{Re}\left[\Psi^*\mathrm{i}\frac{\partial}{\partial r}\Psi\right], \tag{3.4.12}$$

and is obtained from the asymptotic wave function. Recalling (2.3.26),

$$\Psi(r,\theta)_{r\to\infty} = \frac{1}{2\mathrm{i}kr}\sum_{\ell=0}^{\infty}(2\ell+1)C_\ell[\,\mathrm{e}^{\mathrm{i}\delta_\ell}\,\mathrm{e}^{\mathrm{i}kr} - (-1)^\ell\,\mathrm{e}^{-\mathrm{i}\delta_\ell}\,\mathrm{e}^{-\mathrm{i}kr}]P_\ell(\cos\theta), \tag{3.4.13}$$

which with $C_\ell = \mathrm{e}^{\mathrm{i}\delta_\ell}$ and $S_\ell = \mathrm{e}^{2\mathrm{i}\delta_\ell}$ becomes

$$\Psi(r,\theta)_{r\to\infty} = \frac{1}{2\mathrm{i}k}\sum_{\ell=0}^{\infty}(2\ell+1)P_\ell(\cos\theta)\left[S_\ell\frac{\mathrm{e}^{\mathrm{i}kr}}{r} + (-1)^{\ell+1}\frac{e^{-\mathrm{i}kr}}{r}\right]. \quad (3.4.14)$$

We obtain the radial component of the gradient as

$$\frac{\partial}{\partial r}\Psi(r,\theta) = \frac{1}{2\mathrm{i}k}\sum_{\ell=0}^{\infty}(2\ell+1)\frac{P_\ell(\cos\theta)}{r^2}[S_\ell\,\mathrm{e}^{\mathrm{i}kr}(\mathrm{i}kr-1) + (-1)^\ell\,\mathrm{e}^{-\mathrm{i}kr}(\mathrm{i}kr+1)]. \quad (3.4.15)$$

Now forming the product $\left[\Psi^*\mathrm{i}\frac{\partial}{\partial r}\Psi\right]$ which appears in equation (3.4.12) and integrating over a spherical surface at r (the surface element is $2\pi r^2\,\mathrm{d}\cos\theta$), we find, with the help of the orthogonality relation for the Legendre polynomials (2.3.10) that the net ingoing flux is

$$\frac{2\pi\hbar}{\mu}r^2\frac{2}{4k^2r^3}\sum_{\ell=0}^{\infty}(2\ell+1)\times\mathrm{Re}\big\{[S_\ell^*\,\mathrm{e}^{-\mathrm{i}kr} + (-1)^{\ell+1}\,\mathrm{e}^{\mathrm{i}kr}]\cdot\mathrm{i}\cdot[S_\ell\,\mathrm{e}^{\mathrm{i}kr}(\mathrm{i}kr-1) + (-1)^\ell\,\mathrm{e}^{-\mathrm{i}kr}(\mathrm{i}kr+1)]\big\}. \quad (3.4.16)$$

By writing S_ℓ as $\mathrm{e}^{2\mathrm{i}\delta_\ell}$ so that $S_\ell\,\mathrm{e}^{2\mathrm{i}+kr} = \cos 2(\delta_\ell + kr) + \mathrm{i}\sin 2(\delta_\ell + kr)$ the term in curly brackets is easily found to be $kr(1-|S_\ell|^2)$ so the absorption cross-section which is the net ingoing flux divided by the beam current density (equal to the beam velocity $v_\mathrm{p} = \hbar k/\mu$) is finally

$$\sigma_\Lambda = \frac{\mu}{\hbar k}\frac{\pi\hbar}{k\mu}\sum_{\ell=0}^{\infty}(2\ell+1)(1-|S_\ell|^2) = \frac{\pi}{k^2}\sum_{\ell=0}^{\infty}(2\ell+1)(1-|S_\ell|^2), \quad (3.4.17)$$

which of course defines the transmission coefficient as

$$T_\ell = 1 - |S_\ell|^2. \quad (3.4.18)$$

The transmission coefficients are one essential ingredient of the Hauser–Feshbach theory which will be exposed in the following section. They are obviously restricted to the range $0 \le T_\ell \le 1$.

Although the above presentation is valid only for spinless particles, the modifications necessary to take into account spin dependent terms in the optical potential are not complicated. In practice, the important part of the spin dependence results from the (spin–orbit) coupling of the projectile spin to the relative orbital angular momentum. For example, for spin half particles incident on a spinless target there would be two potentials, one corresponding to $J = \ell + 1/2$ and one to $J = \ell - 1/2$. We could then construct an average transmission coefficient by solving these two problems separately and averaging the results over the spin orientation. Alternatively, one may work in terms of the 'channel spin' $\boldsymbol{j} = \boldsymbol{s} + \boldsymbol{S}$ which is the vector sum of the projectile ($\boldsymbol{s}$) and target ($\boldsymbol{S}$) spins. The channel spin

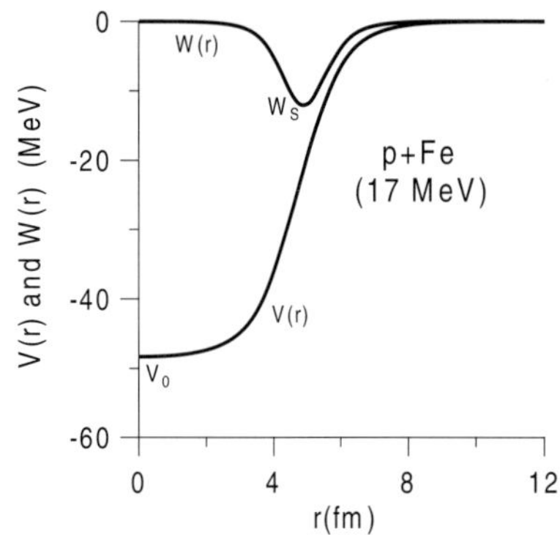

Figure 3.6. Typical optical model potential wells for 17 MeV protons on Fe ($A = 56$). The imaginary potential is of surface form (equation (3.4.21)). The parameters are $V_0 = 48.37$ MeV, $R_V = r_V A^{1/3} = 4.74$ fm ($r_V = 1.24$ fm), $a_V = 0.690$ fm, $W_S = 12.20$ MeV, $R_W = 4.912$ fm, $a_W = 0.456$ fm.

couples to the orbital angular momentum $\boldsymbol{\ell}$ to form the total angular momentum, $\boldsymbol{J}$. One would then write the scattering function as $S_{\ell,j,J}$.

In the sense of being able to fit elastic scattering data the optical model is one of the most successful and widely applied models in nuclear physics. In part, this success is due to the fact that the potential may be simply parametrized. A typical choice associates the (spherically symmetrical) real and imaginary parts of the potential with the so called 'Saxon–Woods' potential wells (see figure 3.6) described by the form factor $f_{\rm SW}(r)$. With this parametrization the real potential is given by

$$V(r) = V_0 f_{\rm SW}(r) = \frac{-V_0}{1 + \exp\left[\frac{(r-R_V)}{a_V}\right]}, \tag{3.4.19}$$

with a similar expression for the imaginary potential. In equation (3.4.19), V_0 is referred to as the depth, R_V as the radius (usually expressed using the 'radius parameter', r_V, where $R_V = r_V A_t^{1/3}$ and A_t is the target mass number) and a_V as the diffuseness parameter which controls the fall-off of the potential at the surface. In some cases (figure 3.6), a 'surface form' is used for the imaginary potential in which case

$$W(r) = 4W_S a_W\left[-\frac{\mathrm{d}}{\mathrm{d}r} f_{\rm SW}(r)\right] = \frac{-4W_S \exp\left[\frac{(r-R_W)}{a_W}\right]}{\left(1 + \exp\left[\frac{(r-R_W)}{a_W}\right]\right)^2}, \tag{3.4.20}$$

where the factor of $4a_W$ is inserted by convention so that the minimum (which occurs at $r = R_W$) is $-W_S$. The six parameters so defined may be freely adjusted so as to obtain a 'best fit' to the data. For particles with spin, spin–orbit

$(\boldsymbol{\ell} \cdot \boldsymbol{s})$ potentials are usually included. For charged particles it is also necessary to include a potential corresponding to the Coulomb repulsion between projectile and target charges. This 'repulsive' potential is usually modelled by considering a point (positive) projectile charge, $Z_\mathrm{p}e$, interacting with a uniformly charged target sphere of radius R_C and charge $Z_\mathrm{t}\,e$ for which

$$\begin{aligned} V_\mathrm{C}(r) &= \frac{Z_\mathrm{p} Z_t e^2}{r}, \ (r \geq R_\mathrm{C}), \\ V_\mathrm{C}(r) &= \frac{Z_\mathrm{p} Z_t e^2}{2R_\mathrm{C}^3}(3R_\mathrm{C}^2 - r^2), \ (r < R_\mathrm{C}). \end{aligned} \tag{3.4.21}$$

The charge radius R_C, is usually fixed to a value somewhat greater than that of the real potential. Examples of optical model fits to data for proton and ^{3}He elastic scattering are given in figures 3.7 and 3.8 [20,21].

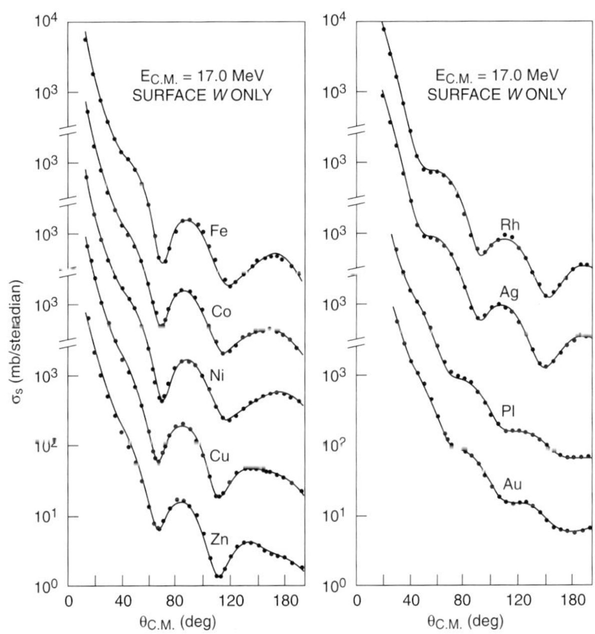

Figure 3.7. Optical model fits (continuous lines) to proton elastic scattering data [20]. The differential cross-sections (points) for nine targets were measured over the full angular range. The optical potential included a spin–orbit term. The quality of the fits is impressive. (Reprinted from *Phys. Rev.*, **131**, Perey F G, p 745. Copyright 1963, by the American Physical Society.)

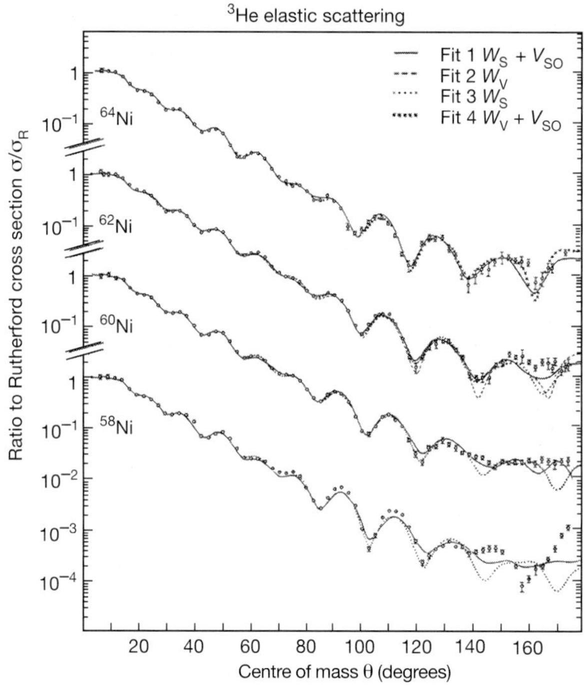

Figure 3.8. As figure 3.7 but for incident ^{3}He ions on Ni targets. Fits using volume or surface imaginary potentials with or without a spin–orbit potential are shown. The differential cross-sections (the symbol $\sigma \equiv \sigma(\theta)$ is meant to refer to $d\sigma/d\Omega$) are shown divided by the 'Rutherford' differential cross-section, $\sigma_R(\theta)$, which correponds to pure Coulomb scattering between point charges. The details of the potentials, which are not important for present purposes, can be found in reference [21]. Once again, except perhaps at extreme backward angles, the data are very well reproduced by the optical model. (Reprinted from *Nucl. Phys.* A, **183**, Cage M E *et al*, p 449. Copyright 1972, with permission from Elsevier Science.)

3.5 The Hauser–Feshbach evaporation model

The aim of the theory proposed by Hauser and Feshbach [22] is identical to that of the Weisskopf evaporation theory inasmuch as it describes reactions which proceed via the formation of a compound nucleus. In principle, the theory provides a proper quantum mechanical treatment of angular momentum. Given that nuclear level densities are strongly angular momentum dependent this is a most important feature. In practice, because of one major assumption, the expression for the specific decay rate is equivalent to that obtained by the intuitive extension of the Weisskopf theory discussed at the end of section 3.3.

Many applications of the Hauser–Feshbach theory have appeared in the literature. Our objective, however, will be to concentrate on those aspects of the theory which are useful for the analysis of nuclei excited to 'not too low' energies where many states may be considered to be included in the spread of energy characteristic of the incident beam. For such excitation energies the compound nucleus often decays by emitting more than one particle (together with gamma rays). One consequence of this 'multistep' emission is that the full quantum mechanical treatment of the set of sequential decay processes is extremely difficult to manage, particularly with respect to directions of emitted particles which are determined by the vector orbital angular momenta appearing in any particular decay chain. Following work by Ericson and Strutinsky [23], the response of experimentalists has been to resort to semi-classical methods to describe the angular momentum related aspects of the process (e.g., angle dependence of energy spectra, angular distributions and correlations of emitted particles) while retaining the basic Hauser–Feshbach formalism to calculate angle and energy integrated emission probabilities. At the same time more and more use has been made of Monte Carlo simulations. These techniques will be explained in the next chapter where we will describe, in some detail, practical aspects of single and multistep Hauser–Feshbach calculations as well as experimental results. In this section we restrict ourselves to a presentation of the model.

It is convenient to begin with the Bohr hypothesis which was written in section 3.2 as $\sigma(a + A \rightarrow B + b) = \sigma(a + A \rightarrow C)P(C \rightarrow B + b)$. We can use the more compact 'channel' form (the formation of C being understood) by representing the channel $a + A$ by the symbol α and $B + b$ by β. We then write

$$\sigma_{\alpha\beta} = \sigma_\alpha P_\beta, \tag{3.5.1}$$

where P_β is the probability for compound nucleus decay to channel β. We will also need the reciprocity theorem (a consequence of the time reversal invariance of the Hamiltonian) which was presented at the end of section 3.3.

Compared with the Weisskopf theory the major modification in the Hauser–Feshbach approach is to replace the Bohr hypothesis by assuming that independence applies to cross-sections corresponding to given values of the total angular momentum and parity. If we first consider the simplest case of entrance and exit channels involving spinless particles, the total angular momentum is the same as the orbital angular momentum and the product of the parities of entrance channel nuclei must also characterize exit channel nuclei. The new independence hypothesis can then be expressed for an arbitrary entrance channel using the transmission coefficients, T_ℓ, as

$$\sigma_{\alpha\beta}(\ell) = \left[\frac{\pi}{k_\alpha^2}(2\ell + 1)T_\ell(\varepsilon_\alpha)\right]P_\beta(\ell), \tag{3.5.2}$$

where, as before, $P_\beta(\ell)$ is the probability of decay to a specific final state. For the inverse reaction involving the same partial wave and the same compound nucleus excitation energy we obviously have

$$\sigma_{\beta\alpha}(\ell) = \left[\frac{\pi}{k_\beta^2}(2\ell+1)T_\ell(\varepsilon_\beta)\right]P_\alpha(\ell). \tag{3.5.3}$$

Dividing these equations and assuming that the reciprocity theorem can be applied to partial cross-sections, we obtain the interesting equality

$$\frac{P_\alpha(\ell)}{T_\ell(\varepsilon_\alpha)} = \frac{P_\beta(\ell)}{T_\ell(\varepsilon_\beta)}. \tag{3.5.4}$$

Equation (3.5.4) states that the ratio of probabilities of decay from the compound nucleus for any pair of channels is equal to the ratio of the corresponding transmission coefficients. Thus, for any decay channel, β,

$$P_\beta(\ell) = \frac{T_\ell(\varepsilon_\beta)}{\sum_\gamma T_{\gamma,\ell}}, \tag{3.5.5}$$

where the sum in the denominator on the right-hand side is intended to represent all accessible exit channels.

Note that, in equation (3.5.5), the energy, ε_β, is fixed by the excitation energy of the compound nucleus, E_C^*, and that of the final state in the residual nucleus, i.e., $\varepsilon_\beta + E_{\mathrm{B}}^* = E_C^* + Q_{C\to b+B}$.

As an obvious generalization of (3.5.5) we may consider that there may be $\Omega(E_{\mathrm{B}}^*)$ states at energy E_{B}^* in which case the decay probability to any one of these states is obtained by multiplying (3.5.5) by $\Omega(E_{\mathrm{B}}^*)$. The denominator may also be written in this way (we replace $T_{\gamma,\ell}$ by $T_{\gamma,\ell}\Omega_G(E_G^*)$). Then, expressing the number of states in a small energy interval, ΔE (over which ε_β can be considered as constant) as $\Omega_{\mathrm{B}}(E_{\mathrm{B}}^*) = \rho_{\mathrm{B}}(E_{\mathrm{B}}^*)\Delta E$ (similarly for $\Omega_G(E_G^*)$) we can finally write the partial cross-section in equation (3.5.2) as

$$\sigma_{\alpha\beta}(\ell) = \sigma_\alpha(\ell)\frac{T_\ell(\varepsilon_\beta)\Omega_{\mathrm{B}}(E_B^*)}{\sum_\gamma T_{\gamma,\ell}\Omega_G(E_G^*)} = \frac{\pi}{k_\alpha^2}(2\ell+1)\frac{T_\ell(\varepsilon_\alpha)T_\ell(\varepsilon_\beta)\rho_{\mathrm{B}}(E_{\mathrm{B}}^*)}{\sum_\gamma T_{\gamma,\ell}\rho_G(E_G^*)}, \tag{3.5.6}$$

in which $\sigma_\alpha(\ell)$ is the partial cross-section for formation of the compound nucleus in channel α. Equation (3.5.6) is the basic form of the Hauser–Feshbach equation for spinless particles. We recall that its derivation is based on the application of the independence hypothesis to two given channels for a single partial wave and on the reciprocity theorem. In practice, it is not very useful since intrinsic angular momenta are almost always involved. In the following generalization, for economy of notation, we use unprimed symbols to represent quantities associated with the entrance channel α and primed symbols for the exit channel β.

For particles with angular momentum the total angular momentum J and, as before, the parity, Π, are conserved throughout the reaction. Assuming the independence hypothesis can be applied for a given J^Π, the cross-section is obtained as a sum of the corresponding contributions. Conservation of parity implies that the product of the parity of the exit channel partial wave, $(-1)^{\ell'}$, and the parities of the emitted particle and the residual nucleus is equal to the parity of the compound nucleus state. Thus, if decay takes place to an isolated final state from a compound nuclear state of definite parity, only either even or odd partial waves can contribute. On the other hand, if the residual nucleus is represented by a state density then it is usually admitted that there are equal numbers of positive and negative parity states so that summing over positive and negative parity states is practically equivalent to ignoring the parity quantum number. This argument is reinforced if positive and negative parity compound nuclear states for given E_C^* and J can be considered to be populated with equal probabilities (this property is usually assumed). One can then consider the parity of the compound nucleus to be undefined so that the parity quantum number may be discarded.

If the transmission coefficients depend only on orbital angular momenta, the partial cross-section to final states of angular momentum S' in the nucleus B is given by

$$\begin{aligned}\sigma_{\alpha\beta}(J,S') &= \sigma_\alpha(J)\frac{\sum\limits_{\ell',j'} T_{\ell'}(\varepsilon_\beta)\rho_{\mathrm{B}}(E_{\mathrm{B}}^*,S')}{\sum\limits_{\gamma,\ell'',j''} T_{\gamma,\ell''}\rho_G(E_G^*)} \\ &= \frac{\pi}{k_\alpha{}^2}\sum_{\ell,j}(2\ell+1)g_\alpha T_\ell(\varepsilon_\alpha)\frac{\sum\limits_{\ell' j'} T_{\ell'}(\varepsilon_\beta)\rho_{\mathrm{B}}(E_{\mathrm{B}}^*,S')}{\sum\limits_{\gamma,\ell'',j''} T_{\gamma,\ell''}\rho_G(E_G^*)} \\ &= \frac{\pi}{k_\alpha{}^2}\frac{(2J+1)}{(2s+1)(2S+1)}\frac{\sum\limits_{\ell j,\ell',j'} T_\ell(\varepsilon_\alpha)T_{\ell'}(\varepsilon_\beta)\rho_{\mathrm{B}}(E_{\mathrm{B}}^*,S')}{\sum\limits_{\gamma,\ell'',j''} T_{\gamma,\ell''}\rho_G(E_G^*)}. \end{aligned} \tag{3.5.7}$$

Equation (3.5.7) requires some explanation. To begin with, the statistical factor, g_α, which multiplies the usual $(2\ell+1)$ weighting to give the cross-section for given total angular momentum, J, is obtained by multiplying the probability, $[(2j+1)/(2s+1)(2S+1)]$, that a given channel spin, j, is formed from projectile (s) and target (S) spins, by the probability, $[(2J+1)/(2j+1)(2\ell+1)]$, that j combines with ℓ to form a given value of J. This factor thus represents an average over entrance channel magnetic substates. Secondly, the sum over the channel spin, j, is over all possible values obtained by coupling the vectors $\mathbf{s}$ and $\mathbf{S}$. It therefore takes values between $|S-s|$ and $S+s$. Similarly, the sum over ℓ for a given j is over all values from $|J-j|$ and $J+j$. The cross-section is depicted as that for a given final angular momentum S' in the residue, B. Obviously 'triangular' conditions which are equivalent to those for ℓ and j apply to ℓ' and j' (j' goes from $|S'-s'|$ to $S'+s'$, where s' is the intrinsic spin of particle b, and ℓ' goes from $|J-j'|$ to $J+j'$). In many cases we may wish to carry out a further sum over S' and thus to calculate the cross-section in the channel β only as a function of the energy of the emitted particle. Finally, we have represented the sum over all possible exit channels simply as $\sum_{\gamma,\ell'' j''} T_{\gamma,\ell''} \rho_G(E_G^*)$ although this sum is meant to include all possible angular momentum and energy conserving final states which can be reached by decay of the compound nucleus (including gamma ray decay).

It may have occurred to the reader that the coupling of the vectors $\boldsymbol{s}$ and $\boldsymbol{S}$ to form $\boldsymbol{j}$ followed by the coupling of $\boldsymbol{j}$ and $\boldsymbol{\ell}$ to form the total angular momentum, $\boldsymbol{J}$, is a matter of choice. One could begin by coupling say, $\boldsymbol{\ell}$ and $\boldsymbol{s}$, to form a vector, $\boldsymbol{I}$, and then couple $\boldsymbol{I}$ and $\boldsymbol{S}$ to form $\boldsymbol{J}$. Indeed such a scheme may be preferable if spin–orbit forces in the optical potential lead to a dependence of the transmission coefficients not only on ℓ, but also on I (section 3.4). In this case one can easily obtain an expression which is equivalent to equation (3.5.7) (a useful excercise). It is usually safe to use average values for the transmission coefficient so that explicit I dependence need not be considered.

In order to investigate the angular dependence of the cross-section we return to equation (3.5.7) using the form appropriate for decay to a specific final state (i.e., suppressing the densities ρ_{B} and ρ_G). The extension to deal with a group of final states with given angular momentum is straightforward. Our aim will be to develop (3.5.7) in such a way that the orientation of the angular momenta of the participants is made explicit (for the entrance channel this implies that we 'expand' the statistical weight, g_α). We do so because, classically, the direction of the emitted particle velocity would be expected to be perpendicular to its orbital angular momentum. At the same time, we generalize this equation by allowing the transmission coefficients to depend on the channel spin and the total angular momentum, J, although this is more formal than useful. Because of the expected symmetry around the beam axis it is convenient to define the angular dependence of the cross-section in terms of the angular distribution of the emitted light particle with respect to the beam axis. We therefore choose this axis as the axis of quantization so that the z component of the entrance channel orbital angular momentum (which is necessarily perpendicular to the beam axis) is $m_\ell = 0$.

Equation (3.5.7) implies that, for a given total angular momentum, J, the angle integrated cross-section is obtained by straightforwardly summing contributions for entrance channel and exit channel orbital angular momenta. This is a rigorous result which is extended (non-rigorously) to the calculation of angular distributions. Thus, in expanding the cross-section so as to explicitly exhibit magnetic quantum numbers, possible interference between partial waves is neglected. The *probability* for a given ℓ (with projection, m_ℓ), to couple to the channel spin $j(m_j)$, to form the compound nucleus with angular momentum, $J(m_J)$, is thus given by the *square* of the Clebsch–Gordan coefficient $(\ell m_\ell, jm_j|Jm_J)$. A similar factor represents the probability that the state $|Jm_J\rangle$ 'decouples' to the exit channel angular momentum ℓ' and the corresponding channel spin j'. The angular distribution for given $\ell' m_{\ell'}$ is simply given by the square of the modulus of the spherical harmonic $Y_{\ell' m_{\ell'}}(\theta, \phi)$ where θ is the polar angle with respect to the beam axis and ϕ, the azimuthal angle (around the beam axis). We thus find that for fixed values of $\ell, m_\ell (= 0)$, j, m_j and J, m_J and fixed values of $j', m_{j'}$, $\ell', m_{\ell'}$ the differential cross-section to a specific final state is

$$\frac{\mathrm{d}\sigma_{\alpha\beta}(\ell, m_\ell = 0, j, m_j, J, m_J, j', m_{j'}, \ell', m_{\ell'})}{\mathrm{d}\Omega} = \frac{\pi}{k_\alpha^2}(2\ell + 1)$$
$$\times (\ell\, 0, j m_j | J m_J)^2 (\ell' m_{\ell'}, j' m_{j'} | J m_J)^2 |Y_{\ell' m_{\ell'}}(\theta, \phi)|^2$$
$$\times \frac{T_{\alpha,\ell,j,J} T_{\beta,\ell',j',J}}{\sum\limits_{\gamma,\ell'',j'',J''} T_{\gamma,\ell'',j'',J''}}, \tag{3.5.8}$$

where, for the sake of brevity, we have suppressed the (energy) arguments of the transmission coefficients and included the channel designations α, β and γ as subscripts.

The cross-section represented by equation (3.5.8) depends only on the polar angle, θ. There is no dependence on the azimuthal angle ($\mathrm{e}^{im\phi}$ in the spherical harmonic $Y_{\ell,m}$). We emphasize that this equation holds if and only if the various contributions can be summed incoherently. Otherwise one would have to construct the square of the modulus of an *amplitude* which would, for example, involve interference between entrance (or exit) channel partial waves.

Before explicitly dealing with sums over the various quantum numbers in equation (3.5.8) we note some restrictions concerning the angular momentum projection quantum numbers. We can freely sum over m_j (all values of m_j are considered to be equi-probable). However because $m_\ell = 0$, $m_J = m_j$. In the exit channel for a given value of m_J we can sum freely over $m_{\ell'}$. However, for each value of $m_{\ell'}$, $m_{j'}$ is fixed by the condition that $m_J = m_{\ell'} + m_{j'}$.

For each entrance channel orbital angular momentum ℓ we know precisely which state is involved ($m_\ell = 0$). In contrast, for the entrance channel spins $\boldsymbol{s}$ and $\boldsymbol{S}$ we do not know which microstate is involved (although, for each reaction there is obviously only one) so that we need to make an average over all such states. As we learned in section 2.6 there is a one-to-one correspondence between states represented in the (s, m_s, S, m_S) and (s, S, j, m_j) bases. For the vectors $\boldsymbol{s}$ and $\boldsymbol{S}$ the total number of entrance channel substates is the product of the number of states for each vector i.e., $(2s + 1)(2S + 1)$, so that the entrance channel average can be performed by summing over j (which goes from $|S - s|$ to $S + s$) and m_j $(-j$ to $+j)$ and dividing by $(2s + 1)(2S + 1)$. For a given ℓ and channel spin j the sum over m_j leads to compound nucleus angular momenta J which vary from $|\ell - j|$ to $\ell + j$. Thus, a free sum over the possible orientations of the channel spin implies a sum over J (not over m_J which is equal to m_j). The situation is not the same in the exit channel. All exit channel states may contribute to the cross-section. Thus in an arbitrary exit channel γ, if the number of possible states were to be doubled, the corresponding statistical weight would also be doubled. It follows that, having calculated the average cross-section to any one exit channel state, we need to sum the result over all such states. For a given J and $m_J(=m_j)$, the sum over exit channel states is over j', ℓ' and $m_{\ell'}$ (or equivalently $m_{j'}$). The final differential cross-section can now be written

$$\frac{d\sigma_{\alpha\beta}}{d\Omega} = \sum_{\ell,j,\ell',j',J,m_j,m_{\ell'}} \frac{1}{(2s+1)(2S+1)} \times \frac{d\sigma_{\alpha\beta}(\ell,0,j,m_j,J,m_J,j',m_{j'},\ell',m_{\ell'})}{d\Omega}. \tag{3.5.9}$$

It is usual to perform the sums over m_j, and $m_{\ell'}$ directly. This is due to the fact that

$$A(\ell,j,\ell',j',J,\theta) = \sum_{m_j,m_{\ell'}} (\ell 0, jm_j|Jm_J)^2(\ell' m_{\ell'}, j'm_{j'}|Jm_J)^2|Y_{\ell' m_{\ell'}}(\theta,\phi)|^2 \tag{3.5.10}$$

can be pre-calculated and inserted into the expression for the cross-section which finally reads

$$\frac{d\sigma_{\alpha\beta}}{d\Omega} = \frac{\pi}{k_\alpha^2} \sum_{\ell,j,\ell',j',J} \frac{(2\ell+1)}{(2s+1)(2S+1)} A(\ell,j,\ell',j',J,\theta) \frac{T_{\alpha,\ell,j,J} T_{\beta,\ell',j',J}}{\sum_{\gamma,\ell'',j'',J''} T_{\gamma,\ell'',j'',J''}}. \tag{3.5.11}$$

Alternative expressions of equation (3.5.11) are given in a number of works [1,4,24]. They are intended to simplify computation. In practical calculations of the decay of excited nuclei, however, their use is limited to rather low energies where one-step emission processes to isolated final states are important. We will further explore angle and energy distributions of emitted particles in the next chapter when we describe the setting up of a practical simulation based on the Hauser–Feshbach formalism.

Equations such as (3.5.7) apply to the emission of charged or uncharged particles. For γ rays, the transmission coefficients must be replaced by $\xi_\ell E_\gamma^{2\ell+1}$. The radiative transition strengths, ξ_ℓ, are usually expressed as ratios to 'standard' strengths referred to as Weisskopf units (W.u.). The Weisskopf transition strength is obtained by considering transitions between two single proton states with total angular momentum $\ell + 1/2$ and $1/2$. The proton wave functions are taken as uniform (constant) functions extending over a nuclear radius, R. Under these circumstances, Weisskopf [25] showed that the width corresponding to (electric) transitions of multipolarity $\ell(\ell > 0)$ is given, *numerically*, by

$$\Gamma_\gamma^{(W)}(\ell) = \frac{4.4(\ell+1)\hbar}{\ell[(2\ell+1)!!]^2}\left[\frac{3}{\ell+3}\right]^2 R^{2\ell}\left[\frac{E_\gamma}{197}\right]^{2\ell+1} \times 10^{21}\ \text{MeV}, \tag{3.5.12}$$

where $\Gamma_\gamma^{(W)}(\ell)$ is the (Weisskopf) gamma ray decay width, R is the nuclear radius in fermis and the double factorial symbol $(2\ell+1)!! = 1 \times 3 \times 5 \cdots \times (2\ell+1)$. Magnetic transitions are expected to follow the same trend but to be slowed by a factor which is roughly estimated as $10(\hbar/m_N c R)^2$ (this factor multiplies the $\Gamma_\gamma^{(W)}(\ell)$ of (3.5.12)) where m_N is the nucleon mass. In [25] Weisskopf stated 'We have published these exceedingly crude estimates only because of the rather unexpected agreement with the experimental material which was pointed out to us by many workers in this field' .

While one can understand the usefulness of a standard unit in terms of which transitions may be described as 'strong' or 'weak', the truth of the matter is that studies of transition strengths show large discrepancies with the widths given by (3.5.12). For example, Pühlhofer, who investigated evaporation residue cross-sections in the reaction $^{19}F + ^{12}C$ at laboratory (^{19}F) energies between 40 and 92 MeV [26] used strengths based on systematic studies of low lying states in light nuclei [27]. These studies indicated values

$$\begin{aligned} \xi_{E1} &= \xi_{M1} = 0.5 \times 10^{-8}, \\ \xi_{E2} &= 0.3 \times 10^{-9}, \end{aligned} \tag{3.5.13}$$

which are, respectively, 0.001, 0.5 and 5 W.u for E1, M1 and E2 transitions. In the decay of light nuclei γ ray emission does not compete significantly with particle emission except for excitation energies near the particle emission thresholds.

The Weisskopf estimate assumes sharp initial and final single particle states. In practice, of course, this situation is found only at low excitation energies. At higher energies many overlapping states appear in any small energy region. We have seen in section 2.5 that the corresponding density of states can be calculated by applying combinatorial methods to a single particle spectrum with equidistant spacing of single particle states. In this model a large number of (neutron and proton) particle–hole configurations are degenerate. For an *electric* transition involving the displacement of a single *proton* from the ground state to a state in which the same proton is excited to N times the s.p. spacing, d, only one of these many degenerate states can be involved. Ignoring selection rules due to angular momentum coupling we would therefore expect a 'degeneracy dilution' of the Weisskopf estimate of the strength of any specific transition. The dilution factor would be expected to be, on average, proportional to the density of excited states so that Γ_γ should vary as $1/\rho_C(E_C^*)$ (i.e., $\Gamma_\gamma \rho_C(E_C^*)$ should be roughly constant). The same argument holds when the lower energy state is not the ground state because the 'upper' states can be built on *all* originally degenerate states at the lower energy by raising a proton by the appropriate integer multiple of d. If we now allow for residual interactions between the single particles, then the real states of the excited nucleus become linear superpositions of the degenerate states. Each 'real' state thus contains a small fraction of the original pure single particle state which we may take to be the inverse of the degeneracy dilution factor.

Let us just make a quick estimate using (3.5.12). We consider electric quadrupole transitions ($\ell = 2$) of energy 1 MeV in a nucleus with radius 5 fm, and by 'plugging in' these numbers we find that $\Gamma_\gamma = 1.48.10^{-11}$ MeV. Now the particle transmission coefficient (say for neutrons) may be expressed in terms of the corresponding width as $T_\ell = 2\pi\Gamma_n(\ell)\rho_C(E_C^*)$ (3.3.35). If we now replace the T_ℓ (for 1 MeV quadrupole γ rays) by ξ_{E2} and use a similar expression for the γ ray width we get $\xi_{E2} = 2\pi\Gamma_\gamma(\ell = 2)\rho_C(E_C^*)$ so that one might expect little or no dependence of ξ_{E2} on the parent nucleus excitation energy. In order to obtain the figures quoted by Pülhofer (i.e., 5 W.u. for $\xi_{\mathrm{E2}} = 3 \times 10^{-10}$) we need to assume $\rho_C(E_C^*) = 0.63\ \mathrm{MeV}^{-1}$ which corresponds to levels near the ground state. The estimate of ξ_{E2} can thus be considered to have been made in a self-consistent manner.

So far we have discussed the Hauser–Feshbach formula simply as a model for evaporation. The use of the optical model which, by fitting cross-sections for elastic scattering using a complex potential provides tranmission coefficients, can be considered as part of the model structure. The justification of the model using reaction theory is complicated and, according to Vogt [1], only partially successful. It is based on the description of any (two-body) reaction using the collision matrix which can be thought of as a generalization of the elastic scattering function $S_{\ell,j,J}$. In terms of the collision matrix, U, an arbitrary reaction cross-section connecting channels α and β is given as

$$\sigma_{\alpha\beta} = \frac{\pi}{k_\alpha^2}\sum_\ell (2\ell+1)\sum_{j,J} g_\alpha \sum_{j',\ell'} |\delta_{\alpha\beta}\delta_{\ell\ell'}\delta_{jj'} - U_{j'\ell'Jj\ell}(\alpha,\beta)|^2. \qquad (3.5.14)$$

The quantity g_α was discussed after equation (3.5.7). It represents the probability of forming a given value of J by coupling the channel spin to the orbital angular momentum. The additional summation indices (which also appeared in equation (3.5.7)) are due to the introduction of the channel spin (e.g., the total angular momentum for given ℓ runs from $J = |\ell - j|$ to $\ell + j$).

Equation (3.5.14) expresses the fact that the definition of a cross-section relies on the relative amplitudes of incoming and outgoing waves in 'asymptotic' regions of space. It is thus not necessary to invoke a potential which represents one possible way of calculating the actual values of particular matrix elements.

An important property of the collision matrix (which follows from the observation that all of the incident flux must end up somewhere) is the so-called unitarity property which states that for any entrance channel specified by $(\alpha;\ \ell,\ j,\ J)$

$$\sum_{\beta,j',\ell'} |U_{j'\ell'Jj\ell}(\alpha,\beta)|^2 = 1. \qquad (3.5.15)$$

The attempt to establish a foundation for the evaporation model begins with the observation that, for reactions which proceed via the formation of the compound nucleus, there are many overlapping compound nuclear states in any small energy region. Tandem accelerator beams have a resolution of the order of a few tens of KeV and cyclotron beams are typically ten times worse. Thus, measured cross-sections and associated matrix elements must usually be considered as energy averages. In this situation it is desirable to express the optical model elastic scattering function $S_{\ell,j,J}$ (and thereby the transmission coefficient) as the average over energies of the corresponding collision matrix element $U_{j\ell Jj\ell}(\alpha,\alpha)$. This can, in fact, be done quite easily although a detailed analysis of the averaging process itself is rather involved (see the article by Vogt [1]). We begin by noting that the partial reaction cross-section for given channel spin and angular momentum can be expressed using unitarity (3.5.15) in terms of the collision matrix element which corresponds to elastic scattering. We can write

$$\sigma_{\mathrm{R}}(\alpha;\ell,j,J) = \frac{\pi(2\ell+1)}{k_\alpha^2} g_\alpha[1-|U_{j\ell Jj\ell}(\alpha,\alpha)|^2], \tag{3.5.16}$$

and of course the elastic scattering cross-section itself is (from (3.5.14))

$$\sigma_{\mathrm{EL}}(\alpha;\ell,j,J) = \frac{\pi(2\ell+1)}{k_\alpha^2} g_\alpha|1-U_{j\ell Jj\ell}(\alpha,\alpha)|^2. \tag{3.5.17}$$

(Note that (3.5.16) refers to the *reaction* and not to the absorption cross-section.) Now applying energy averaging to (3.5.16) and (3.5.17) (and surreptitiously suppressing superfluous subscripts), we obtain

$$\langle\sigma_{\mathrm{R}}(\alpha;\ell,j,J)\rangle = \frac{\pi(2\ell+1)}{k_\alpha^2} g_\alpha[1-\langle|U_{\ell jJ}|^2\rangle], \tag{3.5.18}$$

and

$$\langle\sigma_{\mathrm{EL}}(\alpha;\ell,j,J)\rangle = \frac{\pi(2\ell+1)}{k_\alpha^2} g_\alpha\langle|1-U_{\ell jJ}|^2\rangle, \tag{3.5.19}$$

which latter quantity can be expressed as

$$\langle\sigma_{\mathrm{EL}}(\alpha;\ell,j,J)\rangle = \frac{\pi(2\ell+1)}{k_\alpha^2} g_\alpha\{|1-\langle U_{\ell jJ}\rangle|^2 + \langle|U_{\ell jJ}|^2\rangle - |\langle U_{\ell jJ}\rangle|^2\}. \tag{3.5.20}$$

We may now define an 'optical' or 'shape' elastic (SE) scattering cross-section which indeed allows us to represent the scattering function $S_{\ell,j,J} = \mathrm{e}^{2\mathrm{i}\delta_{\ell,j,J}}$ as an average of collision matrix elements. Thus we retain the first term of (3.5.20) to write

$$\sigma_{\mathrm{SE}}(\alpha;\ell,j,J) = \frac{\pi(2\ell+1)}{k_\alpha^2} g_\alpha|1-\langle U_{\ell jJ}\rangle|^2, \tag{3.5.21}$$

which of course achieves our aim in the sense that $S_{\ell,j,J} = \langle U_{\ell jJ}\rangle$. However, we are now obliged to associate the last two terms in (3.5.20) (proportional to the variance of U) with the *compound* elastic scattering cross-section

$$\sigma_{\mathrm{CE}}(\alpha;\ell,j,J) = \frac{\pi(2\ell+1)}{k_\alpha^2} g_\alpha\{\langle|U_{\ell jJ}|^2\rangle - |\langle U_{\ell jJ}\rangle|^2\} = \frac{\pi}{k_\alpha^2} g_\alpha\sigma^2(U_{\ell jJ}), \tag{3.5.22}$$

which is part of the absorption cross-section (as opposed to the shape elastic scattering cross-section) in the optical model. In this case the absorption cross-section which is the sum of the average reaction cross-section and compound elastic cross-sections is, adding (3.5.18) and (3.5.22)

$$\sigma_{\mathrm{A}}(\alpha; \ell, j, J) = \frac{\pi(2\ell+1)}{k_\alpha^2} g_\alpha [1 - |\langle U_{\ell j J}\rangle|^2], \tag{3.5.23}$$

in which we can again note the association of $S_{\ell,j,J}$ and $\langle U_{\ell j J}\rangle$.

Averaging over resonances is usually carried out by assuming that both the real and imaginary parts of the collision matrix fluctuate over positive and negative values so that their phases may be considered to be random. Fluctuations for different resonances are thus uncorrelated. One consequence is that possible interference between partial waves in the entrance and exit channels is destroyed thus lending support for the construction of the differential cross-section as a sum over individual ℓ waves [28].

In practice the re-emission, from the compound nucleus of a particle in the elastic channel becomes negligible with increasing excitation energy due to competition from the rapidly increasing number of alternative open channels so that $\sigma_{\mathrm{A}} = \sigma_{\mathrm{R}}$ and $\sigma_{\mathrm{EL}} = \sigma_{\mathrm{SE}}$. If this is not the case, it is necessary to calculate and subtract the compound elastic cross-section from the measured cross-section before fitting with the optical model.

The division of the elastic scattering cross-section into compound and 'direct' contributions can, in fact, be made, not only for elastic scattering, but also for non-elastic (reaction) channels. However, this procedure should be thought of only as a first approximation despite the success of corresponding theories. In principle, it is necessary to consider contributions from different stages of the equilibration process leading to the formation of the compound system. Thus, for example, the simple two-component separation cannot account for the pre-equilibrium reaction mechanism (to be discussed in section 3.7).

3.6 Fusion

Fusion is, by definition, the process by which a compound nucleus composed of the sum of projectile and target nucleons is formed. Interest in this topic grew rapidly in the early seventies due to the increasing availability of heavy ion beams and, for about a decade, the problem of understanding heavy ion fusion cross-sections generated a great deal of experimental and theoretical activity.

If the excitation energy in the compound system is sufficiently high (above about 10 MeV) so that compound elastic scattering can be neglected ($\sigma_{\mathrm{A}} = \sigma_{\mathrm{R}}$) the fusion cross-section can be measured by detecting evaporation residues which are the nuclei which remain after emission of light particles from the compound system. In light and medium mass systems ($A < 80$) at excitation energies up to about 2 MeV/nucleon this has proved to be the most direct method. At higher energies it may be necessary to distinguish between evaporation residues and products of deep inelastic scattering (see end of this section). In heavy compound nuclei, fission is a competing decay mode especially if the compound nucleus is formed with high angular momentum so that the compound nucleus cross-section is the sum of the cross-sections corresponding either to evaporation residues or to pairs of fission fragments.

The cross-section corresponding to the fusion process can be written as a sum of partial cross-sections (2.2.17)

$$\sigma_F = \frac{\pi}{k^2} \sum_{0}^{\infty} (2\ell + 1) T_\ell F_\ell, \tag{3.6.1}$$

which defines F_ℓ as the fusion probability for the ℓth partial wave. This seemingly clumsy way of developing σ_F is justified by the fact that we wish to retain the definition of the transmission coefficient T_ℓ defined in scattering theory (quantities such as F_ℓ can then be used to represent any reactions which contribute to the total reaction cross-section).

For light projectiles at energies just above the Coulomb barrier the fusion cross-section is quite close to the total reaction cross-section. For heavier projectiles at higher incident energies the fusion cross-section falls below the reaction cross-section due the growing importance of direct and deep inelastic scattering reactions which take place at the higher angular momenta (impact parameters).

The most simple methods of calculating σ_F use only the conservative real potential supposed to characterize the projectile–target interaction and therefore, do not include the detailed dynamics which one might expect to govern the collision. In such 'static' models there is a finite partial cross-section for fusion for all angular momenta (impact parameters) up to a certain 'critical' value beyond which the fusion probability, $F_\ell = 0$.

The essential physics of the static approach can be understood by considering the model proposed in 1975 by Glas and Mosel [29] which is based on classical trajectories for the projectile–target relative motion. We focus on the radial separation. Assuming spherical symmetry, the radial kinetic energy at a projectile–target separation of r for a trajectory with relative angular momentum $\ell\hbar$ is written (see the radial equation (2.3.19))

$$E_{\text{rad}}(r) = E_{\text{cm}} - V_\ell(r) = E_{\text{cm}} - V(r) - \frac{\ell(\ell+1)\hbar^2}{2\mu r^2}, \tag{3.6.2}$$

where μ is the reduced mass, $V(r)$ is the sum of the attractive nuclear and repulsive Coulomb potentials and the last term represents the centrifugal barrier (which is the energy associated with tangential motion). A typical sketch of the effective potential energy, $V_\ell(r)$ (the negative of the sum of the second and third terms on the right-hand side of equation (3.6.2)) is shown in figure 3.9. For not too high ℓ-values (e.g., $\ell = 20$ in the figure) the combination of the repulsive Coulomb and attractive nuclear forces superimposed on the centrifugal potential produces a characteristic hump referred to as the barrier. The barrier can be roughly characterized by a quadratic form located at a fixed radial separation,

R_B, and thus specified by R_B, the barrier height and the curvature ω_ℓ at the maximum. The (quantum mechanical) penetration probability, *which is equated with the transmission coefficient*, for a parabolic barrier was given by Hill and Wheeler [30],

$$T_\ell(E_{cm}) = \frac{1}{1 + e^{2\pi(V_\ell - E_{cm})/\hbar\omega_\ell}}. \tag{3.6.3}$$

It is usually supposed that R_B is independent of ℓ (see again figure 3.9) so that one may take R_B to be the position of the barrier for $\ell = 0$. Let us first examine the reaction cross-section using such a model. A reaction is assumed to take place provided the energy at the barrier, $E_{rad}(R_B)$, is positive (classically, this implies that the projectile can penetrate to radial separations less than or equal to R_B). Thus, the 'Hill–Wheeler' transmission coefficients are approximated by assuming that T_ℓ falls abruptly to 0 for energies below the barrier so that the corresponding cross-section is, approximately, the sum of all partial cross-sections for which the $E_{rad}(R_B) > 0$ condition is satisfied. We thus obtain

$$\begin{aligned}\sigma_R &= \frac{\pi}{k^2}\sum_0^\infty \frac{2\ell+1}{[1 + e^{2\pi(V_\ell - E_{cm})/\hbar\omega_\ell}]} \\ &\approx \frac{\pi}{k^2}\sum_0^{\ell_B}(2\ell+1) = \frac{\pi}{k^2}(\ell_B+1)^2,\end{aligned} \tag{3.6.4}$$

where ℓ_B is the largest angular momentum quantum number for which $E_{rad}(R_B) \geq 0$. Setting $(\ell_B+1)^2 \approx \ell_B(\ell_B+1)$ and noting that $k^2 = 2\mu E_{cm}/\hbar^2$, equation (3.6.4) can be written (using (3.6.2)) as

$$\sigma_R = \frac{\pi}{k^2}(\ell_B+1)^2 \approx \frac{\pi\hbar^2}{2\mu E_{cm}}\ell_B(\ell_B+1) = \pi R_B^2\left[1 - \frac{V(R_B)}{E_{cm}}\right]. \tag{3.6.5}$$

Since the Coulomb barrier (Coulomb + nuclear potential $= V_\ell(R_B)$ for $\ell = 0$) is positive, equation (3.6.5) predicts that the reaction cross-section rises with increasing energy and tends to a fixed value, πR_B^2, at high incident energies.

Let us now turn to the fusion cross-section. The Glas–Mosel model assumes that fusion takes place whenever the trajectory is such that, at some stage, the projectile and target mass centres are separated by a 'critical' distance R_F [31] which is somewhat smaller than R_B. At this distance the attractive nuclear forces are supposed to be sufficiently strong to pull the projectile and target together. This conjecture allows us to attribute a more precise physical meaning to the quantity F_ℓ by stating that F_ℓ is the probability that the system evolves from R_B to the smaller radial separation R_F. In a static potential model this evolution takes place for all partial waves for which the energy E_{rad} remains positive for r values between R_B and R_F. At low energies (angular momenta) the dip in the potential, V_ℓ, inside the barrier (figure 3.9) is sufficiently pronounced to ensure that R_F is attained for *any* trajectory which surmounts the Coulomb barrier. Thus, $\sigma_F = \sigma_R$. With increasing angular momenta (which come into play progressively

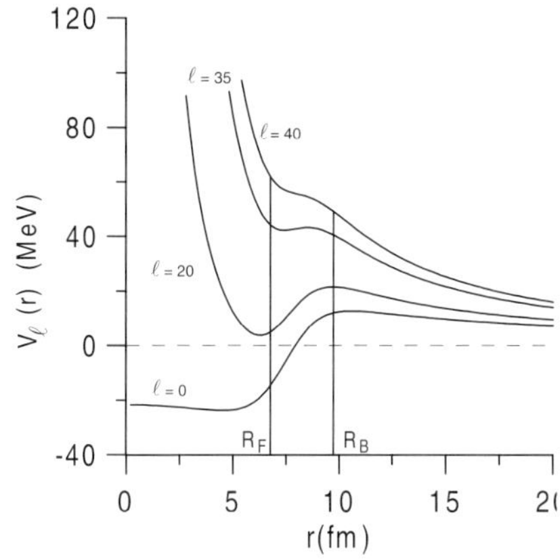

Figure 3.9. Effective potentials, $V_\ell(r)$ (equation (3.6.2)), corresponding to several angular momenta for fusion of $^{16}O + {}^{27}Al$. The Coulomb barrier is indicated by the vertical line at $r = R_B$. The 'pocket' in the potential disappears above $\ell = 35$. Fusion takes place provided that E_{rad} is positive for all $r \geq R_F$.

as the incident energy increases), the potential 'pocket' gradually disappears. Nevertheless, we can still use the reasoning applied to σ_R to directly write the fusion cross-section as

$$\sigma_F \approx \frac{\pi\hbar^2}{2\mu E_{cm}}\ell_F(\ell_F + 1) = \pi R_F^2\left[1 - \frac{V(R_F)}{E_{cm}}\right], \tag{3.6.6}$$

where $V(R_F)$ is the Coulomb + nuclear potential at R_F, and ℓ_F is the largest angular momentum for which the condition, $E_{rad}(r \geq R_F) \geq 0$, holds. The major difference (with σ_R) is of course that the nuclear + Coulomb potential at R_F may be quite different from that at R_B and may even be negative. If this is the case, the fusion cross-section actually diminishes with increasing energy. Plotted as a function of $1/E_{cm}$, we thus expect the fusion cross-section to show a characteristic change of slope which signals that passage by the barrier at R_B is no longer sufficient to ensure fusion. The 'changeover' energy is obviously obtained by solving

$$\pi R_F^2\left[1 - \frac{V(R_F)}{E_{cm}}\right] = \pi R_B^2\left[1 - \frac{V(R_B)}{E_{cm}}\right], \tag{3.6.7}$$

for E_{cm}. An excellent example of this characteristic change of slope which is taken from the review article by Mosel [29] is shown in figure 3.10.

Of course we can criticize the critical distance model and indeed all models which involve only classical trajectories. Even at the radial turning point ($E_{rad} = 0$), the tangential motion would tend to oppose fusion. Furthermore, the fact that the interaction between projectile and target nucleons tends to dissipate the incident energy leads to the conclusion that the simple static (frictionless) models

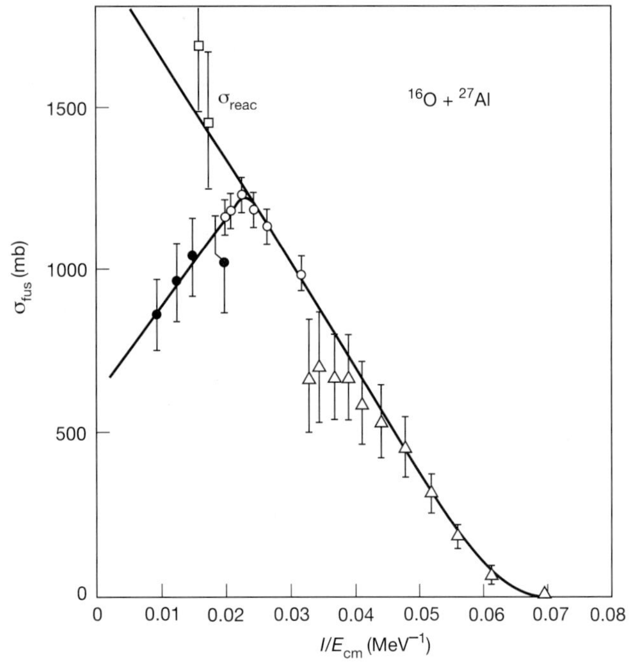

Figure 3.10. Fusion cross-sections for the system $^{16}O + ^{27}Al$ plotted againt the reciprocal of the centre-of-mass energy. The data are from Dauk *et al* 1975 *Nucl. Phys.* A **241** 170 (open triangles), Kozub *et al* 1975 *Phys. Rev.* C **11** 1497 (filled circles) and Back *et al* 1977 *Nucl. Phys.* A **285** 317 (open circles). Note that the reaction cross-section (open squares, Cormier *et al* 1976 *Phys. Rev.* C **13** 682) continues to rise at high energies. The continuous curve represents a fit to the data with the critical distance model [29]. The slight rounding off of the predicted cross-section is due to the use of smoothly varying transmission coefficients in the calculation. (Reprinted from *Heavy Ion Collisions*, vol 2, ed R Bock. Copyright 1980, with permission from Elsevier Science.)

are, in fact, a little too simple. In this context however, we may cite an interesting remark made by Mosel [29] concerning low energy fusion. To first order, frictional forces would be expected to diminish the energy and, at the same time, the angular momentum. Indeed a loss of energy of ΔE should entail a diminution of the square of the angular momentum given by $\Delta E = \hbar^2 \Delta(\ell^2)/2\mu b^2$ where b is the impact parameter. These two changes appear as terms with opposite signs in E_{rad} (the correction is $-\Delta E[1 - b^2/r^2]$) so that the effect of introducing friction in the radial coordinate is somewhat attenuated.

Roughly speaking the static theories are useful for projectile energies up to about 10 MeV/nucleon. At higher energies, there are other physical limitations to fusion which can be framed in terms of a maximum (also referred to as 'critical')

ℓ value. The fusion process is a transition from continuum states to states in the compound nucleus. A compound nucleus cannot be created where no states exist, i.e., beyond the Yrast line (see (2.5.55)). Another limitation arises from studies of the nucleus with large angular momentum pictured as a rotating liquid drop. These calculations (see Cohen, Plasil and Swiatecki [32]) predict that the fission barrier for a compound nucleus is strongly angular momentum dependent. At large angular momenta, the barrier height for any given nucleus decreases and the barrier disappears at a characteristic angular momentum. The compound nucleus thus cannot exist as a quasi-stable entity for angular momenta higher than the critical value $\ell_{\mathrm{cr}}^{(\mathrm{s})}$ ('s' for stability). The fusion cross-section then goes as

$$\sigma_{\mathrm{F}} = \frac{\pi}{k^2}\ell_{\mathrm{cr}}^{(\mathrm{s})}(\ell_{\mathrm{cr}}^{(\mathrm{s})}+1) = \frac{\pi\ell_{\mathrm{cr}}^{(\mathrm{s})}(\ell_{\mathrm{cr}}^{(\mathrm{s})}+1)\hbar^2}{2\mu E_{\mathrm{cm}}}, \tag{3.6.8}$$

and thus falls rapidly with increasing beam energy. This is probably the most important limitation to fusion at high energies.

Models which go beyond the static 'radial' approaches by introducing angular as well as radial coordinates and by including frictional forces were first proposed by Birkelund *et al* [33] and by Bondorf, Sobel and Sperber [34] (see also Seglie *et al* [35]). As an example of a modern calculation we recommend to the reader the work of Fröbrich and Gontchar [36], which includes an excellent account of friction. A report of a recent conference (1994) concerning heavy ion fusion is contained in [37]. Regretably, there is no one *simple* theory which allows an accurate prediction of the fusion probabilities, F_ℓ, over a wide range of projectile energies.

The investigation of heavy ion fusion, or perhaps we should say 'non-fusion', has given birth to studies of a whole new branch of heavy ion collisions which is referred to as deep inelastic scattering. This process concerns the part of the reaction cross-section which, despite considerable loss of relative energy and angular momentum together with exchange of nucleons between projectile and target nuclei, does not lead to fusion. A detailed description of this mechanism falls somewhat outside the scope of this book (see again reference [36]). However, we should just mention that the generalized dynamical collective models do not constitute the only approach to deep inelastic scattering. The relevant literature is extensive and includes application of the random walk process to nucleon and momentum exchange between projectile and target [38] as well as 'microscopic' calculations based on molecular dynamics simulations [39] or more recently on the use of the Boltzmann equation for the one-body distribution function [40]. In agreement with the loss of stability predicted by collective models the calculations show that, as the energy increases, the projectile–target interaction, even at quite small distances and in spite of frictional forces, is incapable of dissipating all the energy of relative motion so that the cross-section for fusion diminishes and may even vanish (at projectile energies above about 50–100 MeV/nucleon).

The deep inelastic process is characterized by the production of excited projectile- and target-like fragments (PLF and TLF) which decay mainly by evaporation after leaving the spatial region in which the collision took place. Readers specifically interested in this subject are referred to the literature cited above, as well as to the reviews of Schröder and Huizenga [41], Lefort [42] and Blocki *et al* [43].

3.7 Pre-equilibrium emission

Despite their sophistication, the description of the collision of two nuclei using classical dynamical calculations, which follow the evolution of collective variables, becomes progressively more difficult at bombarding energies above about 10 MeV/nucleon because of the increasing importance of the prompt emission of light particles (essentially neutrons, hydrogen and helium isotopes) in the initial stages of the reaction. This phenomena is referred to as pre-equilibrium (or precompound) emission. For light projectiles (e.g. protons) at rather low incident energies, where complete fusion accounts for at least 90% of the total reaction cross-section, there is a theoretical basis for understanding this phenomenon which can be linked to the picture of the formation of the compound nucleus proposed by Bohr. Statistical theories starting from this viewpoint were successfully developed by Griffin [44], by Cline and Blann [45], and in a somewhat different way by Harp, Miller and Berne [46] and successors (see, for example, Fabrici *et al* [47]). Pre-equilibrium emission has the effect of producing incomplete 'compound' nuclei which have a distribution of mass, charge, excitation energy and angular momentum rather than the fixed values (or simple distributions in the case of angular momentum) of these quantities which characterize the compound nuclei produced by the complete fusion process.

The simplest theory (and in fact the first to be proposed) is due to Griffin [44] and is usually referred to as the exciton model. The principal aim of the theory is to calculate the energy spectrum of pre-equilibrium nucleons. The basic hypothesis is that the path to equilibrium is governed by nucleon–nucleon collisions which excite (promote) 'particles' to single particle levels above the Fermi surface leaving 'holes' below. In figure 3.11 we show a schematic picture of the process together with two examples involving a two particle–one hole (2p, 1h) state and a three particle–two hole (3p, 2h) state. For a nucleon incident on a target with even numbers of protons and neutrons the initial state can be considered to be a 1p, 0h state, followed, after one collision of the incident nucleon with a target nucleon embedded in the Fermi sea, by a 2p, 1h state (3.11(*a*)). The total number of particles, n_p, and holes n_h which occupy single particle states respectively above and below the Fermi energy, is referred to as the number of excitons $n = (n_p + n_h)$. In both examples shown in figure 3.11(*a*), with $n = 3$ and $n = 5$ excitons, respectively, all particles are bound in the potential well. In contrast, the configurations shown in figure 3.11(*b*) include a particle excited to a single particle state above the energy which constitutes the effective threshold for emission (for a neutron this is the negative of the Q-value which is referred to as the separation energy; for a proton, the separation energy plus the Coulomb barrier). Griffin assumed that such a state will instantaneously decay with emission of the particle to the continuum. When such an emission takes place from a nucleus with total excitation energy E^*, if the kinetic energy of the nucleon is ε and the threshold energy B then the remaining excitation energy is $E^* - B - \varepsilon$.

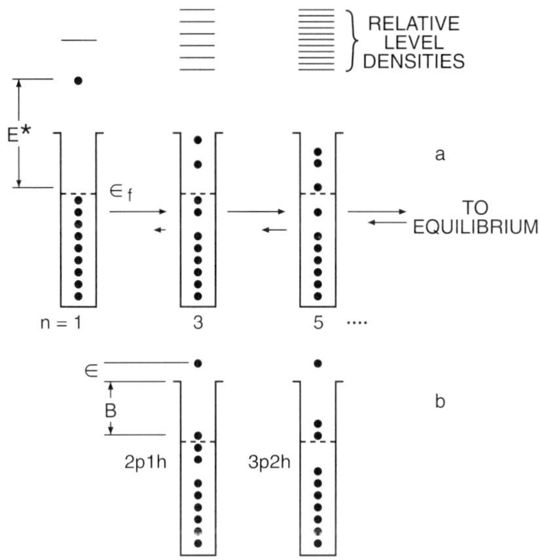

Figure 3.11. Particle hole creation as the equilibrium process proceeds (after Blann [45]). (*a*) The incident nucleon is, of course, in the continuum but succeeding states (2 particle–1 hole and 3 particle–2 hole) are bound. The level density is shown as increasing rapidly with the number of excitons. (*b*) 2 particle–1 hole and 3 particle–2 hole states both with one unbound particle which is assumed to be emitted as a pre-equilibrium nucleon. (Reprinted from Blann M, *Proceedings of the Europhysics Study Conference on Intermediate Processes in Nuclear Reactions*, ed N Cindro, P Kulisic and T Mayer-Kuckuk. Copyright 1973 Springer-Verlag GmbH & Co.)

The most important assumption made by Griffin is that, for a given number of excitons, every particle–hole configuration consistent with the total energy is equally probable. Thus if we consider an n-exciton state with total energy E^* and use the microstate density approximation $\Omega_n(E^*) = \rho_n(E^*)\Delta E^*$, then the probability of occupation of any microstate is

$$\frac{1}{\Omega_n(E^*)} = \frac{1}{\rho_n(E^*)\Delta E^*}. \qquad (3.7.1)$$

We now define a particular *macrostate* as that state in which one particle is separated from the parent nucleus. The density of microstates corresponding to this state is taken to be the convolution (see section 1.3) of the $n-1$ exciton density with the density of phase space states, ρ_{pe}, ('pe' for 'pre-equilibrium') for the residual nucleus + emitted particle.

$$\rho_n(E^*, B) = \int_0^{E^*-B} \rho_{n-1}(E^* - B - \varepsilon)\rho_{\mathrm{pe}}(\varepsilon)\,\mathrm{d}\varepsilon. \tag{3.7.2}$$

If we further restrict the macrostate definition so that the emitted particle has energy between ε and $\varepsilon + \mathrm{d}\varepsilon$ then the corresponding density of microstates is $\rho_{n-1}(E^*-B-\varepsilon)\rho_{\mathrm{pe}}(\varepsilon)\,\mathrm{d}\varepsilon$. Griffin thereby deduced that the probability to observe the macrostate under consideration is simply

$$\frac{\mathrm{d}P_n(\varepsilon)}{\mathrm{d}\varepsilon}\,\mathrm{d}\varepsilon = \frac{\rho_{n-1}(E^* - B - \varepsilon)\rho_{\mathrm{pe}}(\varepsilon)\,\mathrm{d}\varepsilon}{\rho_n(E^*)}. \tag{3.7.3}$$

Note the dual point of view concerning the decaying states. On the one hand, they are simply particle–hole states with an unbound particle (whose single particle energy lies above the threshold). They therefore form a subset of $\rho_n(E^*)$. On the other hand, they can be considered as certainly evolving to microscopic configurations in which the emitted particle is physically separated from the parent nucleus. This point of view is adopted in writing equation (3.7.3) and is similar to that used in the Weisskopf model.

We can complete the theory by writing the full energy spectrum as a weighted sum over exciton numbers. Thus

$$\frac{\mathrm{d}P(\varepsilon)}{\mathrm{d}\varepsilon}\,\mathrm{d}\varepsilon = \sum_n \tau_n \frac{\rho_{n-1}(E^* - B - \varepsilon)\rho_{\mathrm{pe}}(\varepsilon)}{\rho_n(E^*)}\,\mathrm{d}\varepsilon, \tag{3.7.4}$$

where n takes on values $1, 3, 5, \ldots$ if the target nucleus is even, and $2, 4, 6, \ldots$ if it is odd (Griffin assigns a 1p, 0h state to an odd nucleus). The problem is now to estimate the weighting factors τ_n. In his original work Griffin noted that the number of states increases very rapidly with n (this is depicted in figure 3.11(*a*)) so that transitions from states with n excitons to states with $n' = n + 2$ excitons would be expected to predominate compared with transitions to states with $n' = n$ or $n' = n - 2$. He further supposed that pre-equilibrium particle emission is a rare event so that (approximately) all states with given n would certainly evolve to states of $n + 2$. In this case the weighting factors τ_n are proportional to the mean lifetimes of n-exciton states. In fact, Griffin assumed equal lifetimes and thus assigned a uniform weighting in equation (3.7.4). It might reasonably be objected that, at least in the later stages of the process (close to the equilibrium number of particles and holes), that transitions with $n' = n$ or $n - 2$ would be expected to be as frequent as $n' = n + 2$ transitions. This objection is valid. However, the pre-equilibrium emission probability receives a negligible contribution from such states because, for large n, the energy sharing ensures that particles are almost always in single particle states below the threshold for emission. Notwithstanding, if one insists on the notion of *temporal* equilibrium in any one system, one might question the neglect of $n' = n$ transitions which presumably are responsible for the 'local' equilibrium situation which is considered to apply to each exciton number n.

We now turn to the specific evaluation of the state densities. The phase space density of states corresponding to the emitted particle (with mass, m, spin degeneracy, g and kinetic energy, ε) is obtained approximately by assuming an infinite mass for the (stationary) residue as in the Weisskopf theory (see (3.3.7)).

$$\rho_{\mathrm{pe}}(\varepsilon)\,\mathrm{d}\varepsilon = \frac{gV(2\pi m)^{3/2}\varepsilon^{1/2}}{h^3\Gamma(3/2)}\,\mathrm{d}\varepsilon = c\varepsilon^{1/2}\,\mathrm{d}\varepsilon. \tag{3.7.5}$$

From our examination of the Weisskopf model we can see that this is a strong approximation (or perhaps an error). Pre-equilibrium emission does not populate the volume V with particles whose velocities are randomly oriented. The second component, which is the density of states for a given number of excitons, was discussed in section 2.5. For single particle states with uniform spacing d, the number of states for n_{p} particles and n_{h} holes with excitation energy above the ground state κd was obtained as,

$$\Omega_{n_{\mathrm{p}},n_{\mathrm{h}}}(\kappa) = \sum_{k_{\mathrm{p}}=0}^{K} \mathbb{P}_{n_{\mathrm{p}}}(k_{\mathrm{p}})\mathbb{P}_{n_{\mathrm{h}}}(K - k_{\mathrm{p}}), \tag{3.7.6}$$

where

$$K = \kappa - \frac{n_{\mathrm{p}}(n_{\mathrm{p}}+1)}{2} - \frac{n_{\mathrm{h}}(n_{\mathrm{h}}-1)}{2}$$

and $\mathbb{P}_{n_{\mathrm{p}}}(k_{\mathrm{p}})$ is the number of ways of partitioning k_{p} into no more than n_{p} parts. Converting these numbers into densities (dividing by d) and remembering that $n = n_{\mathrm{p}} + n_{\mathrm{h}}$, we can use equations (3.7.4)–(3.7.6) to calculate the energy spectra either for a specific number of excitons, n, or as a sum over exciton numbers. In his original work, Griffin used an approximation to the state density for fixed n which has been recently examined (and criticized) by Böhning [48]. In spite of this difficulty, Griffin was the first to provide a simple account of the emission of pre-equilibrium nucleons. A detailed study of neutron induced reactions using the exciton model has been made by Holub *et al* [49].

Some improvements to the Griffin model were discussed by Cline and Blann [45]. The first improvement was to correct the model by adopting the argument used in the Weisskopf theory which effectively inserts an additional transition probability, which restricts final states of the emitted particles in the volume V to those which are in fact physically realized. This factor is simply the 'Weisskopf' transition rate calculated by considering the inverse reaction in which the emitted pre-equilibrium particle is captured by the $n-1$ exciton residual nucleus. If the inverse cross-section is $\sigma(\varepsilon)$, then the inverse transition rate is $\sigma(\varepsilon)(2\varepsilon/m)^{1/2}/V$. Inserting this factor gets rid of the unknown volume V (which occurs when (3.7.5) is inserted into (3.7.4)) and allows, in principle, an absolute estimation of the emission rate. Thus the specific rate of emission of particles from an n-exciton state is the Weisskopf expression

$$\frac{\mathrm{d}W_n(\varepsilon)}{\mathrm{d}\varepsilon}\,\mathrm{d}\varepsilon = \frac{\rho_{n-1}(E^* - B - \varepsilon)}{\rho_n(E^*)}\,\frac{\sigma(\varepsilon)(2\varepsilon/m)^{1/2}}{V}\,\frac{gV(2\pi m)^{3/2}\varepsilon^{1/2}}{h^3\Gamma(3/2)}\,\mathrm{d}\varepsilon$$

$$= \sigma(\varepsilon)\frac{8\pi gm}{h^3}\frac{\rho_{n-1}(E^* - B - \varepsilon)}{\rho_n(E^*)}\varepsilon\,\mathrm{d}\varepsilon. \qquad (3.7.7)$$

The second improvement is to allow for all energy conserving transitions which may result from a nucleon–nucleon collision i.e., $n' = n-2, n, n+2$. In this case the weighting factors τ_n (equation (3.7.4)) are replaced by the occupation probabilities, $q_n(t)$, for each n-exciton state. These probabilities explicitly introduce the notion of temporal evolution between states with different numbers of excitons. We might expect that the origin of the time variable, t, would be taken to coincide with the occurrence of the first nucleon–nucleon collision. In practice, this origin is used as a parameter.

The functions $q_n(t)$ are obtained from a differential equation. We define the transition probability per unit time from state n to state n' as $\lambda(n, n')$. The set of (so-called) master equations which are used to describe a process taking place on a discrete state space in continuous time (like the Poisson process discussed in section 2.4) then reads

$$\frac{\mathrm{d}q_n(t)}{\mathrm{d}t} = q_{n-2}(t)\lambda(n-2, n) + q_{n+2}(t)\lambda(n+2, n) - q_n(t)[\lambda(n, n-2) + \lambda(n, n+2)], \qquad (3.7.8)$$

where the first two and the last two terms are referred to, respectively, as 'gain' and 'loss' contributions. The transition rates in the Cline–Blann theory were constructed using a result from time dependent perturbation theory (see formula XVII.50 in Messiah [50]) known as Fermi's Golden Rule

$$\lambda(n, n') = \frac{2\pi}{\hbar}|T_{n,n'}|^2\rho'_{n'}(E^*). \qquad (3.7.9)$$

The matrix elements $T_{n,n'}$ were taken as constants, i.e. independent of n. They are, *a priori*, unknown so that only relative values of $\lambda(n, n')$ can be obtained with the consequence that, despite the fact that, in principle, (3.7.7) gives absolute decay rates, an absolute timescale cannot be defined. Furthermore, some modification of the densities (indicated by writing $\rho'_{n'}$ rather then $\rho_{n'}$ in (3.7.9)) was thought to be necessary because, in a single nucleon–nucleon collision, not all final particle–hole states are accessible [51]. Note that equation (3.7.8) incorporates one of Griffin's assumptions, namely, that the rate of pre-equilibrium emission, which should normally be included in the master equation as a loss term, can be neglected. In this sense, particle emission is considered as a small perturbation built on the equilibration process.

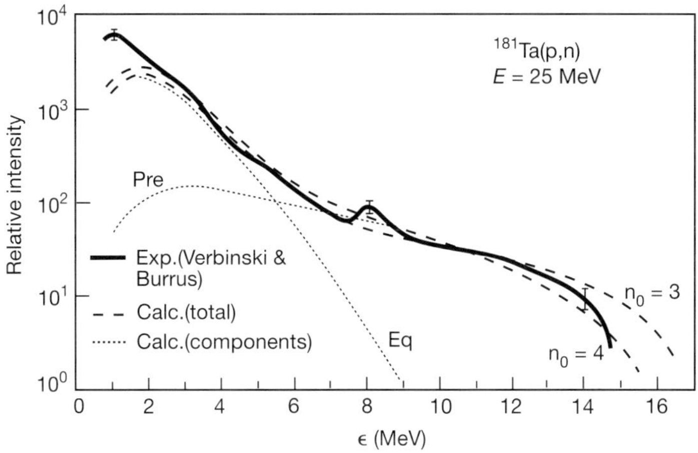

Figure 3.12. Prediction of the Cline–Blann master equation model [45] for the energy spectrum of pre-equilibrium neutrons in the ^{181}Ta (p,n) reaction compared with experimental data (heavy continuous curve). The data are from Verbinski and Burrus 1969 *Phys. Rev.* **177** 1671. Typical error bars are shown at a few energies. The labels n_0 refer to the number of excitons assumed to exist at $t = 0$. The pre-equilibrium and equilibrium components are shown separately for $n_0 = 3$. (Reprinted from *Nucl. Phys.* A, **172**, Cline C K and Blann M, p 225. Copyright 1971, with permission from Elsevier Science.)

Given that the master equation yields the time dependent occupation probabilities $q_n(t)$, the final step is to construct the time dependent energy spectrum (compare with (3.7.4))

$$\frac{\mathrm{d}P(\varepsilon, t)}{\mathrm{d}\varepsilon}\,\mathrm{d}\varepsilon = \sum_{n(\Delta n=2)} q_n(t)\frac{\mathrm{d}W_n(\varepsilon)}{\mathrm{d}\varepsilon}\,\mathrm{d}\varepsilon, \tag{3.7.10}$$

which, of course may be examined at any particular time or integrated over any time interval.

The Cline–Blann model is globally rather successful. A comparison of an experimental energy spectrum with the model prediction is shown in figure 3.12 [45]. The number of excitons used to initiate the equilibration process is not considered to be known in the theory. However, for alpha particle projectiles the best fit to the data is obtained with n_0 ($n(t = 0)$) of the order of five and, for incident nucleons, of the order of three.

In the above models the effect of the Pauli principle is built into the particle–hole densities. An alternative theoretical picture was proposed by Harp, Miller and Berne [46] which is based on the single particle level density. In the HMB model, the effect of the Pauli Principle is incorporated directly into the master equation which determines the evolution of the occupation probabilities of single particle states in a potential well. The modified equation is referred to as the

Boltzmann master equation. Adaptations of this theory (see for example reference [47]) have been extensively used to model the equilibration process. An important feature is that the uniform (or random) occupation of particle–hole states assumed in the exciton model is not guaranteed, especially in the early stages of the reaction.

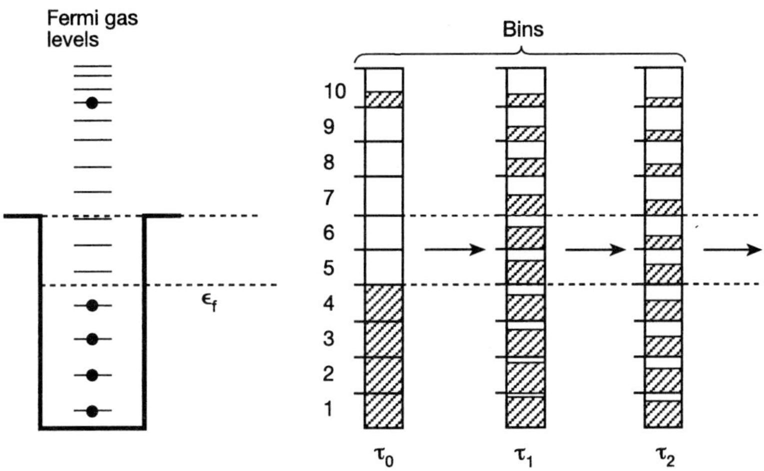

Figure 3.13. Schematic picture of the HMB model [46] for pre-equilibrium emission. The single particle spectrum is shown divided into ten energy bins. Intially, all bins below the Fermi surface are occupied and one bin in the continuum is partially occupied (by the projectile nucleons). The occupation probabilities evolve as the reaction proceeds due to nucleon–nucleon collisions. Continuum bins may emit particles to the exterior with a rate given by the last term of equation (3.7.12). (Reprinted from Blann M, *Proceedings of the Europhysics Study Conference on Intermediate Processes in Nuclear Reactions*, ed N Cindro, P Kulisic and T Mayer-Kuckuk. Copyright 1973, Springer-Verlag GmbH & Co.)

The principle of the calculation is represented in figure 3.13 which is adapted from the review of Blann [45]. The energy of any single particle state will be denoted as ε. The smoothed density of states, $\rho_{\rm sp}(\varepsilon)$ for a single particle in a square well potential is divided into bins such that the ith bin contains g_i states (all states between $\varepsilon_i - \frac{\Delta\varepsilon}{2}$ and $\varepsilon_i + \frac{\Delta\varepsilon}{2}$). Bins below a well depth energy (taken as 48 MeV) cannot emit particles to the exterior and are thus confined to the volume, $V_{\rm p}$, corresponding to the potential, but those above may do so. Transitions rates between bins are governed by a 'Boltzmann' transition probability which, as indicated above, includes blocking factors for occupied bins. The authors define the occupation probability, f_i (they actually use the symbol n_i) corresponding to the ith bin by stating that the number of occupied states in this bin is

$$f_i g_i = f_i \int_{\varepsilon_i - \frac{\Delta\varepsilon}{2}}^{\varepsilon_i + \frac{\Delta\varepsilon}{2}} \rho_{\rm sp}(\varepsilon)\,{\rm d}\varepsilon. \tag{3.7.11}$$

If $f_i = 1$ all states are occupied. The transition rates (transition probabilities per unit time) are written as $\omega_{kl\to ij}$ which is the rate for a process in which a nucleon in one of the g_k states in the kth bin collides with a nucleon in one of the states in the lth bin so as to produce two nucleons in states belonging to the bins i and j. In this process energy must be conserved. The Boltzmann master equation is now written (for simplicity, we do not distinguish between neutrons and protons)

$$\begin{aligned}\frac{\mathrm{d}(g_i f_i)}{\mathrm{d}t} &= \sum_{j,k,l}(g_k f_k)(g_l f_l)\omega_{kl\to ij} g_i(1-f_i)g_j(1-f_j)\delta(\varepsilon_i+\varepsilon_j-\varepsilon_k-\varepsilon_l)\\ &\quad -\sum_{j,k,l}(g_i f_i)(g_j f_j)\omega_{ij\to kl} g_k(1-f_k)g_l(1-f_l)\delta(\varepsilon_i+\varepsilon_j-\varepsilon_k-\varepsilon_l)\\ &\quad -(g_i f_i)\omega_{i\to i'} g_{i'}\delta(\varepsilon_i-48-\varepsilon_{i'}).\end{aligned} \tag{3.7.12}$$

The first term describes transitions from bins k, l to bins i, j. Summed over j, k and l it represents a gain term in the bin i. In the same way we see that the second term is a loss term. In the original work, the first two terms of equation (3.7.12) contained an additional factor of ΔE in the denominator which cancels with an identical factor introduced in the definition of $\omega_{ij\to kl}$. We have simply removed these factors so that the definition of $\omega_{ij\to kl}$ is dimensionally identical with that of $\omega_{i\to i'}$ (inverse time). The last term (contrasting with the Cline–Blann theory) explicitly accounts for 'leakage' to the exterior. The rate $\omega_{i\to i'}$ represents a particle leaving the potential well to find itself outside the volume V_p in the 'exterior' energy bin with energy $\varepsilon_{i'}$. Note that ε_i and $\varepsilon_{i'}$ are positive quantities so that the last delta function in (3.6.12) implies that a particle must have at least 48 MeV to be emitted.

The transition probabilities, $\omega_{ij\to kl}$, are gven in terms of the nucleon–nucleon scattering cross-section, σ_{NN}, and the relative velocity, v_{ij}, of nucleons i and j by

$$\omega_{ij\to kl} = \frac{\sigma_{\mathrm{NN}} v_{ij}}{V_\mathrm{p}\sum_{mn} g_m g_n \delta(\varepsilon_i+\varepsilon_j-\varepsilon_m-\varepsilon_n)}, \tag{3.7.13}$$

which is simply the 'Weisskopf' rate which represents transitions to all energy conserving final states divided by the total number of those states (as indicated above, a factor of ΔE in the numerator has been removed). The transition rate to continuum states in an exterior volume, Ω_{ext}, is constructed in the same way, i.e.,

$$\omega_{i\to i'} = \frac{\sigma_{\mathrm{NUC}} v_{i'}}{g_i \Omega_{\mathrm{ext}}}. \tag{3.7.14}$$

In this case the Weisskopf rate is divided by g_i because it refers to formation of the nucleus (with g_i states) in the inverse process. The cross-section, σ_{NUC} corresponds to the size of the compound nucleus and m_N is the nucleon mass.

The density of external phase space states is the standard one particle expression (equation (1.2.32)) with an extra factor of two due to spin degeneracy

$$\rho(\varepsilon')\,\mathrm{d}\varepsilon' = \frac{4\pi\Omega_{\mathrm{ext}}(2m_\mathrm{N})^{3/2}\varepsilon'^{1/2}}{h^3}\,\mathrm{d}\varepsilon'. \tag{3.7.15}$$

The volume, $\Omega_{\rm ext}$, cancels (as in the Weisskopf model) in equation (3.7.12). In contrast, the nuclear volume,

$$V_{\rm p} = \frac{4\pi}{3}\Big(\frac{\sigma_{\rm NUC}}{\pi}\Big)^{3/2},$$

which replaces $\Omega_{\rm ext}$ in the calculation of the density of single particle states, is obviously a physical parameter of the model.

The 'accessibility' of the kth bin is determined by the blocking factor $(1 - f_k)$. No blocking factors are included for continuum states (so that $f_i g_i$ is replaced simply by g_i). The pre-equilibrium energy spectrum for emission from the ith bin to the i'th exterior bin is given by the time integration of the last term of (3.7.12).

$$\frac{{\rm d}P_i(\varepsilon_{i'})}{{\rm d}\varepsilon_{i'}}\,{\rm d}\varepsilon_{i'} = \int_0^\infty f_i(t) g_i \omega_{i\to i'} g_{i'} \delta[\varepsilon_{i'} - (\varepsilon_i - 48)]\,{\rm d}t, \qquad (3.7.16)$$

in which we have indicated explicitly that the occupation probabilities, f_i, are time dependent. The full energy spectrum is constructed as a sum over i. An example of a 'Harp–Miller–Berne' calculation made by Blann [45] is shown in figure 3.14.

So far we have discussed theories which have been applied mainly to light particle (p, α) induced reactions. Regretably there is no 'standard' statistical theory for pre-equilibrium emission of nucleons and heavier particles (notably alpha particles) which occurs in collisions of heavy ions. Experimental results are often parametrized using a fixed temperature source moving in the laboratory frame of reference with a fixed velocity [52, 53]. The source is usually taken to be isotropic with an energy dependence of the Maxwellian form

$$\frac{{\rm d}^2\sigma}{{\rm d}\varepsilon\,{\rm d}\Omega} = \frac{N_S \varepsilon^{1/2}\,{\rm e}^{-\varepsilon/T}}{2(\pi T)^{3/2}}, \qquad (3.7.17)$$

where N_S is the source multiplicity which is the integral over energy and solid angle of (3.7.17) (one may question the wisdom of using the Maxwellian energy distribution instead of the evaporation formula which goes as $\varepsilon\,{\rm e}^{-\varepsilon/T}$). In section 2.2, we learned how observables for such a source are transformed to the laboratory frame of reference.

Systematics emerging from such analyses suggest that, for nucleon emission, the source velocity is roughly equal to the velocity of the centre of mass corresponding to collisions of individual nucleons of the projectile and target (the nucleon–nucleon centre-of-mass velocity is equal to half the projectile velocity), whereas the temperature, T, is of the order of 5–10 MeV. From (3.7.17) we quickly find that the mean energy of nucleons is $3T/2$ which, for a temperature of 10 MeV, is rather lower than, but of the same order as, the average kinetic energy of nucleons in the nuclear ground state (3/5 of the Fermi energy). Some reduction of the effective Fermi energy which is not incompatible with the observations

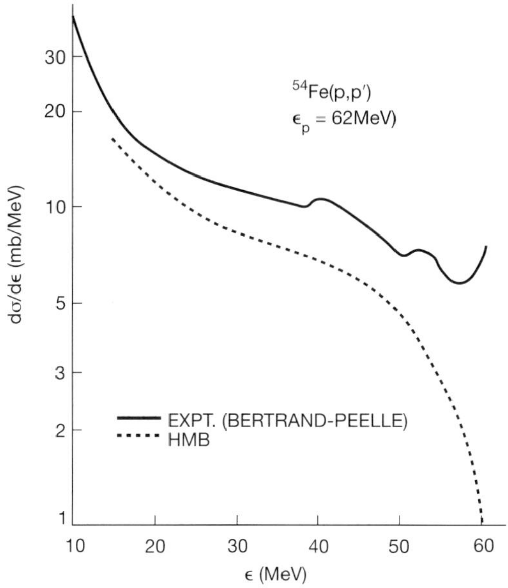

Figure 3.14. Predictions of the HMB model compared with energy spectra of protons in the ^{54}Fe(p, p′) reaction [45]. The experimental data, which do not appear to have been published, are from Bertrand and Peelle (1971, Oak Ridge National Laboratory internal report ORNL4638). The shape of the spectrum is well predicted but the absolute value is low by approximately 50%. (Reprinted from Blann M, *Proceedings of the Europhysics Study Conference on Intermediate Processes in Nuclear Reactions*, ed N Cindro, P Kulisic and T Mayer-Kuckuk. Copyright 1973, Springer-Verlag GmbH & Co.)

would be expected if nucleons emerged from the interior of a low density 'surface overlap' region of projectile and target. Thus it is possible to obtain, if not a theory, then at least a sensible parametrization of the experimental results.

Several alternative theories for pre-equilibrium emission are reviewed in the book by Gadioli and Hodgson [13]. Included in this presentation (one might say panoply) we find, of course, the models discussed above for light projectiles but also microscopic dynamical approaches to heavy ion collisions such as the intra-nucleon cascade model [54], the Landau–Vlasov equation [55] or the nucleon exchange transport theory proposed by Randrup and Vandenbosch [56]. These latter models fall rather outside the scope of the present work. However, they are representative of current theories for pre-equilibium emission in heavy ion induced reactions.

To terminate this section we briefly refer the reader to the considerable efforts deployed, over several decades, by a number of theoreticians whose objective is to construct a fully quantum mechanical description of reactions involving the formation of a compound nucleus. Early theories [57] involved some sort of averaging process applied to the S-matrix which was usually achieved by a direct parametrization of matrix elements. More recent theories, including the widely used FKK (Feshbach, Kerman and Koonin) approach [58], are based on a stochastic distribution of matrix elements of the nuclear Hamiltonian which governs transitions between intermediate states of different complexity and also the coupling to (pre-equilibrium) exit channels. The recent work of Herman *et al* [60] is based on the Gaussian orthogonal ensemble (GOE) which describes the distribution of martix elements. Microstate densities for classes of intermediate states are obtained independently using the single particle spectrum (as in section 2.5). Such theories (at least in the weak coupling limit) can be formulated using the exciton picture in the sense that classes of intermediate states can be referred to by the numbers of particles and holes. The equilibration process can then be decribed in terms of a transport equation governed by averages over the matrix elements [60]. For the moment, numerical applications are limited to rather low energies involving nucleon-induced reactions but attempts to include the direct reaction contributions (unbound intermediate states) are well underway. One important result obtained by Herman *et al* is that there is a rather well-defined distinction, in terms of numbers of excitons, between pre-compound and compound processes.

3.8 Statistical models for fission at low excitation energies

The subject of nuclear fission seems to be characterized by a series of surprising discoveries beginning with the fission phenomenon itself. Indeed the observation of this phenomenon in 1938 by Hahn and Strassman [61] seems to have taken two years before being recognized as such.

A qualitative understanding of the fission mechanism can be obtained by examining the main features of the dynamic evolution when a thermal neutron ($3k_{\mathrm{B}}T/2 = 3.9 \times 10^{-8}$ MeV at 300 K) is captured in a heavy nucleus such as ^{235}U. In this reaction the Q-value for capture (i.e. fusion) provides an excitation energy of 6.54 MeV which is sufficient to provoke a significant (essentially quadruple) deformation in ^{236}U. As this nucleus deforms (at constant volume) the magnitude of the negative contribution of the surface energy to the binding energy increases and thus opposes the deformation. However, at the same time the increasing separation of the nuclear charges (the protons) reduces the Coulomb energy and thus favours further elongation. The competition between these two energy changes has the effect of creating a potential barrier. The top of the barrier corresponds to the 'no return' or saddle point beyond which the system 'descends' to the point of scission into two fragments (figure 3.15). The breakup into two fragments is accompanied by the emission of 2.5 neutrons (on average). In the particular case cited, the barrier is 5.5 MeV and is thus lower than the excitation energy. However, this is not a general rule. With thermal neutrons captured by heavy nuclei the fission barriers are usually of the same order of magnitude as the excitation energies so that neutron capture induced fission may or may not be the dominant decay mechanism. Thus, for example, the capture of a thermal neutron by ^{238}U does not lead directly to fission because the resulting excitation energy, 4.8 MeV, is smaller than the barrier height (5.7 MeV). Rather, the sequence $^{238}\mathrm{U}(\mathrm{n}, \gamma)^{239}\mathrm{U} \rightarrow \beta^- \rightarrow {}^{239}\mathrm{Np} \rightarrow \beta^- \rightarrow {}^{239}\mathrm{Pu}$ takes place resulting in the formation of ^{239}Pu, which, like ^{235}U, fissions upon capture of a thermal neutron. Nuclei such as ^{235}U and ^{239}Pu are referred to as fissile. The existence of such nuclei, together with the kinetic energy of fission fragments (~200 MeV) and the release of fast neutrons (with energies ~2 MeV) which engender a chain reaction, are essential factors for the construction of conventional nuclear power reactors.

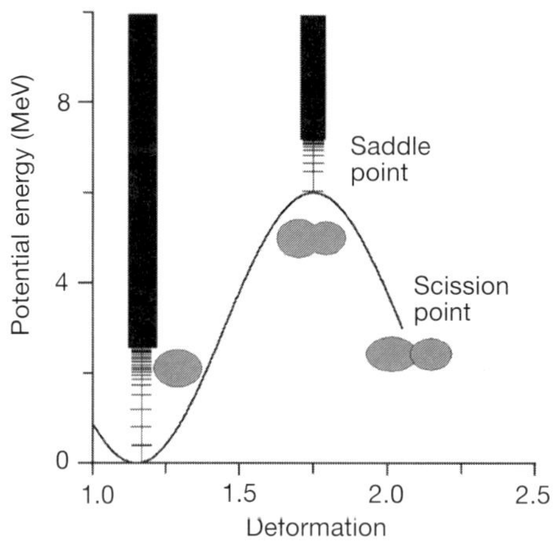

Figure 3.15. Sketch of the potential energy in the Bohr–Wheeler model for low energy fission. The ground state is depicted as slightly deformed. The deformation is defined as the ratio of the long and short axes of the deformed nucleus and is thus equal to one for spherical shapes. The specific decay rate for fission is determined by the density of states at the fission barrier. The energy released in the descent from saddle to scission is assumed to appear as kinetic energy of the fragments (adiabatic limit).

The preceding paragraph enables us to appreciate a major difficulty with models which attempt to describe the fission process which is that the fission barrier which is of the order of a few MeV results from the partial cancellation of two quantities (the surface and Coulomb energies), which are both of the order of several hundred MeV. Thus, nominally small effects such as shell structure may play an important role in determining the characteristics of the barrier. A related difficulty is associated with an essentially exponential dependence of certain observables on the available energy above the fission barrier which obviously entails an enhanced sensitivity to detailed aspects of the structure of the fissioning nucleus.

Understanding the low energy fission process has proved so difficult that, even 60 years after the Bohr–Wheeler liquid drop statistical model [62] which provided a qualitative understanding of fission, there does not seem to exist a well-defined and universally accepted theory. A major element of the debate concerns the importance of the descent from saddle to scission in the determination of the characteristics of the fission products.

The primary objective for theorists is to explain the ratio of the probabilities (widths) for fission and neutron emission. Other important characteristics include the (asymmetric) mass distribution, the kinetic energy of fragments at the scission point, charge distributions, angular distributions, and the variation of the number of post-scission neutrons with the mass asymmetry. Recently, there has been considerable interest in measurements of pre-scission emission of light particles. The 1989 Berlin conference to celebrate the 50th anniversary of the discovery of fission [63], together with the collective work of Wagemans and collaborators concerning fission at low excitation energies [64], together provide a detailed account of the current situation as well as an extensive guide to the literature.

In this section we will concentrate on attempts to understand fission at low excitation energies in terms of statistical mechanics. One problem which occurs in the construction of such models has to do with the coupling between collective variables which describe macroscopic aspects of the fissioning system (such as the elongation), and intrinsic (single particle) degrees of freedom. In the limit of strong coupling, the energy which is liberated due to the decrease in potential energy from saddle to scission would be expected to be distributed between collective and single particle degrees of freedom according to the statistical weights of the corresponding groups of microstates, so that a purely statistical model should be appropriate at all points along the saddle–scission trajectory up to the point of scission. This assumption forms the basis of the theory advocated by Fong [65]. In the opposite limit, liberated energy will appear almost entirely as kinetic energy in collective coordinates, and the population of single particle levels will be always close to the 'ground state' for successive collective configurations. In this case one refers to an adiabatic hypothesis because, beyond the saddle point, the collective degrees of freedom do not exchange energy with the single particle degrees of freedom. For intermediate coupling, statistical theories usually have to be abandoned in favour of detailed dynamical calculations which include an explicit treatment of energy dissipation [36, 66]. However, at the end of this section, we present a 'two temperature' model which attempts to deal with intermediate coupling using an equilibrium statistical model.

Probably the most important statistical theory for fission was provided almost immediately after the discovery of this decay mechanism. The transition state method proposed by Bohr and Wheeler in 1939 [62] applies not only to fission of heavy elements but indeed to any binary division of an excited nucleus. The method makes use of the density of microstates at the saddle point. For this reason it cannot be straightforwardly applied at excitation energies below the barrier. The problem of escape over a potential barrier of a system subjected to random forces (viscocity) was considered in 1940 by Kramers [67] who showed that, for a wide range of viscocity, the outwardly directed probability current associated with the collective (fission) coordinate (the current of phase space points) is dominated by those states close to the barrier. This work thus lends support for the Bohr–Wheeler model at least with respect to the calculation of the width for fission.

Let us begin not with fission but with neutron emission which is the principal competing decay mechanism for heavy nuclei. We do this because the theory, in this case, is very similar to the Weisskopf evaporation theory discussed in section 3.3. Let $\rho_t(E_C^* + Q_n, \Delta t)$ be the density of transition states associated with the time interval Δt so that if $\rho_C(E_C^*)$ is the density corresponding to all microstates of an excited (compound) nucleus the decay probability per unit time (specific decay rate) is simply

$$W_n = \frac{1}{\Delta t}\frac{\rho_t(E_C^* + Q_n, \Delta t)}{\rho_C(E_C^*)}. \tag{3.8.1}$$

The Q-value for neutron emission, which is normally a negative quantity, plays the role of a barrier. It corresponds to the fact that nucleons cannot be considered to be independent particles (as assumed in the simple treatment of evaporation) but experience mutually attractive forces. Otherwise, the notion of the transition state is the same as that introduced in connection with evaporation. Transition microstates are those states which decay in the time interval Δt. We therefore simply need to estimate the number of such states. In fact this can be done directly, without considering the inverse reaction (which was the method employed by Weisskopf). We identify transition states as those states for which a neutron, with a random outwardly directed velocity, lies at the surface of the nucleus in a cylindrical volume element of section ΔS and width ΔR. The geometrical aspects of the problem are then identical with those discussed in connection with evaporation (see figure 3.4(*b*)). We now admit that neutrons may be emitted from all points of the surface of the excited nucleus. Thus, starting from equation (3.3.20) the small 'hole' in the container (which now becomes the parent nucleus) is extended over the full surface of the residual nucleus (i.e., $\Delta S = 4\pi R^2$) and thus becomes a spherical shell of width, ΔR. Furthermore, we note that the microstate density for transition states can be represented as the convolution of the density of states in the residual nucleus (which, for any energy E, we write as $\rho_n(E)$) and the phase space density of states for the emitted neutron. Then, the density of states which correspond to neutrons (mass m_n) escaping in a time interval Δt with energy in the range $(\varepsilon_n, \varepsilon_n + d\varepsilon_n)$ so that the residual nucleus has an excitation energy in the range $(E_C^* + Q_n - \varepsilon_n, E_C^* + Q_n - (\varepsilon_n + d\varepsilon_n))$ is

$$\frac{d\rho_t(E_C^* + Q_n, \Delta t)}{d\varepsilon_n}\, d\varepsilon_n = [\rho_n(E_C^* + Q_n - \varepsilon_n)]\left[4\pi R^2 \frac{2\pi m_n \varepsilon_n\, d\varepsilon_n \Delta t}{h^3}\right]. \tag{3.8.2}$$

This equation is almost identical with the evaporation equation (3.3.20). Apart from the replacement of ΔS by $4\pi R^2$, the density of states in the heat bath asssumed in equation (3.3.20) has been replaced with the density of states for the residual nucleus. Indeed, if we set $Q_n = 0$ (the independent particle approximation) and expand $\rho_n(E_C^* + Q_n - \varepsilon_n)$ we recover the evaporation formula.

The second factor in equation (3.8.2) should, in fact, include a spin degeneracy factor g_n although this factor was not included in the original Bohr–Wheeler theory.

From equation (3.8.1) we obtain the specific decay rate as the density of transition states divided by the density of states of the compound system and by Δt. Thus,

$$W_n = \frac{\Gamma_n}{\hbar} = \frac{4\pi R^2 2\pi m_n}{h^3} \frac{\int_0^{E_C^*+Q_n} \rho_n(E_C^* + Q_n - \varepsilon_n)\varepsilon_n \, d\varepsilon_n}{\rho_C(E_C^*)}$$
$$= \frac{m_n R^2}{\pi \hbar^3 \rho_C(E_C^*)} \int_0^{E_C^*+Q_n} \rho_n(E_C^* + Q_n - \varepsilon_n)\varepsilon_n \, d\varepsilon_n. \tag{3.8.3}$$

By dealing directly with transition states we avoid the introduction of the density 'ρ_{bB}' used by Weisskopf. We thus 'cut out the middle man'. Nevertheless, the result in equation (3.8.3) is identical with that of Weisskopf. Indeed, by associating the quantity πR^2 with the inverse cross-section σ_{bB} introduced in the original derivation we immediately recover the Weisskopf result (to within the trivial inclusion of the spin degeneracy factor g_n). Thus

$$\frac{dW_n}{d\varepsilon_n} d\varepsilon_n = \frac{8\pi m_n[\pi R^2]}{h^3} \frac{\rho_n(E_C^* + Q_n - \varepsilon_n)\varepsilon_n \, d\varepsilon_n}{\rho_C(E_C^*)}. \tag{3.8.4}$$

The transition state formulation is probably to be preferred to the original derivation because of the absence of the intermediate states (the 'bB' states of equation (3.3.2)) and of the consequent need to consider time reversed trajectories.

Bohr and Wheeler also applied the transition state theory to fission above the barrier. For this purpose, transition states were identified as saddle configurations for which the fragment separation (deformation) is increasing. Such states will eventually fission. As for neutron emission, the transition state density is expressed as a product of two factors. The first factor, the density corresponding to internal excitation, will be written as $\rho_f(E_C^* - B_f - \varepsilon)$ where B_f is the fission barrier and ε is the kinetic energy associated with the motion of the fission fragments at the barrier. The second factor is the density associated with the fission fragment kinetic energies. Here, Bohr and Wheeler introduced an important modification. Whereas the neutron was considered to move randomly in three-dimensional space, the relative motion of fission fragments was restricted to correspond to increasing separation of fragments along the deformation axis. There are two consequences. The first is that the factor $\Delta R \Delta S \ 4\pi p^2 \, dp/h^3$ is replaced by the one-dimensional equivalent, $\Delta R \, dp/h$. The second is that the average over angles of the emission probabilities which led to (3.3.17) is not performed so that the fraction of phase space microstates which leave a small region of width,

ΔR, near the saddle point in a time interval, Δt, and therefore decay, is simply $v\Delta t/\Delta R$. With these modifications the specific decay rate for fission is obtained by appropriately modifying equation (3.8.3) (and remembering that $v\,\mathrm{d}p = \mathrm{d}\varepsilon$) as

$$W_\mathrm{f} = \frac{\Gamma_\mathrm{f}}{\hbar} = \frac{1}{h\rho_C(E_C^*)}\int_0^{E_C^*-B_\mathrm{f}} \rho_\mathrm{f}(E_C^* - B_\mathrm{f} - \varepsilon)\,\mathrm{d}\varepsilon. \tag{3.8.5}$$

The integral in equation (3.8.5) simply counts all microstates above the barrier up to the maximum energy $E_C^* - B_\mathrm{f}$. An improvement is possible by noting that transmission over the barrier is not 100% for states with energies near the barrier. This can be taken care of (down to the cut off $\varepsilon = E_C^* - B_\mathrm{f}$) by including a transmission coefficient in (3.8.5). Using the Hill–Wheeler parabolic barrier approximation (3.6.3), and separating out discrete transition states, we finally write

$$W_\mathrm{f} = \frac{1}{h\rho_C(E_C^*)}\left[\sum_i \frac{1}{1+\mathrm{e}^{2\pi(B_\mathrm{f}-\varepsilon_i)/\hbar\omega_\mathrm{f}}} + \int_{\varepsilon_i^{\max}}^{E_C^*-B_\mathrm{f}} \frac{\rho_\mathrm{f}(E_C^*-B_\mathrm{f}-\varepsilon)}{1+\mathrm{e}^{2\pi(B_\mathrm{f}-\varepsilon)/\hbar\omega_\mathrm{f}}}\,\mathrm{d}\varepsilon\right], \tag{3.8.6}$$

in which the ε_i are kinetic energies corresponding to discrete states ($\varepsilon_i^{\max}$ is the energy above which a density is used).

It was soon realized that, despite the unified form of the theory, the application to fission was no easy matter. Thus, for example, the simple liquid drop model predicts only undeformed ground states which are stable with respect to mass asymmetric deformations so that, in contrast with observations, symmetric fission is predicted to be predominant. This difficulty was largely resolved by the inclusion of shell and pairing effects. However, the problem of calculating precise potential energy surfaces (because of shell effects the barrier is, in fact, usually of a 'double hump' form) and fission paths is still the object of much ongoing research.

Another attempt to formulate a statistical theory for fission was made by Fong [66]. In its simplest form, Fong's theory simply attempts to count the number of microstates for each conceivable *scission* configuration (macrostate). The philosophy behind the theory is that the dynamic evolution which leads to fission provides no restriction on the microscopic configurations which may be attained at scission. The damping out of relative motion in the path from saddle to scission is assumed to be strong enough to ensure that the kinetic energy at the scission point can be neglected so that essentially all available energy is shared between the two fission fragments. The theory is thus rather different from that of Bohr and Wheeler. It assumes strong coupling rather than adiabaticity for the saddle to scission path.

Let us discuss the simplest form of the model by supposing that any given scission configuration consists of two spherical touching nuclei. If we can neglect kinetic energy, we would expect the density of microstates to be given by a convolution of the two fragment internal microstate densities. Thus if all the available energy, $E = E_C^* + Q_f - B_C$, were to be concentrated in internal excitation, then, using equation (1.4.3),

$$\rho_{12}(E) = \int_0^E \rho_1(E_1)\rho_2(E - E_1)\, dE_1. \tag{3.8.7}$$

If we assume the simplest form for the internal level densities i.e.,

$$\rho_k(E_k) = c_k\, e^{2\sqrt{a_k E_k}}, \tag{3.8.8}$$

then we can obtain a quite accurate evaluation of (3.8.7) by going beyond the zeroth order approximation of equation (1.3.13). We expand the densities to *second* order around the maximum of the integrand i.e., for $k = 1, 2$

$$\ln[\rho_k(E_k = E_k^{\max} + \varepsilon_k)] = \ln[\rho_k(E_k^{\max})] + \beta_k \varepsilon_k - \frac{\varepsilon_k^2}{2\sigma_k^2}, \tag{3.8.9}$$

where β_k is the usual inverse temperature and

$$\frac{1}{\sigma_k^2} = \left.\frac{-d^2 \ln[\rho_k(E_k)]}{dE_k^2}\right|_{E_k = E_k^{\max}} = \frac{a_k^{1/2}}{2(E_k^{\max})^{3/2}}. \tag{3.8.10}$$

The maximum of the integrand in equation (3.8.7) occurs, as expected, when $\beta_1 = \beta_2$. We quickly find $E_1^{\max} = \frac{a_1 E}{a_1 + a_2}$ (see section 1.4) and thus that $\beta_1 = \beta_2 = \sqrt{\frac{a_1 + a_2}{E}}$. Noting that $\varepsilon_1 + \varepsilon_2 = 0$, we can now rewrite (3.8.7) as

$$\rho_{12}(E) = c_1 c_2\, e^{2\sqrt{(a_1+a_2)E}} \int_{-E_1^{\max}}^{E - E_1^{\max}} e^{-\varepsilon^2/2\sigma^2}\, d\varepsilon, \tag{3.8.11}$$

where

$$\frac{1}{\sigma^2} = \frac{1}{\sigma_1^2} + \frac{1}{\sigma_2^2} = \frac{(a_1 + a_2)^{5/2}}{2 a_1 a_2 E^{3/2}}. \tag{3.8.12}$$

We can evaluate (3.8.11) approximately by stretching the limits of integration to $\pm\infty$. This procedure would be expected to lose validity for very asymmetric splits for which $E_1^{\max}$ (or $E_2^{\max}$) is close to zero. Inserting these limits in (3.8.11), we obtain

$$\rho_{12}(E) = 2\sqrt{\pi}\, c_1 c_2 \frac{(a_1 a_2)^{1/2}}{(a_1 + a_2)^{5/4}} E^{3/4}\, e^{2\sqrt{(a_1+a_2)E}}, \tag{3.8.13}$$

which is one of Fong's basic results. It is interesting to note that the *first* order approximation (evaluated with the proper integration limits) would simply be

$$\rho_{12}(E) = c_1 c_2 E\, e^{2\sqrt{(a_1+a_2)E}}, \tag{3.8.14}$$

so that the energy dependence is not greatly changed by going to second order.

The Fong theory works quite well in the sense of successfully predicting asymmetric fission as well as kinetic energy distributions and prompt neutron distributions. The principal problem is how to prescribe exactly the form of the scission configuration. Depending on this choice, the effects of shell and pairing effects which strongly influence the calculation may vary considerably. Thus, for example, the deformation of the fragments at scission strongly influences the level densities because shell effects are strongly deformation dependent. We illustrate this with a calculation (figure 3.16) for thermal neutron induced fission of ^{235}U by Fong himself [68] using a potential energy surface including shell effects calculated by Mustafa, Mosel and Schmitt [69]. The calculation correctly predicts the most probable mass ratio of 132 to 104 whereas if shell effects are not included the predicted ratio is 140 to 96.

To terminate this section, we briefly discuss a more recent calculation proposed by Wilkins, Steinberg and Chasman [70]. Given an excited undeformed parent nucleus with some excitation energy, E_C^*, we denote the density of states

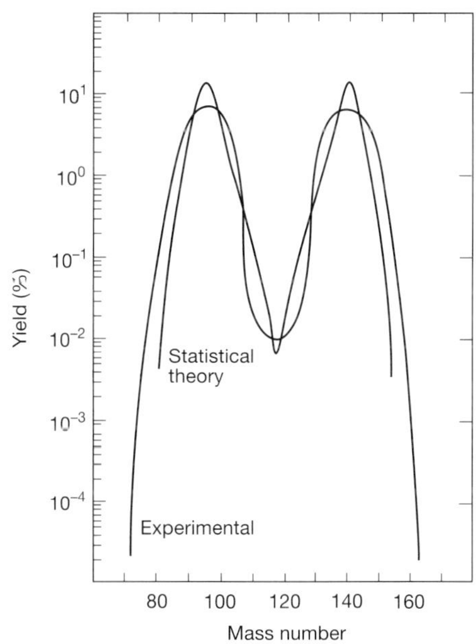

Figure 3.16. Mass distribution resulting from thermal neutron induced fission of ^{235}U. The Fong model reproduces the mass asymmetry and overall shape of the distribution quite well [68] providing that the density of states at fission includes effects of fragment deformations on the shell structure. (Reprinted from *Phys. Rev.* C, **10**, Fong P, p 1122. Copyright 1974, by the American Physical Society.)

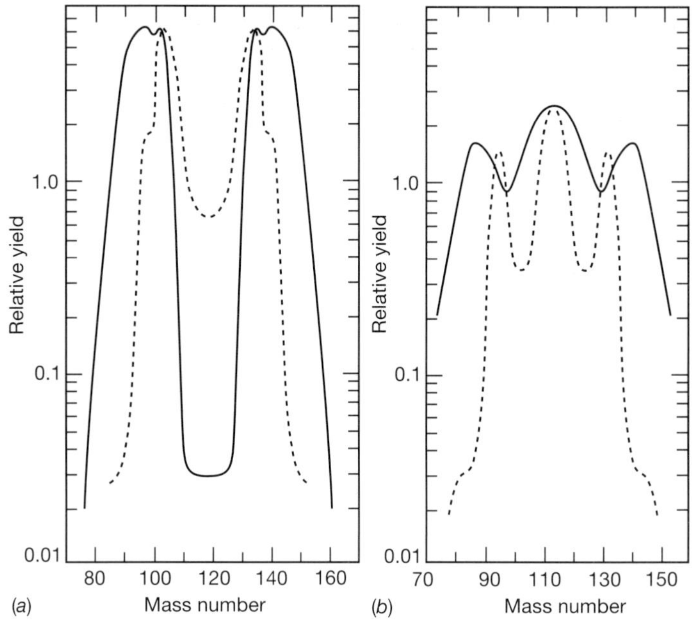

Figure 3.17. Mass distributions for (*a*) thermal neutron induced fission of ^{235}U and (*b*) proton induced fission of ^{226}Ra. The broken curves represent predictions with the scission point statistical model of Wilkins, Steinberg and Chasman [70]. The predictions are qualitatively successful for a wide range of fissioning nuclei. (Reprinted from *Phys. Rev.* C, **14**, Wilkins B D, Steinberg E P and Chasman R R, p 1832. Copyright 1976, by the American Physical Society.)

as $\rho_C(E_C^*)$ and, as in section 1.4, we expand the density $\rho_C(E_C^* - V)$ as

$$\rho_C(E_C^* - V) = \rho_C(E_C^*)\,\mathrm{e}^{-V/T_{\mathrm{coll}}}. \tag{3.8.15}$$

In (3.8.15), T_{coll} is referred to as the 'collective' temperature because V is taken as the difference between the ground state energy of the undeformed parent nucleus and the lowest energy of any given (collectively) deformed configuration. Interest focusses on scission configurations. The (scission) potential energy V is quite sophisticated. It is written in terms of the neutron and proton numbers for the two fragments (N_1, Z_1 and N_2, Z_2) as

$$\begin{aligned} V = {} & V_{\mathrm{LD1}}(N_1, Z_1, \beta_1) + V_{\mathrm{LD2}}(N_2, Z_2, \beta_2) \\ & + S_1(N_1, \beta_1, \tau) + S_1(Z_1, \beta_1, \tau) + S_2(N_2, \beta_2, \tau) + S_2(Z_2, \beta_2, \tau) \\ & + P_1(N_1, \beta_1, \tau) + P_1(Z_1, \beta_1, \tau) + P_2(N_2, \beta_2, \tau) + P_2(Z_2, \beta_2, \tau) \\ & + V_C(N_1, Z_1, \beta_1, N_2, Z_2, \beta_2, d) + V_N(N_1, Z_1, \beta_1, N_2, Z_2, \beta_2, \mathrm{d}), \end{aligned} \tag{3.8.16}$$

where the suffix LD stands for liquid drop, and the parameters β_1 and β_2 refer to the fragment deformations. Subsequent terms after the liquid drop energies refer to shell (S) and pairing (P) corrections. Finally the energy V includes the Coulomb (V_C) and nuclear (V_N) interaction between the two coaxial spheroids whose tips are separated by d. The shell and pairing corrections depend on an intrinsic temperature τ which is said to characterize the excitation of single particle states. The need for a distinct 'single particle' temperature is justified by supposing that the population of nucleons in single particle states does not have time to adjust to the evolution of the potential along the fission coordinate (i.e., intermediate coupling is implicitly assumed). The authors assign what appear to be rather arbitrary collective and intrinsic temperatures (1.0 MeV and 0.75 MeV, respectively). Using fixed values of these parameters and of the separation parameter, d, the model predicts, qualitatively, many of the observed characteristics of fission over a wide range of mass. Calculated mass distributions for thermal neutron induced fission of ^{235}U and proton induced fission of ^{226}Ra are compared with experiment in figure 3.17. An important contribution to the global success of the model appears to be the inclusion of the dependence of shell and pairing corrections on deformation and intrinsic excitation. From their studies, the authors conclude: 'The fact that so many of the observed phenomena of fission are well reproduced in this static scission point model implies that the dynamics of the process in the path from saddle point to scission point either does not play a major role in the determination of the distributions calculated here *or would lead to the same result*'.

We shall be discussing fission again in the next chapter which is mainly concerned with application of the Hauser–Feshbach theory. However, there, the emphasis will be on decay of highly excited nuclei.

References

[1] Vogt E 1968 *Advances in Nuclear Physics* **1** 261 (see also articles in Garg J B (ed) 1972 *Statistical Properties of Nuclei* (New York: Plenum))
[2] Von Witsch W, Von Brentano P, Mayer-Kuckuk T and Richter A 1966 *Nucl. Phys.* **80** 394
[3] Weisskopf V 1937 *Phys. Rev.* **52** 295; Weisskopf V and Ewing D H 1940 *Phys. Rev.* **57** 472
[4] Feshbach H 1960 *Nuclear Spectroscopy, Part B* (NewYork: Academic) p 625.
[5] Bohr N 1936 *Nature* **137** 344
[6] Butler S T 1952 *Phys. Rev.* **88** 685
[7] Bethe H A and Butler S T 1952 *Phys. Rev.* **85** 1045
[8] Ghoshal S N 1950 *Phys. Rev.* **80** 939
[9] John W Jr 1956 *Phys. Rev.* **103** 704
[10] Durham E F and Halbert M L 1965 *Phys. Rev.* B **137** 850
[11] Frenkel I 1936 *Sov. Phys.* **9** 533
[12] Blatt J and Weisskopf V W 1952 *Theoretical Nuclear Physics* (New York: Wiley)
[13] Gadioli E and Hodgson P E 1992 *Pre-Equilibrium Nuclear Reactions* (Oxford: Clarendon); Ait-Tahar S 1986 *DPhil Thesis* University of Oxford

[14] Thomas T D 1964 *Nucl. Phys.* **53** 558
[15] Fernbach S, Serber R and Taylor T B 1949 *Phys. Rev.* **75** 1352
[16] Cook L J, McMillan E M, Peterson J M and Sewell D C 1949 *Phys. Rev.* **75** 7
[17] Feshbach H, Porter C E and Weisskopf V F 1954 *Phys. Rev.* **96** 448
[18] Hodgson P E 1963 *The Optical Model of Elastic Scattering* (Oxford: Oxford University Press)
[19] Fröbrich P and Lipperheide R 1996 *Theory of Nuclear Reactions* (Oxford: Oxford University Press)
[20] Perey F G 1963 *Phys. Rev.* **131** 745
[21] Cage M E *et al* 1972 *Nucl. Phys.* A **183** 449
[22] Hauser W and Feshbach H 1952 *Phys. Rev.* **87** 366
[23] Ericson T and Strutinsky V 1958 *Nucl. Phys.* **8** 284; Ericson T and Strutinsky V 1958–59 *Nucl. Phys.* **9** 689
[24] Thomas T D 1968 *Ann. Rev. Nucl. Sci.* **18** 343
[25] Weisskopf V F 1951 *Phys. Rev.* **83** 1073
[26] Pülhofer F 1977 *Nucl. Phys.* A **280** 267
[27] Endt P M and van der Leun C 1974 *Nucl. Phys.* A **235** 27
[28] Moldhauer P A 1975 *Phys. Rev.* C **11** 426
[29] Glas D and Mosel U 1975 *Nucl. Phys.* A **237** 429; Mosel U 1980 *Heavy Ion Collisions* vol 2 ed R Bock (Amsterdam: North-Holland)
[30] Hill D L and Wheeler J A 1953 *Phys. Rev.* **89** 1102
[31] Galin J, Guerreau D, Lefort M and Tarrago X 1974 *Phys. Rev.* C **9** 1018
[32] Cohen S, Plasil F and Swiatecki W J 1974 *Ann. Phys., NY* **82** 557
[33] Birkelund J R, Tubbs L E, Huizenga J R, De J N and Sperber D 1979 *Phys. Rep.* **56** 107
[34] Bondorf J P, Sobel M and Sperber D 1974 *Phys. Rep.* C **15** 83
[35] Seglie E, Sperber D and Sherman A 1974 *Phys. Rev.* C **11** 1227
[36] Fröbrich P and Gontchar I I 1998 *Phys. Rep.* **292** 131
[37] Stefanini A M, Nebbia G, Lunardi S, Montagnoli G and Vitturi A (ed) 1994 *Heavy Ion Fusion: Exploring the Variety of Nuclear Properties*. (Singapore: World Scientific)
[38] Cole A J 1987 *Phys. Rev.* C **35** 117.
[39] Aichelin J 1991 *Phys. Rev.* **202** 233; Peilert G, Konopka J, Stöcker H, Greiner W, Blann M and Mustafa M G 1992 *Phys. Rev.* C **46** 1457
[40] Bonasera A, Gullminelli F and Molitoris J 1994 *Phys. Rep.* **243** 1
[41] Schröder U and Huizenga J 1983 *Treatise on Heavy Ion Science* vol 2 ed D A Bromley (New York: Plenum)
[42] Lefort M 1980 *Progress in Nuclear and Particle Physics* **4** 197 (see also 1971 Nuclear Structure and Heavy Ion Collisions: *Proc. Int. School of Physics 'Enrico Fermi'* 77 (Amsterdam: North-Holland))
[43] Blocki H, Boneh Y, Nix J, Randrup J, Robel M, Sierk A J and Swiatecki W J 1978 *Ann. Phys.* **149** 374
[44] Griffin J J 1966 *Phys. Rev. Lett.* **17** 478; Griffin J J 1967 *Phys. Lett.* B **24** 5
[45] Cline C and Blann M 1971 *Nucl. Phys.* A **172** 225; Blann M 1975 *Ann. Rev. of Nucl. Sci.* **25** 123; Blann M 1972 *(Lecture Notes in Physics) Proc. of the Europhys. Study Conf. on Intermediate Processes in Nuclear Reactions* ed N Cindro, P Kulisic and T Mayer-Kuckuk (Berlin: Springer)
[46] Harp G D, Miller J M and Berne B J 1968 *Phys. Rev.* **165** 1166

[47] Fabrici E, Gadioli E, Gadioli Erba E, Galmarini M, Fabbri F and Reffo G 1989 *Phys. Rev.* C **40** 2548

[48] Böhning M 1970 *Nucl. Phys.* A **152** 529

[49] Holub E, Pocanic D, Caplar R and Cindro N 1980 *Z. Phys.* A **296** 341

[50] Messiah A 1961 *Quantum Mechanics* (Amsterdam: North-Holland)

[51] Williams F C Jr 1970 *Phys. Lett.* B **31** 184

[52] Awes T C, Poggi G, Gelbke C K, Back B B, Glagola B G, Breur H and Viola V E Jr 1981 *Phys. Rev.* C **24** 89

[53] Fahli A *et al* 1987 *Z. Phys.* A **326** 169

[54] Harp G D and Miller J M 1971 *Phys. Rev.* C **3** 1847

[55] Aichelin J and Bertsch G 1985 *Phys. Rev.* C **31** 1730

[56] Randrup J and Vandenbosch R 1987 *Nucl. Phys.* A **474** 219

[57] Mahaux C and Wiedenmüller H A 1979 *Ann Rev. Nucl. Sci.* **29** 1

[58] Feshbach *et al* 1980 *Ann. Phys.* **125** 429

[59] Nishioka *et al* 1986 *Ann. Phys.* **172** 67; Nishioka *et al* 1988 *Ann. Phys.* **183** 166

[60] Herman M, Reffo G and Wiedenmüller H A 1992 *Nucl. Phys.* A **536** 124

[61] Hahn O and Strassman F 1939 *Naturwissenschaften* **27** 11

[62] Bohr N and Wheeler J A 1939 *Phys. Rev.* **56** 426

[63] Hilscher D, Krappe H J and von Oertzen W (ed) 1989 Fifty Years Research in Nuclear Fission *Nucl. Phys.* A **502**

[64] Wagemans C (ed) 1991 *The Nuclear Fission Process* (Boca Raton, FL: CRC Press)

[65] Fong P 1969 *Statistical Theory of Nuclear Fission* (New York: Gordon and Breach)

[66] Hilscher D and Rossner H 1998 *Ann. Phys., Paris* **17** 471

[67] Kramers H A 1940 *Physica* **7** 284

[68] Fong P 1974 *Phys. Rev.* C **10** 1122

[69] Mustafa M G, Mosel U and Schmitt H W 1973 *Phys. Rev.* C **7** 151

[70] Wilkins B D, Steinberg E P and Chasman R R1976 *Phys. Rev.* C **14** 1832

Chapter 4

Single and multistep evaporation calculations

Lay on Macduff

Macbeth, William Shakespeare

4.1 Introduction

This chapter is concerned with the practical application of the Hauser–Feshbach formalism to the problem of single and multistep evaporation. A rather detailed exposure is presented, partly because light particle evaporation is central to the main theme of the book, and partly because the description paves the way towards an understanding of the more complicated investigations which characterize recent research. The topics to be discussed may be thought of, in some ways, as part of the history of statistical decay processes since they are mainly taken from work carried out in the late 1970s and early 1980s. In this sense, they form an extension of chapter 3. However, the content is largely of a technical nature and can, therefore, also be considered as an introduction to modern simulations. The practical application of the Hauser–Feshbach theory to compound nucleus decay thus provides a bridge between older and more modern ways of comparing theory with experiment. Furthermore, the 'computational' approach to the problem contrasts with the essentially analytical approach provided in the previous chapter and hopefully will lead to a more intuitive appreciation of the theory.

In adopting a sequential evaporation mechanism we suppose that the probability of simultaneous emission of two or more radiations (particles or gamma rays) is negligible. Emitted particles themselves are considered to be in their ground states. Furthermore, it is assumed that the emission of a particle or gamma ray from the compound nucleus produces a 'daughter' nucleus such that the set of all such nuclei represents an equilibrium ensemble. One may, of course, appeal to the ergodic hypothesis to justify the existence of such an ensemble, in which case we may think of each individual daughter nucleus as an equilibrated system prior to decay. Whether we invoke this justification or not, the set of daughter nuclei are considered as compound systems in their own right so that the notion of a sequential cascade follows by induction.

Unfortunately, there is no direct experimental information available concerning the time sequence of particle emission. This can be understood by comparing the sequential emission mechanism with the prompt multifragmentation mechanism in which particles are produced by explosion of a highly excited nucleus. The time scales for these mechanisms would be expected to be quite different but, in both cases, many orders of magnitude smaller than the typical resolution of an experiment (which is of the order of 10^{-10} s). A similar remark applies to spatial configurations. If the reaction products are produced simultaneously we would expect the 'source' to be spatially localized, i.e., comparable with the size of the compound nucleus. For sequential emission it is difficult even to speak of a single source. The emitted particles may be produced over a spatial region which is orders of magnitude larger than the compound nucleus but orders of magnitude smaller than the typical experimental spatial resolution (which is, at best, $\sim 10^{-4}$ m). Experimental information concerning the mechanism is therefore, in this sense, indirect.

An experiment may be set up either to measure one reaction product (inclusive measurement) or more than one reaction product (exclusive measurement) in any particular event. Confidence in the sequential decay mechanism for compound nucleus excitation energies below $\sim$ 2 MeV/nucleon is mainly based on inclusive measurements of the residual nucleus which remains after the compound nucleus decays. The highly successful prediction of evaporation residue yields, angular distributions and energy spectra using Hauser–Feshbach multistep evaporation codes has been taken as strong evidence in support of the picture of sequential evaporation of light particles. Furthermore, accurate predictions of light particle angular and energy distributions have been interpreted as further evidence in favour of the evaporation mechanism.

Until about 1980 exclusive measurements in light and intermediate mass systems were mainly limited to two particle coincidences, involving either two light particles or one particle measured in coincidence with the evaporation residue. Again, these measurements have confirmed the predictions of the multistep Hauser–Feshbach model.

For excitation energies above 2 MeV/nucleon the assumption of simple sequential evaporation is progressively invalidated due to competing emission of larger and larger fragments which may themselves be excited. Although this decay mechanism is a natural extension of evaporation its description goes beyond the Hauser–Feshbach treatment. Models which include this competition will be discussed in the following chapter.

We begin our discussion of evaporation by describing basic Monte Carlo methods. We then go on (sections 4.3 and 4.4) to show, in some detail, how these methods may be applied to the prediction of observables for single and multistep decay. During this part of the presentation it will become apparent that some of the well-established results from theory are subjected to rather severe approximations imposed by the need for a rapidly executing computer program or simply by lack of knowledge of the corresponding input. Some of these approximations are difficult to justify *a priori*. For this reason, the degree of agreement with experiment is considered to provide a measure, not only of the applicability of the statistical model, but also of the success of approximations. Examples of experimental studies are given in section 4.5.

Some dynamical aspects of multistep evaporation can be analysed by analytical methods. This topic will provide an introduction to the notion of event 'topology' and will be discussed at the end of the chapter (section 4.6).

Useful complements to the subject matter considered here may be found in books and review articles. The Monte Carlo method is treated extensively in the book by Hammersley and Handscomb [1]. A broad overview of the application of statistical models to heavy ion reaction studies is provided in the article by Stokstad [2], whereas Oganessian and Lazarev [3] have concentrated on heavy ion induced fission. Finally Mahaux and Weidenmüller [4] have discussed theoretical progress in understanding the success of the Bohr independence hypothesis over a wide range of excitation energy.

4.2 Monte Carlo methods

In many cases it is useful to construct a computer simulation of a physical process. Usually the computer program is intended to simulate the execution of a real experiment so that the corresponding output provides 'data' in the same way as an experiment. In some cases the simulated experiment has no real counterpart in the sense that the corresponding measurement cannot be carried out on a real physical system. This situation is often found, for example, in cosmology or astrophysics where the possibilities to carry out real experiments (e.g., recreate the Big Bang or provoke a supernova) are rather limited. In nuclear physics, the most obvious example concerns computer 'experiments' which simulate projectile–target collisions for fixed impact parameter. In real experiments the impact parameter can sometimes be estimated but almost always with a large associated error bar. Under these circumstances it is probably best to think of a simulation as the 'pseudo experimental' counterpart of a model process. The results of the simulation are determined by a set of input parameters (also called 'source' parameters) which are transformed by the mathematical relations which express the model process into physically intelligible output. Monte Carlo simulation techniques are particularly useful whenever the model process cannot be evaluated analytically or can only be evaluated with great difficulty.

One useful example that springs to mind involves the calculation of multidimensional integrals. In order to familiarize ourselves with the technique which can be applied to this problem, let us consider a simple example involving a one-dimensional function $f(x)$. The source parameters determine $f(x)$ between lower and upper limits (x_0, x_1) and we define the model process as the integral of the function between these limits. Of course there are many numerical techniques available for obtaining an approximation to the required integral. However, we concentrate on the Monte Carlo method. Suppose, for simplicity, that the function is everywhere positive and that the maximum of $f(x)$, between lower and upper limits, is f_m. We begin by locating the maximum, f_m (numerically if necessary) and define a new function $g(x) = f(x)/f_m$ (figure 4.1) which peaks at unity. The evaluation of the area under the curve $g(x)$ is equivalent to knowledge of the integral of $f(x)$.

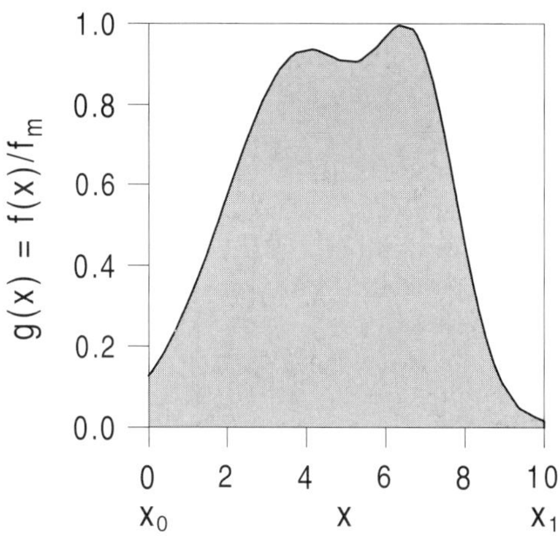

Figure 4.1. Monte Carlo integration: the probability that a number, chosen randomly somewhere inside the box, falls in the shaded area under the curve $g(x) = f(x)/f_m$ is equal to the integral of $g(x)$ divided by the total area of the 'box' $(1 \times (x_1 - x_0))$. The function $f(x)$ was constructed as the sum of two Gaussians whose mean values, standard deviations and maximimum values were 4.0, 2.0, 1.0 and 7.0, 1.0, 0.7, respectively.

Let us now suppose we possess a random number generator capable of producing a very long sequence of uniformly distributed random numbers between 0 and 1. The term 'uniformly' means that the corresponding probability density function $p(x)$ is constant (we remind the reader that the probability density function (PDF) is defined by $\mathrm{Prob}(x' \le x < x' + \Delta x) = \int_{x'}^{x'+\Delta x} p(x)\,\mathrm{d}(x)$). Because the integral of the PDF must be unity, i.e.,

$$\int_0^1 p(x)\,\mathrm{d}x = 1, \tag{4.2.1}$$

we easily see that $p(x) = 1$. We now set up a computer program to approximate the integral of $g(x)$ (equivalently $f(x)$) by carrying out the following operations. (We have tried to describe algorithms without reference to any particular programming language. Nevertheless some FORTRAN constructions have probably leaked into the descriptions.)

Algorithm 4.1.

Step 1 Using the uniform PDF, $p(x)$, generate a random number r between zero and one (uniform random deviate).
Step 2 Using r, generate a new random number. $x_r = x_0 + r(x_1 - x_0)$ which is necessarily uniformly distributed in the interval (x_0, x_1).
Step 3 Generate a new uniform random deviate y_r.
Step 4 If $y_r \le g(x_r)$ increment a counter, i.e., HITS = HITS + 1.
Step 5 Increment a counter which counts the number of times steps 1–4 are executed, TRIALS = TRIALS + 1.
Step 6 Begin again at step 1 unless TRIALS is equal to some preset value, NTRIALS, in which case exit the program outputting HITS and TRIALS (which is equal to NTRIALS).

Note the use of the expression 'uniform random deviate' [5] to designate a random number whose PDF is given by (4.2.1). The pair of random numbers x_r and y_r defines a point which is randomly located in the rectangle whose area is $A = 1 \times (x_1 - x_0)$. If the area under the curve $g(x)$ is $G(x_0, x_1)$, the probability that the random point is located inside this area (i.e. *under* the curve $g(x)$) is simply the ratio of the areas $G(x_0, x_1)$ and A, which is represented by the output of the computer simulation as the ratio HITS/TRIALS. We thus obtain

$$G_{\mathrm{MC}}(x_0, x_1) = A\frac{\mathrm{HITS}}{\mathrm{TRIALS}} = (x_1 - x_0)\frac{\mathrm{HITS}}{\mathrm{TRIALS}}, \tag{4.2.2}$$

in which case the estimation of the required integral of $f(x)$ is

$$F_{\mathrm{MC}}(x_0, x_1) = f_{\mathrm{m}}G_{\mathrm{MC}}(x_0, x_1) = f_{\mathrm{m}}(x_1 - x_0)\frac{\mathrm{HITS}}{\mathrm{TRIAL}}. \tag{4.2.3}$$

(The subscript 'MC' in equations (4.2.2) and (4.2.3) stands for 'Monte Carlo'.) The accuracy of the result depends not only on the statistical accuracy of (4.2.3) but also on the precision with which the maximum can be estimated. If the computer experiment is repeated a large number of times the distribution of values of the output HITS should be represented by a binomial distribution. The expected standard deviation of the mean value of HITS, which is also referred to as the standard error is then

$$\sigma = \sqrt{\frac{\mathrm{HITS}(\mathrm{TRIALS} - \mathrm{HITS})}{\mathrm{TRIALS}}} \tag{4.2.4}$$

(see Hammersley and Handscomb [1]). We can immediately extend the technique to calculate the 'volume' bounded by a closed contour in any number of dimensions. Thus, in a Cartesian system of coordinates, the volume enclosed by a three-dimensional contour $\boldsymbol{C}$ (figure 4.2(*a*)), which has tangent planes situated at x_0 and x_1 (y–z planes), at y_0 and y_1 (z–x planes), and z_0 and z_1 (x–y planes), is obtained if we generalize HITS to denote the number of times a randomly selected point in the range, $x_0 \le x \le x_1$, $y_0 \le y \le y_1$, $z_0 \le z \le z_1$, is found inside the volume enclosed by $\boldsymbol{C}$. The volume is then

$$V_{\mathrm{MC}}(\boldsymbol{C}) = (x_1 - x_0)(y_1 - y_0)(z_1 - z_0)\frac{\mathrm{HITS}}{\mathrm{TRIALS}}. \qquad (4.2.5)$$

We may remark that our procedure for calculating the integral of a function of one variable, $y = f(x)$, is equivalent to defining the rectangular 'box' bounded by the lines at (x_0, x_1) and (y_0, y_1) taking $y_0 = 0$ and $y_1 = f_{\mathrm{m}}$. Futhermore, a volume with a flat base (figure 4.2(*b*)) is generated by a function of two variables, $z = f(x, y)$. It follows that the extension of (4.2.3) to any number of dimensions is straightforward. The integral of a function of n variables requires the construction of an $(n + 1)$-dimensional box.

More important, for the present purposes, is a development arising from the Monte Carlo integration technique which may be used to generate random numbers whose PDF, as a function of some coordinate(s), is determined by a known or at least calculable function. The so-called rejection method, which we now present, is quite easy to understand. Let us again take the function $g(x)$ and suppose we wish to generate a set of random numbers $\{X\}$ in the interval (x_0, x_M) such that for a small interval (Δx) ('small' here means that terms of order $(\Delta x)^2$ and higher can be dropped).

$$\mathrm{Prob}(x \le X < x + \Delta x) = \Delta x g(x). \qquad (4.2.6)$$

In fact, let us start with the specific problem of generating random numbers such that

$$\mathrm{Prob}(x_0 \le X < x_0 + \Delta x) = \Delta x g(x_0). \qquad (4.2.7)$$

Define x_1 so that $\Delta x = (x_1 - x_0)$. Then, from (4.2.2), if we make the notation more precise by writing HITS as $\mathrm{HITS}(x_0, x_1)$ and TRIALS as $\mathrm{TRIALS}(x_0, x_1)$ we find

$$G_{\mathrm{MC}}(x_0, x_1) = (x_1 - x_0)\frac{\mathrm{HITS}(x_0, x_1)}{\mathrm{TRIALS}(x_0, x_1)} \approx \Delta x g(x_0). \qquad (4.2.8)$$

Since $(x_1 - x_0) = \Delta x$, we find immediately that an approximation to $g(x_0)$ is provided by

$$g_{\mathrm{MC}}(x_0) = \frac{\mathrm{HITS}(x_0, x_1)}{\mathrm{TRIALS}(x_0, x_1)}. \qquad (4.2.9)$$

We now go back to the problem stated by (4.2.6) which was to generate the distribution $g_{\mathrm{MC}}(x)$ over a range of values of x from x_0 to x_M which we can represent by the set of M intervals $(x_0, x_1), (x_1, x_2) \ldots (x_{n-1}, x_n) \ldots (x_{M-1}, x_M)$. Since all intervals are equivalent, the problem is solved by the natural generalization of (4.2.9), i.e.,

$$g_{\mathrm{MC}}(x_{n-1}) = \frac{\mathrm{HITS}(x_{n-1}, x_n)}{\mathrm{TRIALS}(x_{n-1}, x_n)}, \qquad (4.2.10)$$

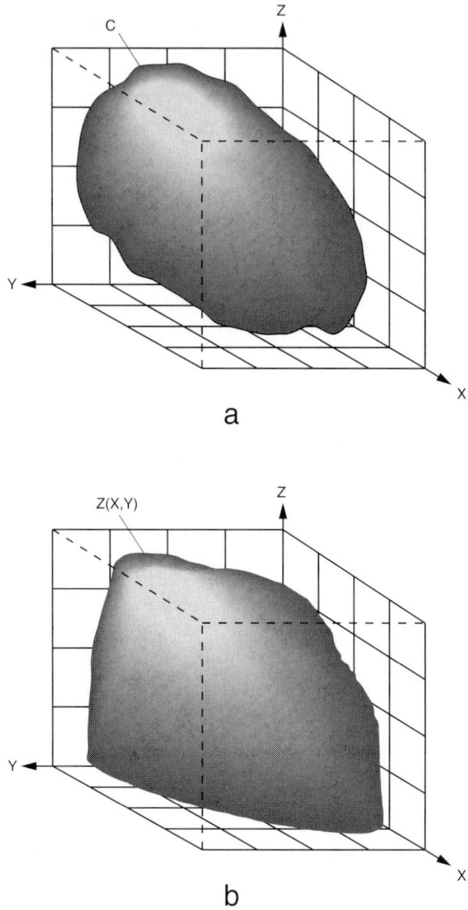

Figure 4.2. Two three-dimensional examples of integration. In (*a*), the volume of the object shown in the figure can be estimated by a Monte Carlo simulation. The MC computer code provides an estimate of the ratio of the number of times a number, which is chosen randomly inside the box formed by the three tangent planes, falls inside the contour ***C*** (see equation (4.2.5)). In (*b*), the volume of the flat-based object is the two-dimensional integral of the function $z(x, y)$.

where $\mathrm{TRIALS}(x_{n-1}, x_n)$ is the number of trials in which a value of x_r (now to be understood as a random number in the range (x_0, x_M)) falls in the interval (x_{n-1}, x_n) and $\mathrm{HITS}(x_{n-1}, x_n)$ is the number of times that, *given this selection*, the random number $y_r \leq g(x_r)$. We thus arrive at the following problem statement and the solution algorithm (remember that the term 'uniform random deviate' refers to a uniformly distributed random number between 0 and 1).

Take $f(x)$ to be a function which is positive in the range (x_0, x_M) with maximum f_m and suppose we wish to simulate $f(x)$ in the range (x_0, x_M).

Algorithm 4.2.

Step 1 Generate a uniform random deviate r_1.
Step 2 Using r_1, generate a new random number, $x_r = x_0 + r_1(x_M - x_0)$ which is neccesarily uniformly distributed in the interval (x_0, x_M). Identify the interval, n, $(1 \leq n \leq M)$ corresponding to the selection x_r and increment a counter, i.e., $\text{TRIALS}(n) = \text{TRIALS}(n) + 1$.
Step 3 Generate a second uniform random deviate y_r.
Step 4 If $y_r \leq g(x_r)(= f(x_r)/f_m)$ then increment a second counter, i.e., $\text{HITS}(n) = \text{HITS}(n) + 1$.
Step 5 Sum number of trials TOT_TRIALS = TOT_TRIALS + 1.
Step 6 If TOT_TRIALS is less than some preset value, NTRIALS, begin again at Step 1.
Otherwise
exit, outputting $\text{HITS}(n)$, $\text{TRIALS}(n)$ for $n = 1, M$.

To within statistical accuracy, and provided the intervals Δx are small enough so that $g(x_{n-1}) \approx g(x_{n-1} + \Delta x)$ the function should be represented by $f_{\text{MC}}(n) = f_m\text{HITS}(n)/\text{TRIALS}(n)$ (one can introduce a sophistication by saying that $f_{\text{MC}}(n)$ represents the function at the midpoint of each interval).

In some cases we may be interested only in relative values of the function. If so, we may use the fact that if all the intervals (x_{n-1}, x_n) are set to have an identical length Δx then the probability that a random number with a uniform PDF in the range (x_0, x_M) falls in any given interval is a constant. Thus relative values are provided simply by $\text{HITS}(n)$. In figure 4.3 we show a typical example involving the simulation of a function, $g(x)$, which is the sum of two Gaussian distributions. The array, HITS has been renormalized either by dividing by the maximum value (figure 4.3(a)) or by requiring the (numerically evaluated) integrals of the original function and the simulated function to be equal (figure 4.3(b)). The second method, which must be used if $f(x)$ is a PDF, is to be preferred especially when the number of trials is limited.

If only relative values are required, it may be thought that we do not need to know the precise value of the maximum. This is not really true. Thus, suppose we overestimated f_m by say 10%. Then the fraction of the rectangular area which lies under $g(x)$ is reduced so that, globally, all values of $\text{HITS}(n)$ are reduced *by the same factor* (curve (a) in figure 4.4). We can then see that although the *relative* values of the simulated function are, on average, unchanged, the efficiency (i.e., the statistical accuracy for a given number of trials) of the generating process is reduced. On the other hand, if the maximum is underestimated (figure 4.4(b)) the function simulated is *not* $g(x)$ but the part of $g(x)$ which lies below unity (the function protrudes outside the 'box'). One must thus make sure that if f_m is estimated numerically (which can be a time-consuming operation) then the

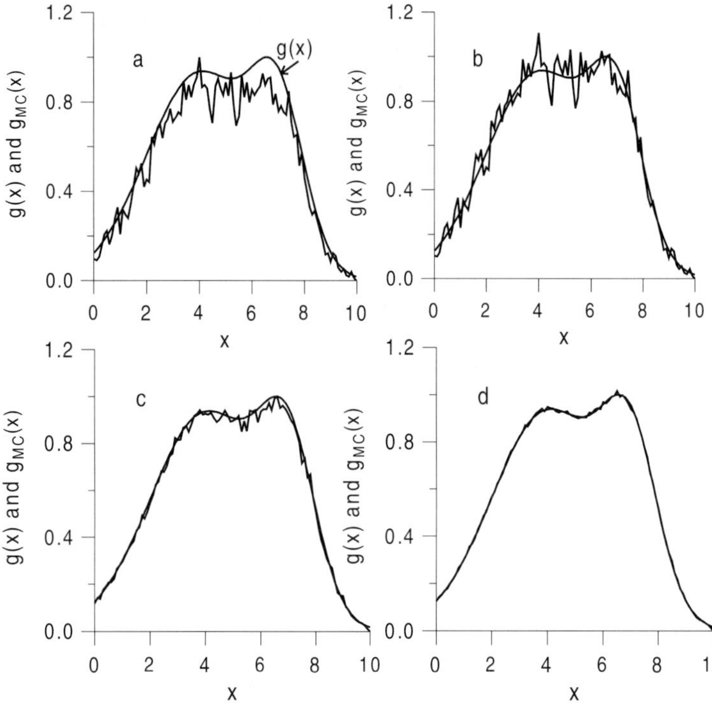

Figure 4.3. The rejection method applied to the the simulation ($g_{\mathrm{MC}}(x)$) of a sum of two Gaussian functions. The original distribution, $g(x)$, is normalized so that it maximizes at 1. (*a*) One thousand events renormalized so that the maximum of the simulated distribution = 1. (*b*), as (*a*), except that the simulation is renormalized to have the same integral as the original function. (*c*) and (*d*), as (*a*) and (*b*), but for 100 000 events.

estimation is always greater than or equal to the exact value. Translated to three dimensions, this means that we can relax our requirement that the limiting planes are actually tangent planes to $\boldsymbol{C}$ and insist only on the weaker condition that the contour $\boldsymbol{C}$ is entirely enclosed in our 'box'.

Functions which are negative over a part of the interval can also be simulated. It is sufficient to (temporarily) redefine the zero. Alternatively, the interval over which the function is defined can be broken into parts corresponding to positive and negative values of the function. Breaking into parts is also recommended for functions which vary strongly over the domain of integration. A further extension follows naturally from the observation that the statistical accuracy of the method is optimized by choosing a bounding box which is not necessarily constructed using a Cartesian system of coordinates, but is more closely adapted to the 'shape' of the function to be simulated. Thus, for a one-dimensional function, $y = f(x)$, it may be possible to generate random numbers corresponding to a known function

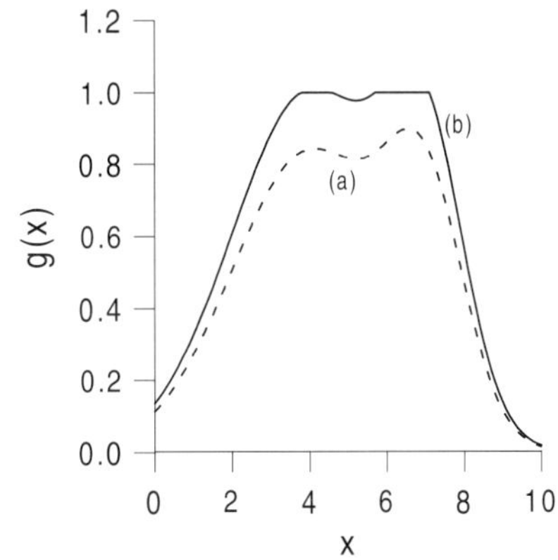

Figure 4.4. Results of the rejection method with a poor estimation of the maximum value of the distribution function to be simulated. The broken curve, (*a*), represents the situation where the maximum value is overestimated (by 10%). Relative values of the distribution function are correctly simulated. The full curve, (*b*), represents the simulation when the maximum value is underestimated (by 8%). In this case, the simulated function is a truncated version of the original distribution.

$y_B = f_B(x)$ which provides an upper bound for y over the required range of x-values. If such a function can be found, the generalization of the results of algorithm 4.2 is provided by $f_B(x_n)\mathrm{HITS}(n)/\mathrm{TRIALS}(n)$.

In all cases considered in this chapter, the function to be simulated represents a probability density function in one or more dimensions and is thus, by definition, positive or zero. Such functions are often simply referred to as probability distributions rather than probability density functions. This can sometimes be misleading since the same term is used to designate the *cumulative* probability distribution function which is the integral of the PDF.

We now discuss an alternative method which is to be preferred whenever the PDF to be simulated can be integrated easily (numerically or analytically). We begin with the definition of the probability density function $p(x)$

$$\mathrm{Prob}(x' \leq x < x' + \Delta x) = \int_{x'}^{x'+\Delta x} p(x)\,\mathrm{d}x, \qquad (4.2.11)$$

and consider a monotonic function of x, $y(x)$. The probability to obtain a value of y in a given small interval of y values is obviously the integral of (4.2.11) over all values of x for which $y(x)$ lies in the specified y interval. If, in some region, x increases monotonically with y, this can be expressed by writing

$$\mathrm{Prob}(y' \leq y < y' + \Delta y) = \int_{y'}^{y'+\Delta y} q(y)\,\mathrm{d}y = \int_{x'}^{x'+\Delta x} p(x)\,\mathrm{d}x, \qquad (4.2.12)$$

and if x decreases monotonically with y by

$$\text{Prob}(y' \leq y < y' + \Delta y) = \int_{y'}^{y'+\Delta y} q(y)\,\mathrm{d}y = \int_{x'}^{x'-\Delta x} p(x)\,\mathrm{d}x, \qquad (4.2.13)$$

in which $q(y)$ is the probability density function (positive or zero) for y, and $x' = x(y')$. We can express both cases, in the limit as $\Delta y \to \mathrm{d}y$ by writing

$$q(y) = p(x)\left|\frac{\mathrm{d}x}{\mathrm{d}y}\right|. \qquad (4.2.14)$$

We now make a specific choice by defining $p(x)$ as the uniform distribution (4.2.1) which, between the limits 0,1, is unity. Then, (4.2.14) can be expressed as

$$q(y)\,\mathrm{d}y = \left|\frac{\mathrm{d}x}{\mathrm{d}y}\right|\mathrm{d}y. \qquad (4.2.15)$$

By making different choices for $y(x)$ (and thereby for $\mathrm{d}x/\mathrm{d}y$) we can use equation (4.2.15) to create new probability distributions using the uniform distribution as a source of random numbers. Thus, for example, if x is a uniform random deviate and $y(x) = -\ln[x]$,

$$\left|\frac{\mathrm{d}x}{\mathrm{d}y}\right| = x = q(y) = \mathrm{e}^{-y}, \qquad (4.2.16)$$

so that y is exponentially distributed. We could also have taken $y(x) = -\ln[1-x]$ in which case $|\mathrm{d}x/\mathrm{d}y| = \mathrm{d}x/\mathrm{d}y = 1 - x$ which again is e^{-y}. The equivalence occurs due to the most important fact that, if x is a uniform random deviate, so is $1 - x$.

Another application which arises from the transformation method may be referred to as the summation method. If $q(y)$ is a PDF with cumulative distribution $Q(y)$ defined over the range y_1 to y_2 ($Q(y_1) = 0$, ($Q(y_2) = 1$) and y is a function of x such that $y(0) = y_1$ then if the PDF, $p(x) = 1$,

$$x' = \int_0^{x'} p(x)\,\mathrm{d}x = \int_{y_1}^{y(x')} q(y)\,\mathrm{d}y = Q(y'), \qquad (4.2.17)$$

where $y' = y(x')$. Random numbers, distributed with a known PDF, $q(y)$, can be obtained by solving (4.2.17) for y'. The equivalence of the PDFs for x and $1-x$ in fact allows us, in applications of (4.2.17), to proceed as if y is aways an increasing function of x (in other words we take integrals of positive PDFs performed in a negative sense as positive). The following algorithm (which returns one number at a time) resumes the procedure.

Algorithm 4.3.

Step 1 Generate a uniform random deviate x'.
Step 2 Integrate $q(y)$ from $y(0)$ to y' where $Q(y') = x'$.
Step 3 Exit returning the value y'.

In some cases we need to deal with functions defined on a discrete space. In this case, we would probably prefer the term 'probability distribution' to the term 'probability density function' (taking care to avoid confusion with the cumulative probability distribution). Thus consider the application of the summation method to a two-dimensional probability distribution defined by a set of values $f_{ij}(i = 1, i_{\max}, j = 1, j_{\max})$ with the obvious condition

$$\sum_{i=1}^{i_{\max}} \sum_{j=1}^{j_{\max}} f_{ij} = 1. \tag{4.2.18}$$

A set of random numbers distributed approximately as f_{ij} is obtained with the following algorithm.

Algorithm 4.4.

Step 1 Generate a uniform random deviate x' between 0 and 1.

Step 2 Evaluate the sum $S(i', j') = \sum_{i=1}^{i'} \sum_{j=1}^{j'} f_{ij}$ adding successive terms until $S(i', j') > x'$.

Step 3 Exit returning the pair of values i', j'.

The above description of Monte Carlo methods, while incomplete, should provide sufficient information to allow the construction of a few simple simulation programs. One useful application concerns the simulation of a source of particles. Let $S(\Omega)\,d\Omega$ represent the probability of emission of particles into the solid angle $d\Omega$. Then, if S is separable in the two angular coordinates θ and ϕ we can write the probability of observing a particle in the solid angle $d\Omega$ at (θ, ϕ) as

$$\frac{d^2 P(\theta, \phi)}{d\theta\, d\phi}\, d\theta\, d\phi = S(\Omega)\, d\Omega = [S_\theta(\theta) \sin\theta\, d\theta][S_\phi(\phi)\, d\phi]. \tag{4.2.19}$$

If S is isotropic we may take $S_\theta = S_\phi =$ const. Then (dropping the constants)

$$\frac{d^2 P(\theta, \phi)}{d\theta\, d\phi}\, d\theta\, d\phi \propto [\, d\cos\theta][\, d\phi], \tag{4.2.20}$$

so that an isotropic source is simulated by selecting values of $\cos\theta$ using a PDF which is uniform in the range $(-1, +1)$, and values of ϕ from a uniform PDF in the range $(0, 2\pi)$. If $S_\theta(\theta) = 1/\sin\theta$ (a distribution which often occurs in practice) the PDF for θ should be uniform in the range $(0, \pi)$.

As further exercises we propose the simulation of the discrete Poisson probability density function defined, in terms of the parameter $s = \langle n \rangle$ for integer $n \geq 0$ by

$$P_n = \frac{s^n\, e^{-s}}{n!}, \tag{4.2.21}$$

and of the Gaussian PDF defined by the mean, x_0, and the standard deviation, σ as

$$g(x) = \frac{1}{\sqrt{2\pi\sigma^2}}\, e^{\frac{-(x-x_0)^2}{2\sigma^2}}. \tag{4.2.22}$$

Both the rejection technique and the summation method may be applied straightforwardly to these problems.

4.3 Monte Carlo treatment of a single evaporation step

The purpose of this section is to show how the Hauser–Feshbach theory may be adapted for computer simulation. The Monte Carlo method is particularly useful for the simulation of multistep decay. Exactly how the method is implemented in this situation will be explained in the next section. Herein, we concentrate on one single evaporation step. Our presentation represents one of several possible strategies and is intended only to provide the reader with a detailed example of the application of the Monte Carlo technique. It corresponds to the author's program 'LANCELOT' which was written in 1979 [6]. Other codes are listed in the review by Stokstad [2]. Let us begin with one particular contribution (for given $\ell' j'$) to the cross-section sum represented in equation (3.5.7). No confusion will arise if we suppress the primes (understanding the numerator to be just one of the terms in the sum in the denominator) so we write this contribution as

$$\sigma_{\alpha\beta} = \sigma_F(E_{\mathrm{i}}^*, J_{\mathrm{i}}) \frac{T_\ell(\beta)\rho_\beta}{\sum_\gamma T_\ell(\gamma)\rho_\gamma}. \tag{4.3.1}$$

In this equation, $\sigma_F(E_{\mathrm{i}}^*, J_{\mathrm{i}})$ (which could also be referred to as σ_α) is the partial cross-section for formation of the compound nucleus, with total angular momentum J_{i}, via the fusion reaction $a + A \rightarrow C$. The excitation energy, E_{i}^*, of the compound nucleus is the sum of the energy in the centre of mass and the fusion Q-value (see (2.2.18)) and is thus fixed by the parameters of the entrance channel. The total angular momentum, on the other hand, must be selected from some distribution as we shall shortly see.

Starting with the compound nucleus at excitation energy E_{i}^* and angular momentum J_{i} (i for 'initial') our objective will be to construct a discrete probability distribution whose elements are the probabilities for particles or gamma rays to be emitted, and which will therefore allow us to make a Monte Carlo selection of the type of emission (γ, n, p, $\ldots \alpha, \ldots$). Thus, we will progressively integrate over other variables. During the course of this integration we will obtain PDFs which will allow us to select the angular momentum of the emitted particle (or the multipolarity of the emitted gamma ray), the energy associated with the emission and the final angular momentum in the residual nucleus.

We begin by making use of the independence hypothesis, i.e., of the form of equation (4.3.1). If $\sigma_F(E_{\mathrm{i}}^*, J_{\mathrm{i}})$ is known, or can be modelled, we need only to calculate the various decay probabilities corresponding to the exit channels β.

However, in order to begin the simulation of a single evaporation step we need to select a particular value for the compound nucleus angular momentum, J_{i}. To apply the Monte Carlo technique to the selection of J_{i} we create a PDF

$$q(J_{\mathrm{i}}) = \frac{\sigma_F(J_{\mathrm{i}})}{\sum_{J_{\mathrm{i}}} \sigma_F(J_{\mathrm{i}})} \tag{4.3.2}$$

(in which we have temporarily suppressed the excitation energy E_{i}^*). It would be possible, in principle, to model the partial fusion cross-sections, $\sigma_F(J_{\mathrm{i}})$, using the critical distance theory for fusion outlined in section 3.6. However, given the imprecision of the theory, especially at high energies, many authors have preferred to employ a simple parametrization for the fusion cross-section (see figure 4.5) of the form

$$\sigma_F(J_{\mathrm{i}}) = \frac{\pi}{k^2} \sum_{J_{\mathrm{i}}=J_{\mathrm{min}}}^{\infty} \frac{2J_{\mathrm{i}}+1}{1+\mathrm{e}^{(J_{\mathrm{i}}-J_{\mathrm{cr}})/\Delta J}}, \tag{4.3.3}$$

where the sum goes over integral or half integral values of J_{i} (J_{min} is 0 or 1/2) according to whether the sum of the projectile and target spins (the channel spin) is even or odd. In the limit of small ΔJ the fusion cross-section is approximately

$$\sigma_F \approx \frac{\pi}{k^2}(J_{\mathrm{cr}}+1)^2, \tag{4.3.4}$$

so that fusion theories are useful in the sense that the cut off angular momentum J_{cr} can be estimated from the fusion cross-section. The parameter ΔJ may also be estimated using theories for fusion. It is of the the order of 1 or 2 (in units of $\hbar$). With (4.3.3), the PDF for J_{i} is obviously

$$q(j_{\mathrm{i}}) = \frac{\frac{2J_{\mathrm{i}}+1}{1+\mathrm{e}^{(J_{\mathrm{i}}-J_{\mathrm{cr}})/\Delta J}}}{\sum\limits_{J_{\mathrm{i}}=J_{\mathrm{min}}}^{\infty} \frac{2J_{\mathrm{i}}+1}{1+\mathrm{e}^{(J_{\mathrm{i}}-J_{\mathrm{cr}})/\Delta J}}}, \tag{4.3.5}$$

so that a random selection of J_{i} can be made using the summation technique (algorithm 4.4) explained in the preceding section.

Turning now to the exit channel we can see, at least schematically, that if we calculate the relative probabilities for all possible decay channels for any value of J_{i}, we can explicitly construct the denominator of (4.3.1). It follows that we need only to calculate the quantity

$$P_\beta = T_\ell(\beta)\rho_B, \tag{4.3.6}$$

for all decay channels. The question then naturally arises as to what we mean by *all* decay channels. In most calculations, particle emission is truncated above ^{6}Li because the large (negative) Q-values for emission of heavier particles diminishes their 'competitivity'. Gamma ray emission is included only for multipolarities $\ell = 1, 2$. Higher multipolarities are progressively much weaker. Leaving aside

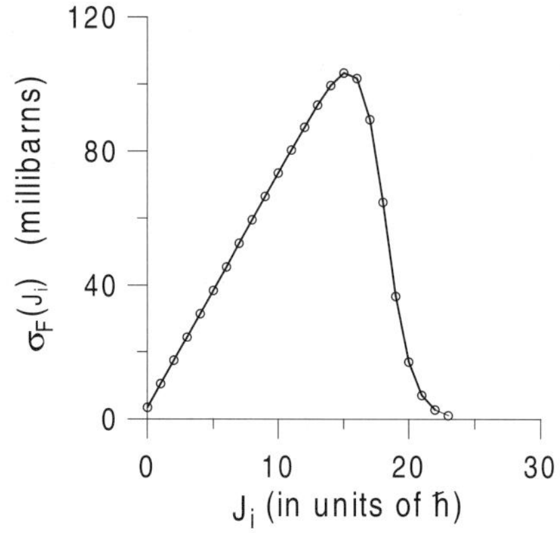

Figure 4.5. Partial cross-sections for fusion of ^{20}Ne + ^{12}C at 65.5 MeV (Ne) incident energy. In this example $\pi/k^2 = 3.5$ mb. The distribution is given by equation (4.3.3) with $J_{\rm cr} = 18$ and $\Delta J = 1$.

events for which the probabilities are known (or supposed) to be small is standard practice in simulations. The cut-off depends on the sensitivity of the experiment to such events.

Equation (4.3.6) refers to a particular value of the relative orbital angular momentum ℓ for a particle b emitted from the compound nucleus C, to leave a residual nucleus B. This equation is the starting point for practical Hauser–Feshbach calculations in which the final states are represented by a density. Other parameters are implicitly included by denoting the exit channel as β. Thus $T_\ell(\beta)$ is the transmission coefficient for the inverse reaction in channel $\beta(b + B \to C)$. We suppose (in line with most reported research) that the transmission coefficient does not depend on the spins of b and B nor on the total angular momentum $J_{\rm i}$.

We now expand the notation by explicitly citing certain parameters implicit in β. Before doing so, a precision concerning the energy of the emitted particle is necessary. We define ε_β as the summed kinetic energy of nuclei b and B. In the coordinate system fixed in the compound system, which, in the present situation, coincides with the centre of mass, b and B *share* the energy ε_β in such a way that the sum of their momenta is zero (we will come back to this point at the end of this section). Remember that this is not the convention used in the original Weisskopf model. In that formalism the residual nucleus was considered to be stationary in the centre of mass so that $\varepsilon_\beta = \varepsilon_b$.

Having made this precision we write the relative probability for a compound nucleus with excitation energy $E_{\rm i}^*$ and angular momentum, $J_{\rm i}$, to emit a particle b with energy ε_β, spin s_b, and relative orbital angular momentum ℓ to form a residual nucleus B with excitation energy $E_{\rm f}^*$ and angular momentum $J_{\rm f}$ (f for 'final') as

$$P_\beta(E_{\rm i}^*, J_{\rm i}, s_b, \ell, \varepsilon_\beta, E_{\rm f}^*, J_{\rm f}) = T_\ell(\varepsilon_\beta)\rho_B(E_{\rm f}^*, J_{\rm f}). \tag{4.3.7}$$

We recall (section 3.5) that, by a suitable adaptation of the transmission coefficient, we can include gamma ray emission in this equation. We will make this more explicit below but for the moment we concentrate on the application of this equation to particle emission. Suppose for the moment that the emitted particle spin $s_b = 0$ (as it is for α particle emission). Then the angular momentum of the residual nucleus, J_f, is the vector sum of the compound nucleus angular momentum J_i and the orbital angular momentum ℓ. Thus for given ℓ we can sum over J_f and write

$$P_\beta(E_i^*, J_i, s_b = 0, \ell, \varepsilon_\beta, E_f^*) = \sum_{J_f=|J_i-\ell|}^{J_i+\ell} T_\ell(\varepsilon_\beta)\rho_B(E_f^*, J_f)$$

$$= T_\ell(\varepsilon_\beta) \sum_{J_f=|J_i-\ell|}^{J_i+\ell} \rho_B(E_f^*, J_f), \qquad (4.3.8)$$

where we assumed that the transmission coefficient is independent of J_f. From (4.3.8) we can immediately proceed by summing over orbital angular momenta. We obtain (notice that the number of arguments on the left-hand side is getting smaller)

$$P_\beta(E_i^*, J_i, s_b = 0, \varepsilon_\beta, E_f^*) = \sum_{\ell=0}^{\ell_{max}} T_\ell(\varepsilon_\beta) \sum_{J_f=|J_i-\ell|}^{J_i+\ell} \rho_B(E_f^*, J_f). \qquad (4.3.9)$$

The limit ℓ_{max} is fixed by the condition that, for all conceivable kinetic energies ε_β, $T_\ell(\varepsilon_\beta) = 0$ for $\ell > \ell_{max}$ (ℓ_{max} is referred to as ℓ_{cr} in specific theories for fusion (section 3.6)). We now relax the imposed condition $s_b = 0$. In this case the angular momentum in the residual nucleus, J_f, is the vector sum of the three angular momenta $\boldsymbol{J}_i$, $\boldsymbol{\ell}$ and $\boldsymbol{s}_b$. Let $\boldsymbol{J}$ be the vector sum of $\boldsymbol{J}_i$ and $\boldsymbol{\ell}$ so that $\boldsymbol{J}_f$ is the vector sum of $\boldsymbol{J}$ and $\boldsymbol{s}_b$.Then the modified form of (4.3.9) reads

$$P_\beta(E_i^*, J_i, s_b, \varepsilon_\beta, E_f^*) = \sum_{\ell=0}^{\ell_{max}} T_\ell(\varepsilon_\beta) \sum_{J=|J_i-\ell|}^{J_i+\ell} \sum_{J_f=|J-s_b|}^{J+s_b} \rho_B(E_f^*, J_f). \qquad (4.3.10)$$

We have yet to integrate over energy. Before doing so, however, we introduce a development [6] which enables the last two sums in (4.3.10) to be simplified and thereby carried out with a minimum investment in computer time. The simplification is based on the fact that the state density, $\rho_B(E_f^*, J_f)$, can be written (equation (2.5.47)) as

$$\rho_B(E_f^*, J_f) = W_B(E_f^*, M_f = J_f) - W_B(E_f^*, M_f = J_f + 1), \qquad (4.3.11)$$

where $W(E_f^*, M_f)$ is the density of states with angular momentum projection quantum number M_f. Using this equation let us carry out the last two sums in (4.3.10) by interchanging the order of summation so as to successively sum the contributions with $J_f = |J - s_b|$, $J_f = |J - s_b| + 1$, etc. Consider the first term in this double sum (with $J_f = |J - s_b|$) which is the sum of terms with J from $|J_i - \ell|$ to $J_i + \ell$

$$P_\beta(E_i^*, J_i, s_b, \varepsilon_\beta, E_f^*; J_f = |J - s_b|) = \sum_{\ell=0}^{\ell_{max}} T_\ell(\varepsilon_\beta)$$
$$\times [W_B(J_f = \big||J_i - \ell| - s_b\big|) - W_B(J_f = \big||J_i - \ell| - s_b\big| + 1)$$
$$+ W_B(J_f = \big||J_i - \ell| + 1 - s_b\big|) - W_B(J_f = \big||J_i - \ell| + 1 - s_b\big| + 1)$$
$$+ W_B(J_f = \big||J_i - \ell| + 2 - s_b\big|) - W_B(J_f = \big||J_i - \ell| + 2 - s_b\big| + 1)$$
$$+$$
$$+$$
$$+ W_B(J_f = |J_i + \ell - s_b|) - W_B(J_f = |J_i + \ell - s_b| + 1)]. \tag{4.3.12}$$

We immediately see that all terms except the first positive contribution and the last negative contibution cancel so that (4.3.12) amounts to

$$W_B(J_f = \big||J_i - \ell| - s_b\big|) - W_B(J_f = |J_i + \ell - s_b| + 1). \tag{4.3.13}$$

The same cancellation occurs when we consider all successive contributions with $J_f = |J - s_b| + 1 \ldots J + s_b$ so that finally we can replace equation (4.3.10) by

$$P_\beta(E_i^*, J_i, s_b, \varepsilon_\beta, E_f^*) = \sum_{\ell=0}^{\ell_{max}} T_\ell(\varepsilon_\beta)\left[\sum_{K_1=||J_i-\ell|-s_b|}^{|J_i-\ell|+s_b} W_B(E_f^*, K_1) - \sum_{K_2=|J_i+\ell-s_b|+1}^{J_i+\ell+s_b+1} W_B(E_f^*, K_2)\right]. \tag{4.3.14}$$

We now come to the (continuous) energy variable. Relative to the residue B, the particle b is emitted with an energy ε_β which ranges from 0 up to a maximum value given by the (usually negative) Q-value corresponding to the emission of b from the compound nucleus, i.e, $\varepsilon_{max} = E_i^* + Q_{C\to bB}$. Furthermore, the excitation energy in the residual nucleus is obtained from the compound nucleus excitation energy as $E_f^* = E_i^* + Q_{C\to bB} - \varepsilon_\beta$. Thus

$$P_\beta(E_i^*, J_i, s_b) = \sum_{\ell=0}^{\ell_{max}} S_\beta(\ell) = \sum_{\ell=0}^{\ell_{max}} \int_0^{\varepsilon_{max}} T_\ell(\varepsilon_\beta)$$
$$\times \left[\sum_{K_1=||J_i-\ell|-s_b|}^{|J_i-\ell|+s_b} W_B(E_f^*, K_1) - \sum_{K_2=|J_i+\ell-s_b|+1}^{J_i+\ell+s_b+1} W_B(E_f^*, K_2)\right]$$
$$\times \delta(E_f^* - (E_i^* + Q_{C\to bB} - \varepsilon_\beta))\,d\varepsilon_\beta, \tag{4.3.15}$$

where we introduced (defined) the summand $S_\beta(\ell)$. Equation (4.3.15) represents the relative probability for the emission of particle b from the compound nucleus

$(E_{\rm i}^*, J_{\rm i})$. Its evaluation involves a double summation and a numerical integration. Furthermore, these operations must be carried out for all competing decay channels (particles and gamma rays).

We now turn to the practical evaluation of (4.3.15). We may begin with the integration over the emitted particle energy. One simplification is obtained by taking the state densities of each residual nucleus to be continuous functions even at very low energies where individual states are resolved. The low energy parametrization of the state densities involves detailed inspection of many nuclear level schemes (a task that was actually carried out in [6]). The alternative, which was implemented in a recent work [7], is to split the integral over energy into a sum over discrete states and an integral over the energy region where a continuous state density function can be used. The summations over discrete states must then be performed explicitly. Using continuous densities, a further approximation used in the LILITA program (written by Gomez del Campo) [8] was to parametrize the energy dependence of the density in order to an provide an analytical evaluation of the energy integrals.

A considerable advantage springs from the fact that, at least in its analytical form (see section 2.5) the density is a separable function of energy and angular momentum. The advantage is apparent if we denote the positive and negative contributions in (4.3.15) as $S_\beta^+(\ell)$, and $S_\beta^-(\ell)$ so that

$$S_\beta(\ell) = S_\beta^+(\ell) - S_\beta^-(\ell), \tag{4.3.16}$$

where $S_\beta^+(\ell) =$

$$\int_0^{\varepsilon_{\max}} T_\ell(\varepsilon_\beta)\left[\sum_{K_1=||J_{\rm i}-\ell|-s_b|}^{|J_{\rm i}-\ell|+s_b} W_B(E_{\rm f}^*, K_1)\delta(E_{\rm f}^* - E_{\rm i}^* - Q_{C\to bB} + \varepsilon_\beta)\right]{\rm d}\varepsilon_\beta, \tag{4.3.17}$$

and $S_\beta^-(\ell) =$

$$\int_0^{\varepsilon_{\max}} T_\ell(\varepsilon_\beta)\left[\sum_{K_2=|J_{\rm i}+\ell-s_b|+1}^{J_{\rm i}+\ell+s_b+1} W_B(E_{\rm f}^*, K_2)\delta(E_{\rm f}^* - E_{\rm i}^* - Q_{C\to bB} + \varepsilon_\beta)\right]{\rm d}\varepsilon_\beta. \tag{4.3.18}$$

Writing the state density $W_B(E_{\rm f}^*, K)$ as a product of energy dependent and angular momentum projection dependent factors (2.5.48),

$$W_B(E_{\rm f}^*, K) = \rho_B(E_{\rm f}^*)\frac{1}{\sqrt{2\pi\sigma^2}}\,{\rm e}^{-K^2/2\sigma^2}, \tag{4.3.19}$$

and denoting the Gaussian distribution ${\rm e}^{-K^2/2\sigma^2}/\sqrt{2\pi\sigma^2}$ by $g(K, \sigma)$, we can write the sum S^+ as the product of two independent factors,

$$S_\beta^+(\ell) = \left[\int_0^{\varepsilon_{\max}} T_\ell(\varepsilon_\beta)\rho_B(E_{\rm i}^* + Q_{C\to bB} - \varepsilon_\beta)\,{\rm d}\varepsilon_\beta\right]\sum_{K_1=||J_{\rm i}-\ell|-s_b|}^{|J_{\rm i}-\ell|+s_b} g(K_1, \sigma), \tag{4.3.20}$$

with a similar expression for $S_\beta^-(\ell)$. The energy integrand is the same in both cases and the corresponding integral can therefore be carried out (albeit numerically) just once. This usually involves another approximation which does relatively little to lighten the computation but does simplify the program organization. The transmission coefficients $T_\ell(\varepsilon_\beta)$ should, in principle, be obtained from Optical Model studies of elastic scattering. Unfortunately, such studies are not available for all possible emitted particles at all energies. Furthermore we recall that, in principle, the transmission coefficient refers to the inverse reaction $b + B \to C$ in which the nucleus B is normally excited. Faced with these difficulties, most authors have had recourse to the low energy theory for fusion discussed in section 3.6 in which the transmission coefficients are expressed in the Hill–Wheeler form [9] as

$$T_\ell(\varepsilon_\beta) = \frac{1}{1 + e^{2\pi(V_\ell - \varepsilon_\beta)/\hbar\omega_\ell}}. \tag{4.3.21}$$

The parameters V_ℓ and ω_ℓ for any emitted particle and residual nucleus (b, B) are obtained by comparing (4.3.21) with 'real' transmission coefficients obtained from Optical Model fits to elastic scattering at precise energies (as available). Figure 4.6 shows such a comparison for alpha particles incident on the nucleus ^{28}Si. While it may seem regretable to have to abandon the precision of a significant part of the theory of compound nucleus decay, one can certainly appreciate the practical value of a simple analytical expression for transmission coefficients.

Gamma ray 'transmission coefficients' were introduced and discussed in section 3.5. They take the form $\xi_\ell \varepsilon_\gamma^{2\ell+1}$ where ε_γ is the energy of the gamma ray and the parameters ξ_ℓ are taken from systematic studies of gamma ray widths for low lying levels. For light nuclei, typical values for dipole and quadrupole strengths are $\xi_1 = 5 \times 10^{-9}$ and $\xi_2 = 3 \times 10^{-10}$.

Having carried out the integration over energies we can directly perform the sums over K_1 and K_2 in order to obtain $S_\beta(\ell)$ for each exit channel β. We can then sample this distribution using the summation method as in algorithm 4.4. A particular type of particle together with its angular momentum (or gamma ray with its multipolarity) is selected in the same Monte Carlo step by adding successive terms in the cumulative probability distribution

$$p_{\beta'}(\ell') = \frac{\sum\limits_{\text{exit chan}(\beta)} \sum\limits_{\ell} S_\beta(\ell)}{\sum\limits_{\text{all }\beta,\ell} S_\beta(\ell)}, \tag{4.3.22}$$

until $p_{\beta'}(\ell')$ (which is bounded in the interval (0,1)) is greater than a uniform random deviate ($p(x) = 1.0, 0 \le x < 1$).

Having selected the type of particle and the angular momentum we can use the cumulative probability distribution function

$$q(\varepsilon_{\beta'}) = \frac{\int_0^{\varepsilon_{\beta'}} T_\ell(\varepsilon_\beta)\rho_B(E_i^* + Q_{C\to bB} - \varepsilon_\beta)\,\mathrm{d}\varepsilon_\beta}{\int_0^{\varepsilon_{\max}} T_\ell(\varepsilon_\beta)\rho_B(E_i^* + Q_{C\to bB} - \varepsilon_\beta)\,\mathrm{d}\varepsilon_\beta}, \tag{4.3.23}$$

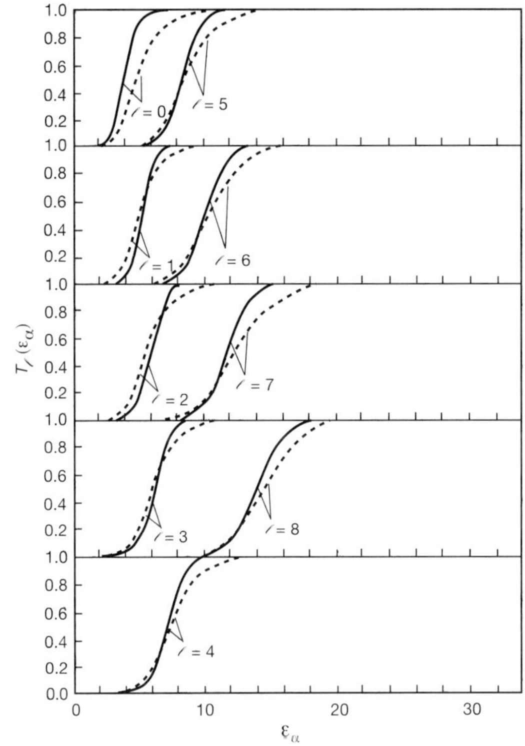

Figure 4.6. Transmission coefficients for elastic scattering of alpha particles from ^{28}Si calculated with optical model parameters fitted to elastic scattering data (broken curve). The 'true' transmission coefficients are compared with approximate values (solid curve) obtained from the Hill–Wheeler parabolic barrier approximation (4.3.21). The real potential parameters used for this calculation were $V = 165.0$ MeV, $r_0 = 1.45$ fm ($R_V = r_0 A^{1/3}$), $a_V = 0.52$ fm.

common to $S^+$and S^- to select the emitted particle energy, ε'_β. This step can also be carried out using the rejection method. Because $T_\ell(\varepsilon_\beta) \to 1$ for large values of ε_β the high energy tail of the energy spectrum is governed by the density $\rho_B(E_\mathrm{i}^* + Q_{C\to bB} - \varepsilon_\beta)$ which falls off (approximately exponentially) with increasing ε_β.

It remains to select the angular momentum in the residual nucleus. This may be achieved quite simply by returning to equation (4.3.10) from which we extract the term

$$F(b, \ell, \varepsilon_\beta) = \sum_{J=|J_\mathrm{i}-\ell|}^{J_\mathrm{i}+\ell} \sum_{J_\mathrm{f}=|J-s_b|}^{J+s_b} \rho_B(E_\mathrm{f}^*, J_\mathrm{f}). \tag{4.3.24}$$

(Remember that, at this stage, the characteristics of the emitted radiation, b, ℓ and E_f^* (equivalently ε_β) have already been fixed. For a given emitted particle (gamma ray), b, orbital angular momentum (multipolarity), ℓ, and energy ε_β we can construct this sum, term by term, separating and summing all terms with a given J_f to form the partial sums $f(b, \ell, \varepsilon_\beta, J_\mathrm{f})$. Then

$$g(J_\mathrm{f}) = \frac{f(b, \ell, \varepsilon_\beta, J_\mathrm{f})}{F(b, l, \varepsilon_\beta)}, \tag{4.3.25}$$

is a probability density function for the discrete variable J_f. A particular value for this quantity can thus be selected again using the simple Monte Carlo summation technique.

Let us summarize. By writing out the probability distribution for a single (Hauser–Feshbach) evaporation step we have shown how to apply Monte Carlo methods to make a single selection of the initial compound nucleus angular momentum, J_i, the type of emitted particle (gamma ray), b, its orbital angular momentum (multipolarity), ℓ, its energy, ε_β (or ε_γ), and the final angular momentum in the residual nucleus J_f. The initial compound nucleus excitation energy E_i^* is fixed by the entrance channel and the residual nucleus excitation energy, E_f^*, is fixed by the selection of ε_β. For gamma ray emission the Q-value is 0 because the final nucleus is, of course, the same as the initial nucleus. At this stage, we can repeat the whole procedure beginning with new selections of J_i, thereby generating a series of events specified by b, E_i^*, J_i, ℓ, ε_β (or E_f^*) and J_f.

Before turning to the treatment of multistep emission, we have two more tasks to perform. The first concerns the direction of the emitted particle. The distribution of possible directions is of course crucial for comparison with experiment. For a single emission step it is possible to derive either a quantum mechanical [10] or a classical [11] probability density function corresponding to emitted particle directions (the quantum mechanical distribution was treated in section 3.5). However, the application of these distributions to multistep emission using analytical methods is rather involved. For this reason, it is preferable to apply the simple Monte Carlo techniques exposed in section 4.2. The Monte Carlo method which we describe here is based on the classical treatment.

We begin by creating a coordinate system which is firmly fixed in the laboratory with the z-axis (denoted by Z_lab) pointing towards the roof, and the y-axis (Y_lab) aligned along the beam direction (figure 4.7). The relative orbital angular momentum, ℓ_α, in the entrance channel is perpendicular to its linear momentum and is thus a vector randomly oriented in a plane perpendicular to the beam direction (we use the subscript 'α' to distinguish the entrance channel orbital angular momentum from the orbital angular momentum characterizing the decay process which we have written as 'ℓ'). This will therefore be true, on average, for the initial compound nucleus angular momentum which is $\boldsymbol{J}_\mathrm{i} = \boldsymbol{\ell}_\alpha + \boldsymbol{s}_a + \boldsymbol{S}_A$. Restraining $\boldsymbol{J}_\mathrm{i}$ to lie exactly in the plane perpendicular to the beam direction might seem to represent a rather drastic approximation but we should remember that ground state spins, s_a and S_A, which respectively characterize the projectile and target nuclei in the formation of the compound nucleus ($a + A \to C$) are usually close to zero so that the dispersion around the mean direction of $\boldsymbol{J}_\mathrm{i}$ is quite small.

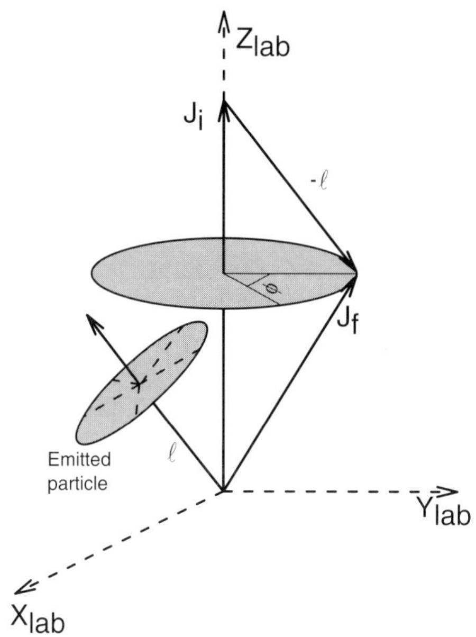

Figure 4.7. Classical angular momentum coupling and particle emission. Intrinsic spin is not taken into account except through its influence on J_i and J_f. An initial rotation $\boldsymbol{R}_0$ is performed so that $\boldsymbol{J}_\mathrm{i}$ lies along the laboratory z-axis as shown in the figure. The orientation, ϕ, of the angular momentum triangle ($\boldsymbol{J}_\mathrm{i}, \boldsymbol{\ell}, \boldsymbol{J}_\mathrm{f}$) is then randomly selected. Finally, the direction of the emitted particle is randomly selected in the plane perpendicular to $\boldsymbol{\ell}$. The final emitted particle direction is found by applying the inverse rotation $\boldsymbol{R}_0^{-1}$.

If $\boldsymbol{J}_\mathrm{i}$ is randomly oriented perpendicular to the beam direction, we can generate a random number uniformly distributed between the limits $(0, 2\pi)$ to select a particular direction for this vector. We can now imagine a '$\boldsymbol{J}_\mathrm{i}$' Cartesian coordinate system in which the z-axis coincides with $\boldsymbol{J}_\mathrm{i}$, the x-axis is fixed to lie in the X_lab–Z_lab plane and the y-axis coincides with Y_lab. We now rotate the J_i system around Y_lab (the beam axis) so that it coincides with the z-axis of the 'lab' system (see again figure 4.7). The advantage of this rotation, which we refer to as $\boldsymbol{R}_0$, is that it enables us to work in a unique (lab) system. We possess the moduli of the vectors $\boldsymbol{\ell}$ and $\boldsymbol{J}_\mathrm{f}$ so that we can create the triangle with sides J_i, ℓ and J_f. The plane represented by this triangle is constrained only by the condition that $\boldsymbol{J}_\mathrm{i}$ coincides with Z_lab and thus, consistently with the statistical approach, may be arbitrarily rotated around the Z_lab axis. A particular orientation can thus be selected again using a uniformly distributed random number, ϕ, in the range $(0, 2\pi)$. The relative orbital angular momentum vector, $\boldsymbol{\ell}$, is thereby fixed in our coordinate system. We now use the fact that, classically, the angular momentum of the emitted particle is perpendicular to its linear momentum. Equivalently, we say that the particle is emitted in a random direction perpendicular to ℓ. This direction may thus be selected using, once more, a random number uniformly distributed in the range $(0, 2\pi)$. In fact, the simplest way to locate the particle direction in the laboratory coordinate system is to form the rotation matrix $\boldsymbol{R}_\ell$ which takes $\boldsymbol{\ell}$ into $\boldsymbol{J}_\mathrm{i}$ (which we have already rotated to coincide with Z_lab). Then, having selected the particle direction randomly in the X_lab–Y_lab plane, the inverse rotation is performed so that the particle direction is correctly expressed relative to the '$\boldsymbol{J}_\mathrm{i}$' coordinate system.

If we are interested only in the distribution of emission angles for this first emitted particle, we could repeat the selection process thereby generating a series of events. However, in each case we would need to correct each selection for the 'unjustified' rotation $\boldsymbol{R}_0$. In order to re-instate the correct orientation it is sufficient to perform the inverse rotation, $\boldsymbol{R}_0^{-1}$ on all selected vectors including the direction of the emitted particle. In the particular case of single step emission the set of rotations $\boldsymbol{R}_0^{-1}$ (or $\boldsymbol{R}_0$) corresponds to a uniformly random angle $(0, 2\pi)$ in the X_lab–Z_lab plane so the problem is easily solved (for multistep emission, as we will see in the next section the operation is a little more complex). Given this random rotation it is easy to see that the distribution of directions of the first emitted particle is uniform with respect to rotations around the beam axis. In addition, because of the random rotation of the $(J_\mathrm{i}, \ell, J_\mathrm{f})$ triangle around $\boldsymbol{J}_\mathrm{i}$ it will be clear that the distribution of directions of ℓ, and thereby of the emitted particle, is symmetric with respect to the X_lab–Z_lab plane. A more important statement is that these characteristics are maintained for the inclusive observation of all particles in an evaporation cascade because the final angular momentum, $\boldsymbol{J}_\mathrm{f}$, which becomes the initial angular momentum, $\boldsymbol{J}_\mathrm{i}$, for the next evaporation step is, *on average*, a vector which remains perpendicular to, and randomly oriented around, the beam direction.

We have not quite finished. Our final task concerns the conservation of momentum. The energy ε_β which was selected previously is the energy of the emitted particle, b, relative to the residual nucleus B. However, we need the corresponding energy in the (b, B) centre-of-mass system. Recalling that the total vector momentum in the centre-of-mass system is zero, we write

$$\boldsymbol{p}_b + \boldsymbol{P}_B = 0. \tag{4.3.26}$$

If the energy of the emitted particle in the centre-of-mass system is ε_b, and that for the residual nucleus is ε_B we also have

$$\varepsilon_\beta = \varepsilon_b + \varepsilon_B. \tag{4.3.27}$$

Taking the masses of b and B to be M_b and M_B respectively and combining (4.3.26) and (4.3.27), we obtain

$$\varepsilon_b = \frac{p_b^2}{2M_b} = \varepsilon_\beta \frac{M_B}{M_b + M_B}. \tag{4.3.28}$$

The emitted particle thus appears in the laboratory system of coordinates with momentum components given, in terms of the momentum, $p_b = \sqrt{2M_b\varepsilon_b}$, by

$$\begin{aligned} p_{b,x} &= p_b \sin\theta_b \cos\phi_b, \\ p_{b,y} &= p_b \sin\theta_b \sin\phi_b + M_b V_{\mathrm{CM}}, \\ p_{b,z} &= p_b \cos\theta_b, \end{aligned} \tag{4.3.29}$$

where V_{CM} is the velocity of the the centre of mass (in this case of the compound nucleus) in the laboratory system. It is important to note a slight inconvenience associated with (4.3.29) which arises because experiments are almost always conducted using a system of coordinates with the z-axis fixed along the beam direction. Let us denote such a system using primes. Then comparison with experiment requires an additional rotation which is simply achieved by writing

$$\begin{aligned} p_{b,z'} &= p_{b,y}, \\ p_{b,x'} &= p_{b,z}, \\ p_{b,y'} &= p_{b,x}. \end{aligned} \tag{4.3.30}$$

To terminate this section we will briefly discuss the observation of gamma rays. The angular distribution of gamma rays from an initial state with angular momentum (quantum number) J_{i} to a final state with angular momentum J_{f} depends on the multipolarity of the gamma ray ℓ and on its projection on an arbitrary axis of quantization which is usually referred to as the z-axis. We may take this axis to coincide with $\boldsymbol{J}_{\mathrm{i}}$. For stretched transitions ($\ell_z = -\ell$) which remove the maximum angular momentum from the compound nucleus we find that $J_{\mathrm{f}} = J_{\mathrm{i}} - \ell$. In this case, if θ_b is the (polar) angle between the direction of the emitted gamma ray and the z-axis, the angular distributions are

$$W(\theta_b) = \tfrac{3}{4}(1 + \cos^2\theta_b), \tag{4.3.31}$$

for dipole radiation and

$$W(\theta_b) = \tfrac{5}{4}(1 - \cos^4\theta_b), \tag{4.3.32}$$

for quadrupole gamma rays [2]. This makes the study of dipole–quadrupole mixtures, which are typical of evaporation, rather difficult because the association of the two terms tends to produce rather featureless angular distributions. Non-stretched transitions also help to smooth out the observed distributions. This remark applies to decay of compound nuclei where an additional complication arises from the fact that the initial angular momentum is randomly oriented perpendicular to the beam axis, but is also pertinent when the emitting system is an excited target nucleus produced by an inelastic collision and is thus made to rotate about an axis perpendicular to the reaction plane [12]. Furthermore, in this latter situation, the description of the inelastic scattering reaction mechanism complicates the analysis of the data. We shall see in section 4.5 that studies of evaporation in heavy nuclei where gamma ray emission is important have concentrated on the number (multiplicity) of emitted gammas rather than on angular distributions.

4.4 Multistep (sequential) evaporation

The extension of the Monte Carlo method to deal with multistep emission is quite straightforward. Thus, suppose we proceed to step 2 of the evaporation cascade. Following the first evaporation, one requires the mass and charge of the residue, B, the vector momentum of this nucleus in the laboratory frame (equation (4.3.30)), the excitation energy, E_{f}^*, and the final vector angular momentum $\boldsymbol{J}_{\mathrm{f}}$. These quantities are considered, respectively, as the new compound nucleus, its vector momentum, the initial excitation energy, E_{i}^*, and the initial angular momentum, $\boldsymbol{J}_{\mathrm{i}}$, for the second step of the evaporation cascade. We can therefore easily simulate a particular sequence of emitted particles and gamma rays and, by repeating the simulation a large number of times, obtain the relative yields and the kinematic properties of the various emitted particles and the residual nuclei. We saw, in the previous section, that the (Monte Carlo) selection of the particle direction in the laboratory is created along the cascade using only the moduli of the angular momentum vectors. A typical evaporation cascade (^{32}S $\rightarrow \alpha$, p, n $\rightarrow$ ^{26}Al) which contains sufficient information for the calculation of particle directions and energies is therefore represented by a table of the form

	b	A	Z	E_{i}^* (MeV)	J_{i} ($\hbar$)	ℓ ($\hbar$)	ε_β (MeV)	J_{f} ($\hbar$)	E_{f}^* (MeV)
Compound	α	32	16	50	17	8	5.0	9	35
1st daughter	p	28	14	35	9	4	2.5	5.5	21
2nd daughter	n	27	13	21	5.5	2	3.0	3	4
Residue (g.s.)	-	26	13	0	5	-	-	-	-

The table represents one possible realization (with slightly rounded energies) of a decay chain. The cascade is represented in figure 4.8. The emission process terminates when all probabilities for all competing particle exit channels are found to be zero. This happens when the excitation energy falls below the lowest particle emission threshold (in ^{26}Al this is the proton emission threshold at 6.31 MeV). The ensuing gamma ray emission then de-excites the residue which is left in its ground state. No account of this emission is included in the description of the process, partly because the formalism is not sufficiently detailed to deal with individual levels (the ground state of ^{26}Al is actually 5+) and precise gamma ray emission strengths, and partly because the majority of experiments are usually concerned with observation of charged particles. Furthermore, the momentum associated with emitted gamma rays can usually be neglected. This is not to say that gamma rays are uninteresting. The study of low lying levels in nuclei using gamma ray spectroscopy forms a vast area of study in nuclear physics providing detailed information on the occupation of single particle levels and/or on the collective characteristics of excited states.

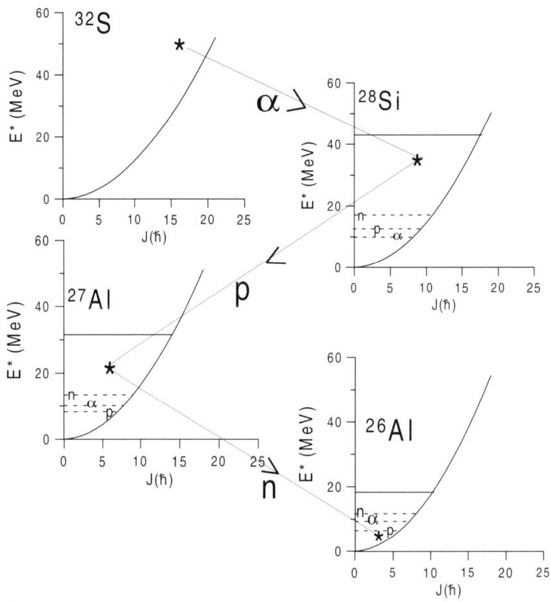

Figure 4.8. Multistep evaporation (evaporation cascade) from the compound nucleus ^{32}S. The continuous horizontal lines indicate the maximum excitation energy E^*_{max} which could be reached in each nucleus, and the n, p and α thresholds are shown as dashed horizontal lines. The cascade stops when the point reached in the E^*–J plane lies below the lowest particle emission threshold, in which case the evaporation residue (^{26}Al) gamma decays to the ground state. For heavy nuclei, gamma ray decay competes significantly even above the particle emission thresholds.

Obtaining the set of directions of emitted particles is quite straightforward but requires careful 'book-keeping'. The reader will recall that in the preceding section the angular momentum of the compound system, $\boldsymbol{J}_{\mathrm{i}}$, which was chosen to have a random direction in the plane perpendicular to the beam axis, was rotated in this plane so as to coincide with the laboratory Z_{lab} axis. We referred to this rotation as the rotation matrix $\boldsymbol{R}_0$. The direction of the emitted particle was then obtained by calculating its direction in the '$\boldsymbol{J}_{\mathrm{i}}$' frame and applying the inverse rotation $\boldsymbol{R}_0^{-1}$. For the second step in the evaporation cascade we have the same problem. The vector $\boldsymbol{J}_{\mathrm{f}}$ which was fixed in the laboratory frame becomes the vector $\boldsymbol{J}_{\mathrm{i}}$ for the second evaporation step. Let $\boldsymbol{R}_1$ represent the rotation which takes this vector into Z_{lab}. Then, after applying $\boldsymbol{R}_1$ the problem of calculating the particle direction is identical to that for the first evaporation step. However, to establish the true direction we now have to transform the particle direction by the 'cumulative' rotation matrix which is $\boldsymbol{R}_0^{-1}\boldsymbol{R}_1^{-1}$. This process may, of course, be repeated at each step of the evaporation cascade. We thus calculate the momentum components $(p_{b,x'}, p_{b,y'}, p_{b,z'})$ given by equation (4.3.30) for each emitted particle. The vector momentum of the residual nucleus can then be easily computed by observing that the total (conserved) momentum is just that of the compound nucleus.

With this remark, we complete the detailed description of the use of Monte Carlo techniques in applications of the Hauser–Feshbach model to single step and multistep evaporation. We emphasize that we have described one possible strategy and that different authors have made use of different approximations and somewhat different level densities. This is to be regreted because it renders the comparison of the various analyses somewhat ambiguous. Nevertheless, as we shall see in the next section, the success of the various codes provides quite convincing evidence for the overall pertinence of the statistical sequential decay model.

4.5 Comparison with experiment

It will be obvious from the results presented in this section that the multistep Hauser–Feshbach model is rather successful in reproducing experimental measurements for compound nuclear excitation energies up to about 2 MeV/nucleon. Especially for evaporation from light compound nuclei this success is such that the model has been considered as a tool which can be used to extract source parameters, such as the fusion cross-section (which, via the parameter $J_{\rm CR}$, is used to characterize the initial distribution of angular momentum in the compound nucleus (4.3.3)). Regretably, we cannot attempt a comprehensive review of the large body of published experimental work. This would require a volume in itself. Thus, we will simply present a few examples in the hope that the interested reader will be able to use the cited papers as a basis for further study.

As remarked by Stokstad [2], experimental results fall into natural categories. For light and medium-weight nuclei, gamma ray emission is not important. For light compound nuclei with large angular momenta, alpha particle emission predominates (see figure 4.9) because heavier particles remove relatively large amounts of angular momentum (leading to a higher density of states in the final nucleus). Thus, initial states close to the Yrast line tend to decay by emitting a cascade of alpha particles whereas, away from the Yrast line, proton and neutron emission are more important [13]. On the other hand, in heavy nuclei gamma ray emission is more important, and charged particle emission is somewhat suppressed (the larger Coulomb barriers lead to lower final state densities) so that neutron emission dominates the evaporation cascade. In such nuclei, a further complication arises from the fact that fission may effectively compete with neutron emission and thus must be included in the panoply of possible decay channels.

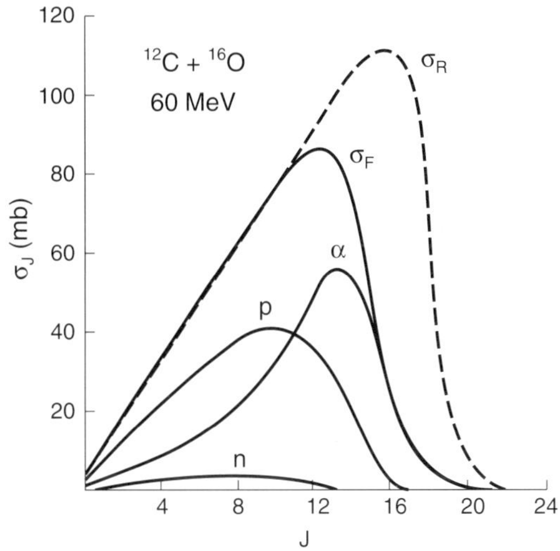

Figure 4.9. First step emission probs for ^{28}Si (produced in the 60 MeV ^{16}O + ^{12}C fusion reaction) displayed as a function of the compound nucleus angular momentum [13]. Nucleons are preferentially emitted from low angular momentum states whereas heavy particles such as alpha particles are emitted from states with large angular mometum. The suppression of neutron emission in this particular case is a Q-value effect. Note the distinction between the reaction cross-section (broken curve labelled σ_R) and the fusion cross-section (σ_F). (Reprinted from *Nucl. Phys.* A, **263**, Weidinger A *et al*, p 511. Copyright 1976, with permission from Elsevier Science.)

Let us begin with light systems. Figure 4.10 shows a calculation made by Pühlhofer [14] using the CASCADE code for yields of residual nuclei in the ^{16}O + ^{27}Al and ^{19}F + ^{27}Al reactions at several incident energies (shown in the figure) leading to compound nuclei with excitation energies below 2 MeV/nucleon. The predominant yields are correctly predicted to within about 20%. A similar calculation was made by the author and collaborators [5] for the ^{20}Ne + ^{12}C reaction at 66.5 MeV and the ^{16}O + ^{16}O reaction at 55.0 MeV. In this case, angular distributions and energy spectra were also calculated using the methods explained in the preceding sections (figures 4.11 and 4.12). The good overall agreement between predictions and measurements is taken as evidence for the validity of the assumption of sequential emission and for the parametrization of the level densities and transmission coefficients. Particularly interesting is the double humped structure in the ^{27}Si energy distribution (also visible for ^{26}Al) which is due to the fact that alpha particle emission, with its associated Coulomb barrier, followed by proton emission is the predominant decay chain for this residue (this characteristic shape was mentioned at the end of section 2.2).

As a study of light particle correlations we can cite the work of Kuang-Hsi *et*

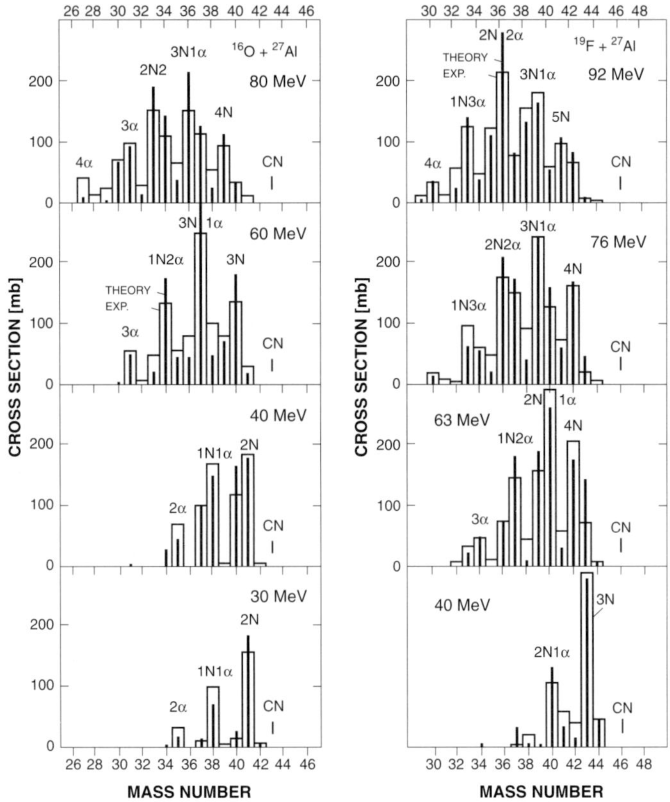

Figure 4.10. Predicted and measured evaporation residue cross-sections [14] in the fusion reactions $^{16}O + ^{27}Al$ and $^{19}F + ^{27}Al$. The regular structure in the relative yields is due to the fact that evaporation is dominated by mass 1 (nucleons) and mass 4 (α particles). Can you explain why? (Reprinted from *Nucl. Phys.* A, **280**, Pühlhofer F *et al*, p 267. Copyright 1977, with permission from Elsevier Science.)

al [15]. Results for p–p, p–α and α–α azimuthal angle correlations measured in the plane perpendicular to the beam are shown in figure 4.13. The α–α correlation is particularly strong and can be qualitatively understood by considering the case when the two α particles are emitted at polar angles of $\theta_1 = \theta_2 = 90°$ in the centre-of-mass system. If the angular momentum of successive α particles is anti-aligned with the direction of the compound nucleus angular momentum, $\boldsymbol{J}_\mathrm{i}$ (this would be expected if the point E_i^*, J_i is close to the Yrast line), the two α particles would be expected to be emitted in the same plane perpendicular to $\boldsymbol{J}_\mathrm{i}$. This plane intersects the measurement plane (perpendicular to the beam) at azimuthal angles $\phi = 0°$ and $\phi = 180°$ so that the correlation would be expected to be strong at $\phi_1 - \phi_2 = 0°$ and $\phi_1 - \phi_2 = 180°$ and weak at $\phi_1 - \phi_2 = 90°$.

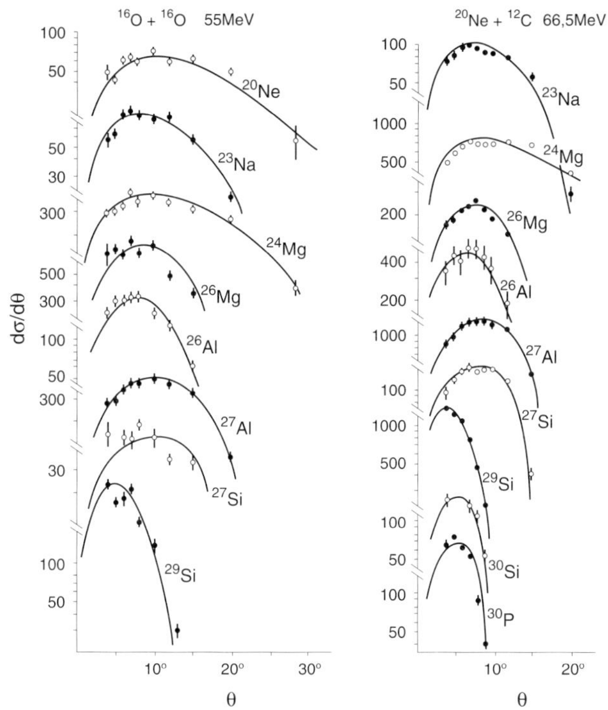

Figure 4.11. Angular distributions of evaporation residues (plotted as $\mathrm{d}\sigma/\mathrm{d}\theta$) for the fusion reactions ^{16}O+^{16}O and ^{20}Ne+^{12}C. The predictions obtained with the LANCELOT code [6] are shown as continuous lines. The overall description of the data is highly satisfactory. Notice the broad distribution for ^{24} Mg which is due to the 2α emission from initial states with large angular momentum. In contrast, the production of ^{23}Na involves lower α energies (there must be energy left for emission of the proton). (Reprinted from *Nucl. Phys.* A, **341**, Cole A J *et al*, p 284. Copyright 1980, with permission from Elsevier Science.)

Even though, in the experiment, the observations were not made at centre-of-mass angles $\theta_1 = \theta_2 = 90°$ (the authors estimate average angles of $\theta_1 = \theta_2 = 110°$) the overall form of the correlation is preserved.

Light particle-evaporation residue energy–energy correlations were studied by Ost *et al* [16] in the reaction ^{28}Si + ^{12}C. The objective of the study was to search for 'non-statistical' contributions to the observed cross-sections. Detailed comparisons of the data with multistep Hauser–Feshbach statistical model calculations showed no significant discrepancy.

Let us now consider briefly studies of evaporation from heavy nuclei. In this case, as is shown in figure 4.14 [17], the predominant decay modes involve the emission of neutrons and gamma rays. One thus speaks of residues formed

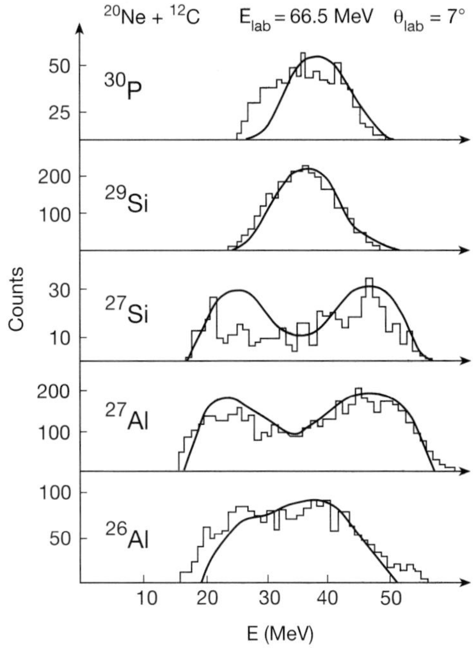

Figure 4.12. Typical energy spectra for evaporation residues measured at small angles in the laboratory and compared with LANCELOT predictions [5]. The 'double bump' structure observed for the ^{27}Si and ^{26}Al is due to the fact that the recoil imparted to these residues is dominated by the emission of an alpha particle (see text). (Reprinted from *Nucl. Phys.* A, **341**, Cole A J *et al*, p 284. Copyright 1980, with permission from Elsevier Science.)

by xn emission. Near the Yrast line (where the level densities diminish to zero) quadrupole gamma ray transitions, which may be likened to transitions near the particle emission threshold, are responsible for the evacuation of angular momentum. This decay mode involves a series of discrete transitions between states with similar (collective band) structure. Inter-band transitions also occur due to dipole transitions. The effect of these 'Yrast cascades' is well illustrated in a calculation (figure 4.15) carried out by Hillis *et al* [18] for decay from an ^{164}Er nucleus excited to 53.8 MeV. Neutron and statistical gamma ray emission dominate as long as the population in the (E_f^*, J_f) space, which becomes the (E_i^*, J_i) space for the following emission step, is not too close to the Yrast line. Otherwise the collective gamma ray cascade takes over leading to the formation of a ground state residue. One consequence is that the (E_f^*, J_f) population corresponding to neutron emission from high J_i values in the compound nucleus tends to 'hit' the Yrast line after evaporation of only one or two neutrons so that the yields of the corresponding residues can be attributed to large angular momentum in the

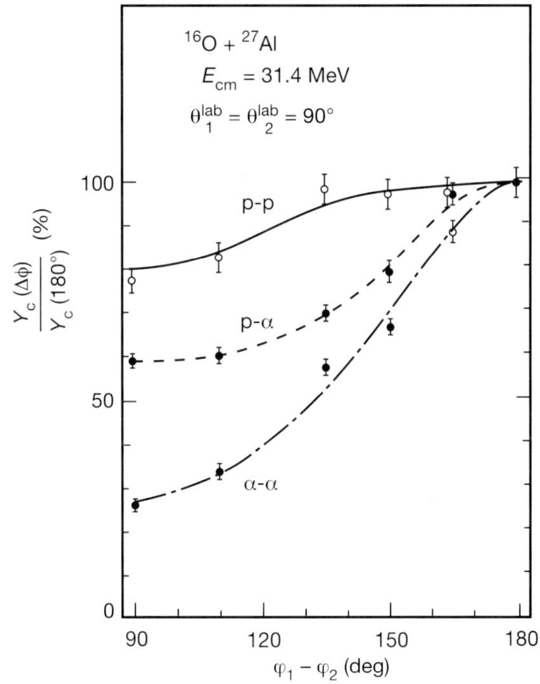

Figure 4.13. Evaporated particle angular correlations measured in the plane perpendiclar to the beam following the fusion reaction $^{16}O + ^{27}Al$ at an incident oxygen energy of 50 MeV [15]. The strongly anisotropic α, α correlation is due to the fact that α particles tend to be emitted with large angular momentum which is almost anti-aligned with the angular momentum of the parent nucleus. (Reprinted from *Nucl. Phys.* A, **316**, Kuang-Hsi T *et al*, p 189. Copyright 1979, with permission from Elsevier Science.)

compound system. By extending this reasoning and assigning the removal of an average angular momentum to each gamma ray, we see that measurement of the characteristics of the 'Yrast' gamma ray multiplicity distribution provides direct infomation on the 'pre-gamma' angular momentum distribution for each residue in the xn chain. More important is the observation that if the same exercise is carried out as a function of the gamma ray energy, then the collective cascade can be used to provide some estimate of the moment of inertia which characterizes successive rotational states near the Yrast line [19,20]. The interested reader is referred to the review of Garret, Hagemann and Herskind [21] which describes studies of gamma ray 'continuum' spectra.

Another major aspect of the statistical decay of heavy systems concerns fission. This would be expected because, as we have seen in chapter 3, fission barriers tend to be small (or even non-existent) in heavy systems. The use of the simple Bohr–Wheeler treatment of fission is usually justified for excited systems

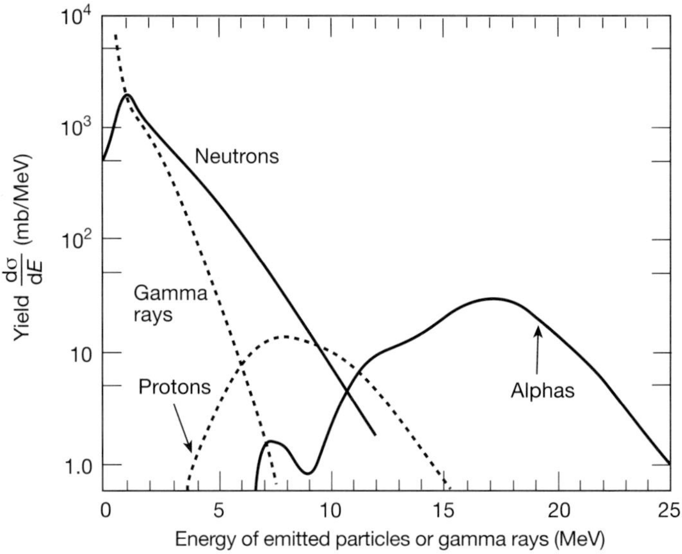

Figure 4.14. Calculated yields of gamma rays, protons, neutrons and alpha particles following the fusion of 90 MeV ^{16}O projectiles with ^{140}Ce [17]. In heavy nuclei, charged particle emission is strongly suppressed due to the large Coulomb barriers. (Reprinted from *Phys. Rev.*, **157**, Grover J R and Gilat J, p 802. Copyright 1967, by the American Physical Society.)

by invoking the gradual disappearance (with increasing excitation energy) of shell and paring corrections which are essential for a successful description of fission at low excitation energies. The ratio of the widths for fission and neutron evaporation, Γ_f/Γ_n, is calculated by assuming the relative probability for fission to be determined by the state density at the saddle point, i.e., at the top of the fission barrier (section 3.8). The corresponding probability for neutron emission is of course determined mainly by the state density in the appropriate daughter nucleus as in evaporation from light nuclei. The study of fission–neutron competition thus concentrates on the difference between the Q-value for neutron emission and the fission barrier height, and on the influence of deformation on the state densities, which in turn may be considered to depend on the effective single particle level density (section 2.5). This latter influence is often expressed in terms of the level density parameter $a = g_0\pi^2/6$ (section 2.5) so that one speaks of the ratio a_f/a_n. Deformation dependent shell and pairing effects on the single particle spacing may also be taken into account although, as mentioned above, one would expect their influence to decline with increasing excitation energy.

A most important angular momentum dependent effect arises due to the centrifugal force which tends to favour fission. At high angular momentum, the fission saddle point is found at smaller deformation. At the same time, the

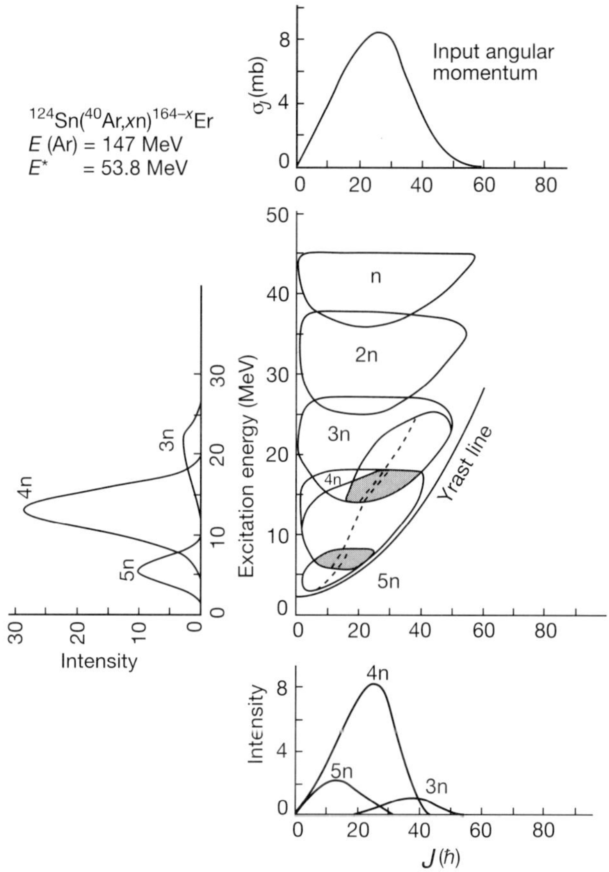

Figure 4.15. Predicted populations [18] in the E^*–J plane following the fusion reacton ^{40}Ar(147 MeV) + ^{124}Sn. The intial angular momentum distribution has the typical triangular form (see figure 4.5). The centre panel represents the E^*–J plane for different nuclei in the decay chain (the Yrast line is not unique as shown in the figure). The shaded area in the populations for $3n$ and $4n$ emission corresponds to the region where gamma ray emission competes with neutron emission. To the right of the broken curve there is a large probability that the system cools by emitting a series of gamma rays (Yrast cascade). Projections of E^*–J populations leading to residues corresponding to $3n$, $4n$ and $5n$ reactions are shown on the left and bottom panels. (Reprinted from *Nucl. Phys.* A, **325**, Hillis D L *et al*, p 216. Copyright 1979, with permission from Elsevier Science.)

deformation of the lowest energy state at the same angular momentum increases because, with increasing deformation, the higher associated moment of inertia lowers the collective rotational energy. At some angular momentum the lowest energy state is identical with the saddle configuration so that the fission barrier

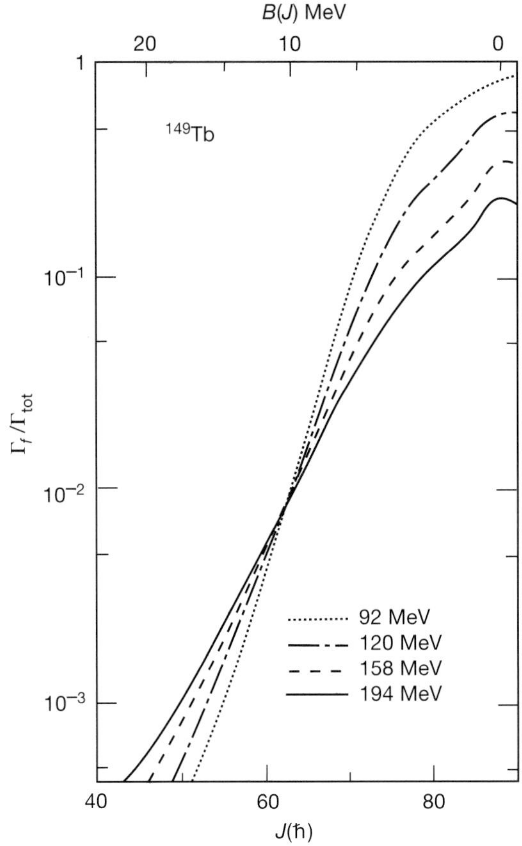

Figure 4.16. Ratio of fission and total decay widths as a function of angular momentum in the nucleus ^{149}Tb. The top horizontal scale indicates the height of the fission barrier [22]. As can be seen, the barriers are much more strongly affected by angular momentum than by excitation energy. (Reprinted from *Phys. Rev.* C, **17**, Beckerman M and Blann M, p 1615. Copyright 1978, by the American Physical Society.)

vanishes. In this situation it is not only neccessary to follow the evolution of the barrier height with angular momentum but also to consider the dependence of the level density on the angular momentum of the fissioning system. The lowering of the fission barrier with increasing angular momentum is illustrated in figure 4.16 which is taken from the work of Beckerman and Blann [22] (see also Plasil and Blann [23]). The fission barrier in ^{149}Tb decreases from 28.5 MeV at $0\hbar$ and vanishes at about $90\hbar$. The corresponding fission width (expressed as the ratio to the total width) increases almost exponentially. Because the average angular momentum of the compound system increases with the beam energy, one would expect a strong energy dependence of the fission probability. As shown in

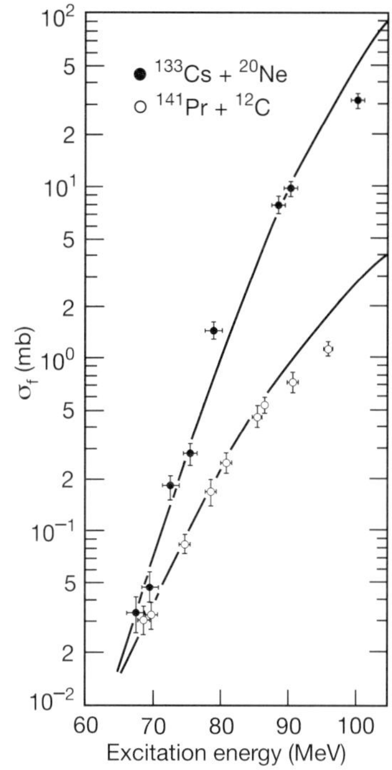

Figure 4.17. Cross-sections for fission as a function of the excitation energy in ^{153}Tb [23]. The difference in the ^{12}C + ^{141}Pr and ^{20}Ne + ^{133}Cs entrance channels is due to the difference in the angular momentum distribution produced in the compound nucleus. The solid curves represent statistical model predictions. (Reprinted from *Phys. Rev. Lett.*, **45**, Plasil F *et al*, p 333. Copyright 1980, by the American Physical Society.)

figure 4.17, this is indeed what is observed [24].

The preceding examples are quite impressive. Indeed, it is fair to say that the multistep Hauser–Feshbach theory is one of the most successful statistical theories of nuclear decay. One may further remark that the theory could be rigorously extended with little effort to include cases where evaporated particles are, in fact, excited nuclei. To the author's knowledge, this has not yet been attempted probably due to anticipated limitations in computing time.

4.6 Recoil correlations in sequential emission

The problem of momentum correlations in sequential emission can be formulated without reference to any detailed theory. The only feature which is essential concerns the mechanism itself. Thus, one may hope to emphasize 'generic' differences between sequential and non-sequential decay processes and thereby to use measurements of momentum (or velocity) correlations to identify the decay mechanism. Having identified the mechanism, detailed predictions of correlations can, of course, be used to test a particular statistical theory. In this section we will investigate 'recoil' correlations which characterize multistep evaporation [25]. Correlations which are characteristic of other decay mechanisms will be discussed in the following chapter.

We consider the sequential emission of a number of particles from a nucleus whose mass and charge steadily diminish up to the formation of a final evaporation residue. The multistep evaporation process is such that the emitted particles themselves are not subject to further decay. The correlations to be discussed arise because of the recoil momentum imparted to each 'residual' nucleus along the evaporation chain by each evaporated particle. They will be formulated in terms of the evaporation energy (referred to as ε_β in section 4.3) which, at any evaporation step, is the sum of the energies of the emitted particle and the recoiling residue measured in a coordinate frame fixed in the emitter. For the first evaporation step this frame is identical with the centre-of-mass frame which characterizes the projectile–target interaction.

Let us denote the mass and the centre-of-mass momentum of the residue following the emission of the kth particle respectively as M_k and $\boldsymbol{P}_k$ ($\boldsymbol{P}_0$, which is the momentum of the compound nucleus in the centre-of-mass system, is, by definition, zero). After evaporation of the first particle, with mass m_1, the residue mass is M_1. This nucleus is not stationary in the centre-of-mass system because the emission of the first particle imparts a recoil 'kick' to the residue. Thus we can write

$$\boldsymbol{P}_1 = -\boldsymbol{p}_1, \tag{4.6.1}$$

where $\boldsymbol{p}_1$ is the particle momentum measured in the frame of the emitter (in this case, the compound nucleus). Now consider the emission of the second evaporated particle, with mass m_2, and let us denote its momentum in a coordinate system fixed in M_1 as $\boldsymbol{p}_2$. This particle gives a recoil of $-\boldsymbol{p}_2$ to the new residue (mass M_2) whose centre-of-mass momentum thus becomes

$$\boldsymbol{P}_2 = \frac{M_2}{M_1}\boldsymbol{P}_1 - \boldsymbol{p}_2. \tag{4.6.2}$$

We now take the square on both sides of (4.6.2) and average over the angle θ, between $\frac{M_2}{M_1}\boldsymbol{P}_1$ and $-\boldsymbol{p}_2$. Only the dot product contains the angle θ so that we obtain

$$\begin{aligned}\langle P_2^2\rangle_\theta &= \frac{M_2^2}{M_1^2}p_1^2 + p_2^2 - 2\frac{M_2}{M_1}\langle \boldsymbol{P}_1 \cdot \boldsymbol{p}_2\rangle_\theta \\ &= \frac{M_2^2}{M_1^2}P_1^2 + p_2^2 - 2\frac{M_2}{M_1}P_1 p_2\langle\cos\theta\rangle. \end{aligned}\tag{4.6.3}$$

The important point is that the direction of the recoil vector $-\boldsymbol{p}_2$ is, on average, uncorrelated with that of the recoiling residue so that, in (4.6.3), the average value

$\langle \cos\theta \rangle = 0$. It follows that, by taking averages over angles, the average energy of the residue which is proportional to the square of its momentum, is simply the sum of the two contributions. Thus

$$\frac{\langle P_2^2 \rangle_\theta}{2M_2} = \frac{M_2}{M_1}\frac{P_1^2}{2M_1} + \frac{p_2^2}{2M_2}. \tag{4.6.4}$$

Let us refer to the energy corresponding to the emission of the kth particle in the coordinate system fixed in the emitter as ε_k. As explained in section 4.3 (see equation (4.3.27)), the energy ε_k is shared between the emitted particle whose energy we may denote as e_k and the residue (energy E_k) in inverse proportion to the masses m_k and M_k. For the second emitted particle we can write

$$e_2 = \frac{p_2^2}{2m_2} = \varepsilon_2 \frac{M_2}{M_1}, \tag{4.6.5}$$

where $M_1 = m_2 + M_2$ is the residue mass before emission of the second particle. We can use (4.6.5) to write (4.6.4) in terms of energy. Thus the kinetic energy (averaged over angles) is

$$E_2 = E_1 \frac{M_2}{M_1} + e_2 \frac{m_2}{M_2} = \varepsilon_1 \frac{M_2}{M_1}\frac{m_1}{M_0} + \varepsilon_2 \frac{m_2}{M_1}. \tag{4.6.6}$$

Equation (4.6.6) is in a form which is suitable for iteration. Thus, the centre-of-mass residue energy after three particle emissions is (using (4.6.6) for E_2)

$$\begin{aligned} E_3 &= E_2 \frac{M_3}{M_2} + \varepsilon_3 \frac{m_3}{M_2} \\ &= \varepsilon_1 \frac{m_1 M_3}{M_0 M_1} + \varepsilon_2 \frac{m_2 M_3}{M_1 M_2} + \varepsilon_3 \frac{m_3 M_3}{M_2 M_3}, \end{aligned} \tag{4.6.7}$$

and, after emission of n particles, we therefore find

$$E_n = M_n \sum_{k=1}^{n} \varepsilon_k \frac{m_k}{M_{k-1} M_k}. \tag{4.6.8}$$

The energy represented in (4.6.8) was obtained by averaging over all emission angles at each evaporation step. The only way that this can be accomplished in experiments is by averaging over events. Thus, for comparison with experiment, we can think of (4.6.8), and indeed the preceding equations, as referring to averages over a large number of events. However, in averaging over events, we also average over the probability distributions of the energies of the emitted particles (the ε_k spectra). One consequence is that ε_k should be replaced by the average energy of the kth emitted particle. To emphasize this point, we should write (4.6.8) as an average over events which we simply refer to by using the brackets $\langle\rangle$.

$$\langle E_n \rangle = M_n \sum_{k=1}^{n} \langle \varepsilon_k \rangle \frac{m_k}{M_{k-1} M_k}. \tag{4.6.9}$$

We can express the same result in terms of momentum

$$\langle P_n^2 \rangle = 2M_n^2 \sum_{k=1}^{n} \langle \varepsilon_k \rangle \frac{m_k}{M_{k-1} M_k}. \tag{4.6.10}$$

Since the average energies $\langle \varepsilon_k \rangle$ in (4.6.10) are not directly measurable, we will seek to replace them with the energies (or squares of momenta) of the light emitted particles. The kth light particle is emitted from the $(k-1)$th residue. The average energy in the centre of mass is thus

$$\langle e_k \rangle = \langle E_{k-1} \rangle \frac{m_k}{M_{k-1}} + \langle \varepsilon_k \rangle \frac{M_k}{M_{k-1}}. \tag{4.6.11}$$

Multiplying by $2m_k$ and rearranging terms, (4.6.11) yields

$$2m_k \langle \varepsilon_k \rangle = \langle p_k^2 \rangle \frac{M_{k-1}}{M_k} - \langle P_{k-1}^2 \rangle \frac{m_k^2}{M_k M_{k-1}}. \tag{4.6.12}$$

The quantity $2m_k \langle \varepsilon_k \rangle$ given in equation (4.6.12) can thus be eliminated from (4.6.10). We obtain

$$\langle P_n^2 \rangle = M_n^2 \sum_{k=1}^{n} \left[\frac{\langle p_k^2 \rangle}{M_k^2} - \frac{m_k^2 \langle P_{k-1}^2 \rangle}{M_{k-1}^2 M_k^2} \right]. \tag{4.6.13}$$

Equation (4.6.13) represents a generic characteristic of sequential emission which is expressed as an iterative calculation of $\langle P_n^2 \rangle$ in terms of the mean square momenta of emitted light particles. The seed for the iteration is obtained by observing that, when $n = 1$, $\langle P_0^2 \rangle$ is the mean square momentum of the compound nucleus in the centre of mass which is identically zero. We notice that in the limit where all evaporated particle masses are much smaller than the compound nucleus mass, the second term can be neglected and the first term gives

$$\langle P_n^2 \rangle \approx M_n^2 \sum_{k=1}^{n} \frac{\langle p_k^2 \rangle}{M_k^2} \approx \sum_{k=1}^{n} \langle p_k^2 \rangle, \tag{4.6.14}$$

which is just the result expected for addition of randomly oriented momenta. The main effect of the recoil is therefore to be found in the second term on the right-hand side of (4.6.13). It will be apparent that all contributions are negative. This is a consequence of the fact that the direction of the residue anywhere along the evaporation chain is, on average, anti-correlated with the direction of each of the emitted particles (see figure 4.18).

The application of equation (4.6.13) is not quite as straightforward as it may seem. The principal difficulty is that use of this equation requires an assumption concerning the *order* in which the light particles are emitted. A less important problem arises because neutrons are not usually detected so that a correction for this unseen contribution is necessary.

We thus terminate our account of evaporation. Part of the next chapter will be devoted to other types of correlation which are characteristic of sequential binary decay and of the explosive decay mechanism referred to as multifragmentation.

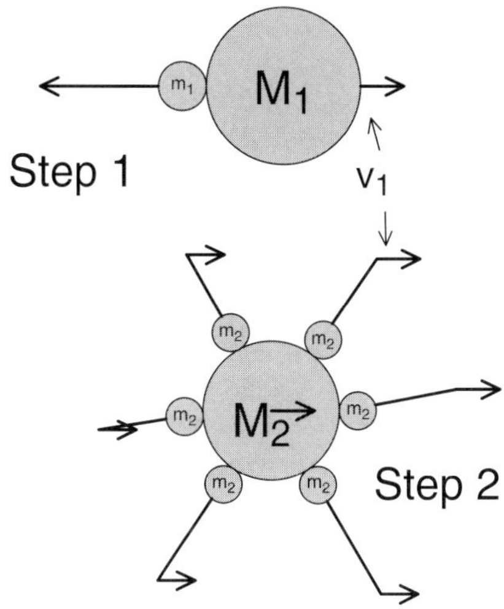

Figure 4.18. Recoil velocities in the evaporation process. The velocity v_1, which is imparted to the first residue, is, *on averages* unchanged by subsequent emissions. Thus, on the second and succeeding steps all products are characterized by an average velocity v_1 in the original compound nucleus (centre-of-mass) system.

References

[1] Hammersley J M and Handscomb D C 1964 *Monte Carlo Methods* (London: Methuen)

[2] Stokstad R G 1985 *Treatise on Heavy Ion Science* vol 3, ed D A Bromley (New York: Plenum)

[3] Oganessian Y Ts and Lazarev Y A 1985 *Treatise on Heavy Ion Science* vol 4, ed D A Bromley (New York: Plenum)

[4] Mahaux C and Weidenmüller H A 1979 *Ann. Rev. Nucl. Part. Sci.* **29** 1

[5] Press W H, Teukolsky S A, Vetterling W T and Flannery B P 1986, 1992 *Numerical Recipes* (New York: Cambridge University Press) Ch. 7

[6] Cole A J, Longequeue N, Menet J, Lucas J J, Ost R and Viano J B 1980 *Nucl. Phys.* A **341** 284

[7] Matuse T, Beck C, Nouicer R and Mahboub D 1997 *Phys. Rev.* C **55** 1380

[8] Gomez del Campo G, Stokstad R, Biggerstaff J A, Dayras R A, Snell H and Stelson P 1979 *Phys. Rev.* C **19** 2170

[9] Hill D L and Wheeler J A 1953 *Phys. Rev.* **89** 1102

[10] Thomas T D 1968 *Ann Rev. Nuc. Sci.* **18** 343

[11] Ericson T and Strutinsky V 1958 *Nucl. Phys.* **8** 284; 1958–59 *Nucl. Phys.* **9** 689

[12] Dayras R A, Stokstad R G, Hensley D C, Halbert M L, Sarantites D G, Westerberg L and Barker J H 1980 *Phys. Rev.* C **22** 1485

[13] Weidinger A, Busch F, Gaul G, Trautmann W and Zipper W 1976 *Nucl. Phys.* A **263** 511

[14] Pühlhofer F 1977 *Nucl. Phys.* A **280** 267

[15] Kuang-Hsi T, Dossing T, Gaarde C and Larsen J S 1979 *Nucl. Phys.* A **316** 189
[16] Ost R, Cole A J, Clover M R, Fulton B R and Sikora B 1980 *Nucl. Phys.* A **342** 185
[17] Grover J R and Gilat J 1967 *Phys. Rev.* **157** 802
[18] Hillis D L, Garret J D, Christensen O, Fernandez B, Hagemann G B, Herskind B, Back B B and Folkman F 1979 *Nucl. Phys.* A **325** 216
[19] Newton J O, Lee I Y, Simon R S, Aleonard M M, El Masri Y, Stephens F S and Diamond R M 1977 *Phys. Rev. Lett.* **38** 810
[20] Deleplanque M A, Lee I Y, Stephens F S, Diamond R M and Aleonard M M 1978 *Phys. Rev. Lett.* **40** 629
[21] Garret J D, Hagemann G B and Herskind B 1986 *Ann. Rev. Nucl. Sci.* **36** 419
[22] Beckerman M and Blann M 1978 *Phys. Rev.* C **17** 1615
[23] Plasil F and Blann M 1975 *Phys. Rev.* C **11** 508
[24] Plasil F, Ferguson R L, Hahn R L, Obenshain F E, Pleasonton F and Young G R 1980 *Phys. Rev. Lett.* **45** 333
[25] Cole A J *et al* 1993 *Phys. Rev.* C **47** 1251

PART 3

SEQUENTIAL BINARY DECAY AND MULTIFRAGMENTATION

Chapter 5

Multidetectors, sequential binary decay and the characterization of multifragment decay processes

"Of shoes and ships and sealing wax,
And cabbages and kings,
And why the sea is boiling hot,"

The Walrus and the Carpenter, Lewis Carroll

5.1 Introduction

We arrive now at a new turning point in our account of statistical decay. We shall henceforth be dealing with decay mechanisms which are more complex than the particle evaporation mechanism considered in the preceding chapter. The evolution of the decay mechanism is associated with increasing compound nucleus excitation energy and angular momentum. These 'modern' studies have been spurred on by the increasing availability of higher and higher energy heavy ion beams. Going beyond typical beam energies for single stage cyclotrons and linear accelerators (10–20 MeV/nucleon) has been possible using two coupled cyclotrons or single superconducting devices (typically 15 times lighter than equivalent conventional magnets). The development of high performance ion sources, notably the electron cyclotron resonance (ECR) source by Geller and collaborators [1], has also been a major contributing factor. The principal advantage of this device is that the energy input to the electron-ion plasma is supplied in the form of microwaves. The Lamour precession frequency of the electrons is in resonance with the microwaves at certain (magnetic) regions inside the source. The source gas is thus ionized by energetic electrons leading to formation of a magnetically confined plasma. Fragile and rapidly burned cathodes and heating filaments characteristic of older designs are thus absent resulting in increased longevity and robustness. The ECR ion source facilitates the attainment of high energies because it produces highly charged ions which may be extracted and injected into an accelerator. Current beam energies used in studies of the decay of 'hot' nuclei range up to hundreds of MeV per nucleon. However, the range of energy with which we will be concerned is limited to below 500 MeV/nucleon and, for most purposes, below 100 MeV/nucleon.

Before discussing the evolution of the decay mechanism, a word concerning terminology is appropriate. We use the term 'fragment' to refer to any reaction product (excepting gamma rays). Examination of the literature reveals that fragments of mass approximately in the range 6–30u are often referred to as intermediate mass fragments (IMFs) whereas the term 'complex fragments' can be used to refer to all fragments beyond alpha particles. The word 'particles' is conventionally reserved for fragments with mass ≤ 4. The appearance of events including more than one complex fragment has led to the use of the term 'multifragmentation' to refer to the corresponding decay process. However, with increasing excitation energies, a new kind of process thought to be due to the simultaneous disintegration of the parent nucleus has been observed and referred to as 'prompt multifragmentation'. To avoid possible confusion we will use this term or the shortened form 'multifragmentation' only to refer to this latter kind of event. multifragmentation will be treated in detail in the following two chapters.

With increasing excitation energy and angular momentum the compound nucleus decays by emitting not only longer and longer chains of light particles but also by the 'evaporation' of heavier and heavier fragments on a progressively shorter timescale. This competition occurs despite the fact that the emission of heavier fragments is associated with more negative Q-values. Part of the explanation has to do with temperature. It will be recalled from the Weisskopf theory (equation (3.3.12)) that the transition rate for emission of a given fragment includes a factor $e^{Q_{C\to bB}/T}$ so that the dependence on the Q-value is exponential. However, at high excitation energies (high temperatures) differences in Q-values for emitted particles or fragments have less influence on the relative emission probabilities.

The decreasing influence of ground state Q-value differences is not the only reason why the emission of complex fragments may be favoured. We saw, in chapter 1, that the density of states corresponding to internal excitation in a binary division maximizes when the excitation energies are such that the two fragments have equal temperatures. For light emitted particles this maximization is inoperable because the emitted particles may be considered to possess only ground states (the few known excited states are at rather high energies so that corresponding Q-values lead to strongly diminished emission probabilities). However, for heavier fragments excited states are available. The resulting increase in the number of microstates corresponding to internal excitation is thus another factor which offsets the Q-value difference and, thereby, increases the competitivity of large fragment production.

We need also to consider the influence of moments of inertia associated with binary divisions (the term includes all processes from nucleon evaporation to symmetric fission) starting from compound systems with non-negligible angular momentum. The moment of inertia about the centre of mass for two point particles (fragments) is μR^2 where μ is the reduced mass and R, the separation. We saw, in our discussion of fusion (section 3.6), that the energy associated with relative angular momentum of two fragments is $\ell(\ell+1)\hbar^2/2\mu R^2$. The smaller this energy, the greater will be the energy available for internal excitation of the fragments. Furthermore, the density of microstates associated with the emission of any fragment is largely dominated by the internal degrees of freedom. It is easily seen that the reduced mass maximizes for division of the parent nucleus into two equal mass fragments. It follows that large compound nucleus angular momentum favours more symmetric splits and, in this sense, offsets Q-value differences.

While on the subject of angular momentum we may mention a difficulty for would-be modellers of binary division. In many studies, it is assumed that two fragments separate along a fission coordinate which passes through their mass centres (i.e. back to back). Thus, the relative motion of a complex fragment with respect to the residual nucleus is considered to be one-dimensional as in the Bohr–Wheeler theory of fission. This choice obviously influences the density of microstates corresponding to the relative kinetic energy. The system angular momentum manifests itself as a rigid rotation of the saddle or scission configuration (see figure 5.1) as if the two fragments were stuck together at the touching point and the corresponding rotational energy is included as a contribution to the fission barrier. This problem is a little delicate. The Bohr–Wheeler theory contrasts the emission of large fission fragments and neutrons at low excitation energies (implicitly low angular momentum) so the assumption of (respectively) one-dimensional and isotropic three-dimensional relative motion seems justified. However, if the emitted fragment is a light complex fragment, such as a lithium or carbon nucleus, the choice may not be obvious. Artificial restrictions on the number of degrees of freedom associated with the relative motion which limit the relative orbital angular momenta may be unwise.

Emission of excited complex fragments implies, of course, that such fragments may further decay. Given that these fragments are 'cooler', light particle emission is strongly favoured. However, it is difficult to establish any hard and fast rule. Consequently, one must consider that, with increasing excitation energy, the sequential binary division mechanism (figure 5.2(b)), which may be thought of as a generalization of sequential evaporation (figure 5.2(a)), gradually replaces this latter mechanism. Given that fragments may be formed at excitation energies below their particle emission thresholds and thus gamma decay to their ground states, or that evaporation following one or more binary divisions leads to such fragments, a typical measured event may include a number of complex fragments as well as light particles.

The investigation of these high multiplicity events has led to the construction of 4π multidetectors which are capable in principle of simultaneously detecting all charged fragments in a collision. These devices have provoked significant

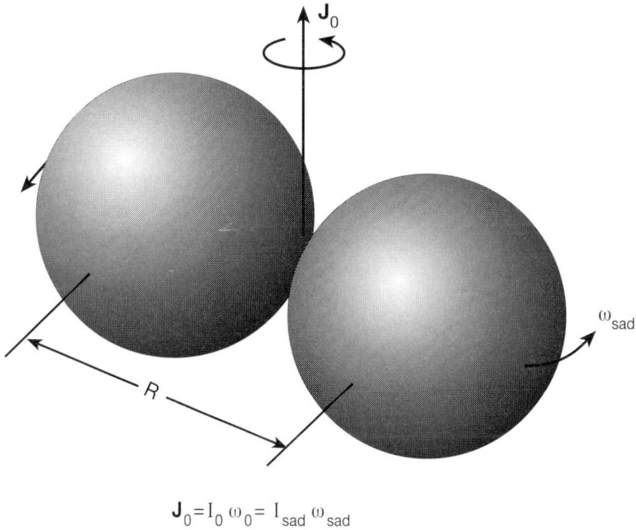

Figure 5.1. Binary division of an excited nucleus into two fragments. The scission configuration is pictured as two touching spheres rigidly rotating, with an angular velocity equal to that of the compound nucleus multiplied by the ratio of the moments of inertia of the compound nucleus and the scission configuration.

changes in the methodology of experiments. We thus begin this chapter with a brief 'detour' (section 5.2) in order to describe multidetectors and their operation. Evolution of techniques of data reduction and analysis has also become necessary, simply because of the huge amount of complicated (and correlated) information which is obtained in a single experimental 'run'. The reader is referred to articles by Désesquelles and collaborators [2] for a description of new methods pertaining to comparison of data with simulation models.

In subsequent sections (5.3–5.5), we will present three theoretical approaches to the problem of the emission of complex fragments. The thermodynamical approach of Friedman and Lynch [3] provides the basis for a study of the influence of cooling and expansion of the evaporation residue on emission probabilities. A second model (GEMINI [4]) is exposed in some detail in section 5.4. If emitted complex fragments are themselves excited, the Weisskopf and Hauser–Feshbach models (at least in their original forms) cannot be used to calculate the decay widths. GEMINI is based on a generalization of the transition state concept (used in the Bohr–Wheeler model) proposed by Moretto [5]. It may be referred to as a 'static' model in the sense that the time variable does not appear explicitly. In contrast, the model proposed in reference [6] by Barbagallo, Richert and Wagner (section 5.5) imposes a dynamical expansion on the system with binary division probabilities obtained from a natural extension of the Weisskopf transition rate. Despite some superficial resemblance, the model assumptions

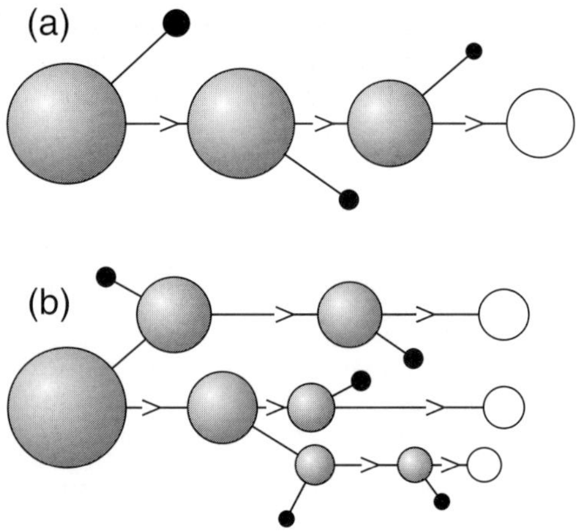

Figure 5.2. Sequential evaporation (*a*) and sequential binary decay (*b*). In (*a*) the light particles (black spheres) are emitted sequentially to form a single complex fragment residue (open circle). In (*b*) the initial compound system may divide into two excited complex fragments. Either or both fragments may be considered as new compound nuclei and thus may emit light particles or again experience binary dvision into two complex fragments. Thus, the final state may be composed of light particles and *several* 'cold' complex fragments, none of which can be formally identified as the 'residue'.

differ significantly from those made by Friedman and Lynch.

One of the major difficulties at high excitation energies occurs with attempts to make a clear cut distinction between sequential binary decay and multifragmentation. Indeed, in the limit of very short times for the production of several fragments by sequential binary decay, one can imagine that this distinction becomes increasingly problematic (and perhaps redundant). The last part of the present chapter (section 5.6) will be concerned with this distinction and, more specifically, with characteristics of fragment correlations. The methods described aim at presenting images of event sets which are of help in identifying the decay mechanism. Of course, this latter discussion has little to do with statistical mechanics as such. However, information of this kind is useful in deciding what kind of restrictions may be imposed in the construction of appropriate statistical models.

5.2 Multidetectors

In this section we briefly describe the typical configuration and operation of charged particle multidetectors. Technical details are not included but can be found in the cited references. Several such detectors were built from about 1985 onwards [7–14].

Most multidetectors are composed of a large number ($\sim$200) of fixed detection devices covering a significant proportion of the full (4π) solid angle. They thus contrast sharply with earlier detection systems which consisted of a few detection devices (e.g. Si telescopes) attached to rotatable arms. Such devices subtended typical solid angles of one hundredth of a steradian.

The main considerations in the design of multidetectors are concerned with reconciling the need to detect all or nearly all fragments (often with a wide range of velocities) produced in a single event while avoiding multiple hits (two or more fragments impinging simultaneously on the same detector). A typical configuration, the AMPHORA multidetector [8], is shown in figure 5.3(*a*). AMPHORA is composed of 140 CsI(Tl) scintillators which cover 82% of 4π. The dectectors are arranged to form a forward wall (48 units) and a backward ball. The missing solid angle mostly corresponds to beam entrance and exit ports and to the solid angle obscured by the target support assembly.

A light particle impinging on any one CsI crystal in the array gives rise to a light output with characteristic rapid and slow components. Online analysis of the corresponding electric pulse produced by a photomultiplier provides a pair of digital signals which may be used to identify the mass and charge of the incident particle. Heavier fragments may be identified in charge by measuring the energy loss (light output) in thin plastic scintillators (typically 100 μm) placed on the front faces of the CsI crystals. Low energy heavy fragments are stopped in such scintillators and thus not identified so that for some experiments it is necessary to install thin entrance window ionization chambers in front of the CsI crystals. The same basic principles are used in the MSU miniball (188 detectors, figure 5.3(*b*)) [9] and the Oak Ridge Dwarf ball + wall (105 detectors) [10].

The INDRA detector at GANIL in France [11] is currently one of the most advanced devices in operation (figure 5.4). It is composed of 336 independent CsI crystals. Very low energy thresholds, especially important at backward angles, are achieved by the use of thin entrance window ionization chambers. At forward angles identification is facilitated by including high resolution thin Si solid state detectors in front of the CsI crystals.

The ALADIN spectrometer [12], installed at the SIS facility in GSI, Darmstadt, is designed to detect reaction products emitted at small angles with respect to the beam direction (approximately $+/-4.5^\circ$ in the horizontal and vertical directions). The full detection system associated with the spectrometer includes three separate sections (figure 5.5): the target chamber which contains a Si–CsI detector array (used for detecting light particles at angles between 7° and 40°), the bending magnet, and the forward angle detection system. The latter system includes a multiple sampling ionization chamber (MUSIC), which can be used as a tracking device as well as for charge identification of heavy fragments. The time of flight (TOF) wall, which is formed of 80 vertical scintillator strips, is used for charge identification of light fragments (typically $Z \leq 8$) and for measurement of event multiplicities and fragment velocities. The time origin (the start signal) for the time of flight measurement is taken from thin scintillator detectors placed in

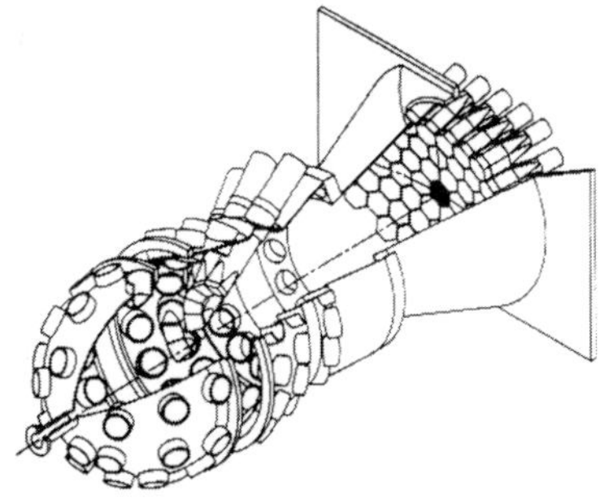

(a)

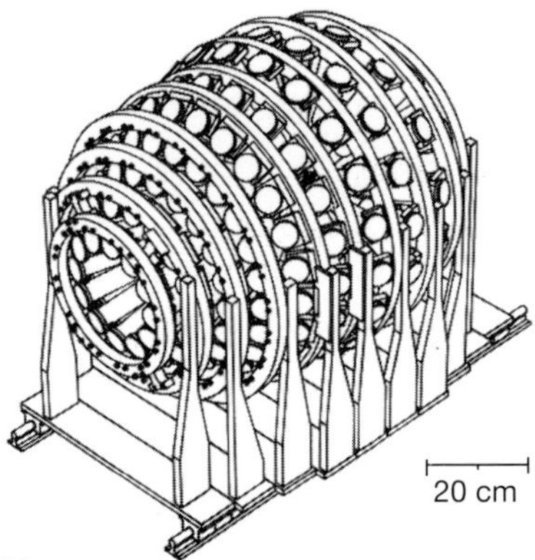

(b)

Figure 5.3. (*a*) The Grenoble multidetector AMPHORA [8]. The target is situated in the centre of the ball on the left. Small angles are covered by the forward wall. The wall detectors and the first two rings of the ball detectors are equipped with plastic scintillator foils for charge identification of complex fragments. (Reprinted from *Nucl. Instrum. Meth. Phys. Res.* A, **281**, Drain D *et al*, p 52. Copyright 1989, with permission from Elsevier Science.) (*b*) The Michigan State University miniball multidetector [9] which covers almost 90% of the full 4π solid angle. The 188 Cs(I) detectors are arranged in 11 coaxial rings. All detectors are equipped with 40 μm plastic scintillators. (Reprinted from *Nucl. Instrum. Meth. Phys. Res.* A, **295**, De Souza R T *et al*, p 365. Copyright 1990, with permission from Elsevier Science.)

the beam upstream from the target.

The FOPI [13] detector, also at GSI, deserves special mention because it includes magnetic analysis in time projection chambers as well as energy loss measurements. Because the detected fragments have to pass through the walls of the tracking chambers, the FOPI configuration is better adapted for higher energy beams typical of the SIS facility at GSI.

Finally (the list is not exhaustive) we would like to mention the Indiana

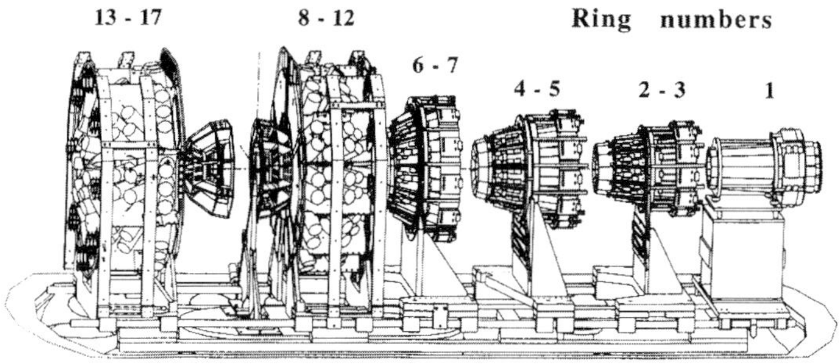

Figure 5.4. The INDRA multidetector [11]. Sets of detector rings are mounted on distinct segments which may be separated for ease of access as shown in the figure. The 336 detectors cover 90% of 4π. Low energy thresholds for complex fragment detection are obtained using thin walled ionization chambers operated with low pressure C_3F_8 gas. (Reprinted from *Nucl. Instrum. Meth. Phys. Res.* A, **357**, Pouthas J *et al*, p 418. Copyright 1995, with permission from Elsevier Science.)

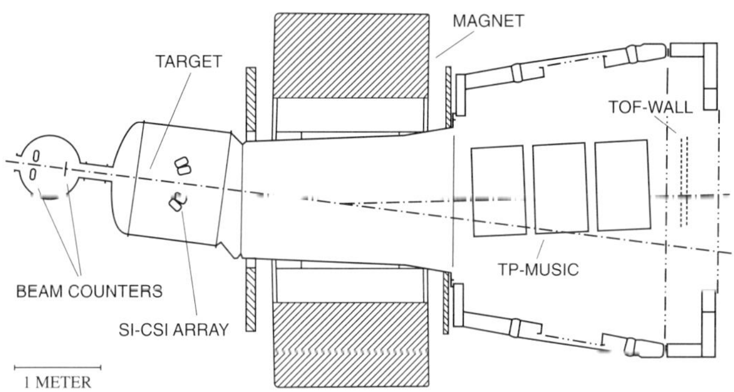

Figure 5.5. The ALADIN forward angle detection set-up operated at the SIS facility in GSI, Darmstadt [12]. The beam is incident on the target shown at the left-hand side of the figure. The time projection (TP) chamber is used in reconstruction of the trajectories of complex reaction products. The multiple sampling ionization chamber (MUSIC) provides charge identification. The time of flight (TOF) wall of plastic scintillator detectors is shown on the right.

silicon sphere array (ISIS, [14]), which uses 162 triple element detectors disposed in a spherical configuration. Each detector is composed of a gas (C_3F_8) ionization chamber, a silicon wafer and a CsI 'stopping' detector. This device is specifically adapted to measurements of the decay products of excited nuclei formed in light particle (e.g. anti-proton) induced reactions for which the centre-of-mass velocity of the recoiling compound nucleus is small.

The quantity of raw information acquired in a single experimental run is enormous (of the order of 10 gigabytes for 10^8 events in a typical run with AMPHORA) and requires several months of arduous data reduction in order to produce information which can be compared directly to theoretical models. Thus, for example, it is important to calibrate all detectors using beams of heavy ions specifically produced for this purpose. It is also necessary to include corrections for differences in detector efficiencies so that absolute angular distributions and energy spectra may be obtained. The various charges (or masses) of detected fragments may be identified using two-dimensional spectra in which specific fragments appear as part of (more or less) separated clouds (figure 5.6). Each detected fragment corresponds to a single point in such a spectrum. Thus, the fragment identification procedure involves the construction of 'masks' which divide the observed spectrum into specific charge (and/or mass) regions.

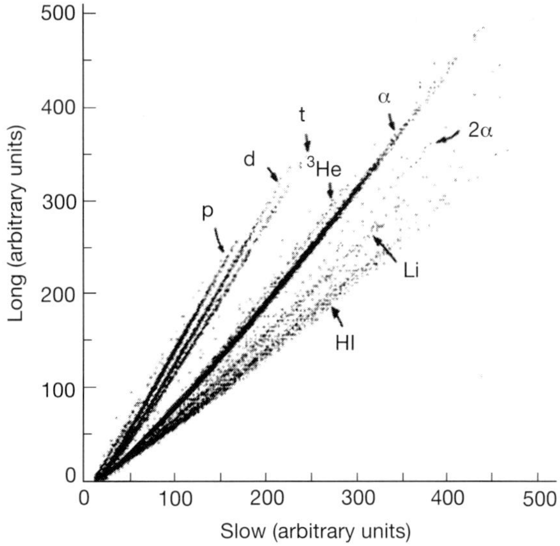

Figure 5.6. Identification plot obtained in one of the Oak Ridge Dwarf ball detectors [10]. The plot concerns light particle identification which is achieved by plotting the so called 'slow' (1 μs) and long (7 μs) light outputs from the CsI scintillator on x and y axes, respectively. (Reprinted from *Nucl. Instrum. Meth. Phys. Res.* A, **294**, Stracener D *et al*, p 485. Copyright 1990, with permission from Elsevier Science.)

Once the raw data has been reduced and converted to a physically intelligible form, various gating procedures (see section 5.6) are used to select events corresponding to different ranges of projectile–target impact parameter and, if possible, to identify the corresponding decay mechanisms.

Some understanding of the response of a multidetector composed of $N_{\rm d}$ detectors to an event with charged fragment multiplicity K can be achieved by considering the detectors to be identical and to be 100% efficient. If we assume the fragments to be randomly distributed over the detectors, the response can be formulated using combinatorial methods. Let each detector subtend a solid angle $\Delta\Omega$ and therefore be represented by a geometrical efficiency of $\delta = \Delta\Omega/4\pi$. The probability that a single fragment will be detected is then given by the global efficiency of the device, i.e., $N_d\delta$, and the probability that k out of K fragments will 'hit' the detector system is the binomial probability,

$$\begin{aligned} P(k, K; N_d) &= \text{Prob [that the detectors are hit by } k \text{ out of } K \text{ fragments]} \\ &= K!\frac{(N_d\delta)^k}{k!}\frac{(1 - N_d\delta)^{K-k}}{(K-k)!}. \end{aligned} \tag{5.2.1}$$

We now assume that if any single detector is hit by more than one fragment it will fire once and reject the other incident radiations due to dead time in the electronic circuits. It is then legitimate to seek the probability of p (apparent) hits which is usually written $P^K_{N_d,p}$. A derivation of this distribution has been given by Van der Werf [15]. If $p = K$ the event was perfectly detected. If $p = K - 1$, then either one fragment missed the detection system and the others were perfectly detected, or all fragments hit the system and there was one double hit. We write

$$P^K_{N_d,p} = \sum_{k=p}^{K} P(k, K; N_d)Q(p \mid k), \tag{5.2.2}$$

where $Q(p \mid k)$ is the conditional probability that if k fragments hit the detection system then p detectors will fire (i.e., the event with true multiplicity K will appear as an event with multiplicity p even though $k \geq p$ hits actually occurred). To obtain Q we exploit a standard result from combinatorial analysis. Stirling's number of the second kind, $S^{(p)}_k$ is defined as the number of ways of partitioning k elements into p non-empty subsets [16]. A non-empty subset in this context is a detector which has been hit by one or more fragments. If we choose the p detectors at random from a total of N_d then there are N_d ways of choosing the first detector, $N_d - 1$ choices for the second... $N_d - (p - 1)$ for the pth so that the number of ways of choosing p detectors from N_d is simply $N_d!/(N_d - p)!$. It follows that the number of ways that p detectors, chosen randomly from N_d, will fire when we know that the detection system as a whole was hit by k fragments is $N_d!/(N_d - p)! \times S^{(p)}_k$. This is a number, not a probability. However, the total number of ways for k fragments to hit N_d detectors is simply $N_d{}^k$ (N_d possible choices for each of the k fragments) so that we can write the probability $Q(p \mid k)$ as

$$Q(p \mid k) = \frac{1}{N_d{}^k}\frac{N_d!}{(N_d - p)!}S^{(p)}_k, \tag{5.2.3}$$

and thus finally construct $P^K_{N_d,p}$ using equation (5.2.2). Given that $S^{(p)}_k$ can be written explicitly as

$$S^{(p)}_k = \frac{1}{p!}\sum_{j=0}^{p}(-1)^{p-j}\frac{p!}{j!(p-j)!}j^k, \tag{5.2.4}$$

we can use equations (5.2.1), (5.2.3) and (5.2.4) to write

$$p^K_{N_\mathrm{d},p} = \sum_{k=0}^{K} K! \frac{(N_\mathrm{d}\delta)^k}{k!} \frac{(1-N_\mathrm{d}\delta)^{K-k}}{(K-k)!} \frac{1}{N_\mathrm{d}{}^k} \frac{N_\mathrm{d}!}{(N_\mathrm{d}-p)!} \times \frac{1}{p!} \sum_{j=0}^{p} (-1)^{p-j} \frac{p!}{j!(p-j)!} j^k, \tag{5.2.5}$$

where we have been able to extend the sum over k down to 0 (cf (5.2.2)) because $S_k^{(p)}$ is 0 for $k < p$ (one or more of the p subsets is empty). This looks complicated but in fact may be simplified considerably by changing the order of summation. Cancelling the factor $N_\mathrm{d}{}^k$ and grouping the resulting factor δ^k and the factor j^k we rewrite (5.2.5) as

$$P^K_{N_\mathrm{d},p} = \sum_{j=0}^{p} \frac{(-1)^{p-j}}{j!(p-j)!} \frac{N_\mathrm{d}!}{(N_\mathrm{d}-p)!} \left[K! \sum_{k=0}^{K} \frac{(j\delta)^k}{k!} \frac{(1-N_\mathrm{d}\delta)^{K-k}}{(K-k)!} \right]. \tag{5.2.6}$$

Happily, for all concerned the last term in square brackets is the binominal expansion of $[1-(N_\mathrm{d}-j)\delta]^K$ so that (5.2.6) can finally be expressed as

$$P^K_{N_\mathrm{d},p} = \frac{N_\mathrm{d}!}{(N_\mathrm{d}-p)!} \sum_{j=0}^{p} \frac{(-1)^{p-j}}{j!(p-j)!} [1-(N_\mathrm{d}-j)\delta]^K. \tag{5.2.7}$$

The probability distribution $P^K_{N_\mathrm{d},p}$ can be used to determine such quantities as the average number of detectors which fire for an event of multiplicity K or simply the probability that K detectors fire (all fragments detected, no double hits) which is $P^K_{N_\mathrm{d},K}$.

Equation (5.2.7), while useful for design purposes and for planning experiments, results from a somewhat idealized view of the detection system and is therefore not very useful for data analysis. In practice, the detection efficiency for a given event depends not only on the fragment masses, charges and energies, but also on correlations in the set of fragment directions and inhomogeneities in the multidetector (beam pipe, target support etc). Thus, for example, in comparing model predictions with observations it is best to abandon estimates such as (5.2.7) in favour of detailed computer simulations of the detector response. Such programs are often referred to as 'filters' and used in association with Monte Carlo versions of theoretical model predictions (event generators). The filtered output informs the experimentalist as to how the model prediction would appear, viewed through the multidetector. The influence of detector filters have been discussed by Peter *et al* [17] and by Brandan *et al* [18]. These studies showed that imperfections in the detection systems may strongly distort certain global variables or spectra (section 5.6) constructed from the detected event set. The degree of distortion depends on the specific reaction (projectile–target mass asymmetry, beam energy). The work of the pre-cited authors thus emphasizes that the (computer) construction of an accurate filter is a crucial stage in the comparison of data with theoretical predictions.

In writing this short section I am strongly aware that enormous amounts of work necessary for the construction and operation of complex devices such as multidetectors often go without proper acknowledgement. This neglect occurs on the one hand because the nature of the task cannot be simply described, and on the other because such a description would serve little purpose. In this domain, students learn, not from books, but directly from colleagues and supervisors. I trust that I will be forgiven for injecting a personal note by paying tribute to the long hours spent by researchers and students alike in setting up, running and analysing experiments. Without this effort there would be little incentive to pursue the study of nuclear decay to Bohr's 'explosive' limit.

5.3 The sequential decay model of Friedman and Lynch

Turning away from experimental procedures we come back, in this section, to decay mechanisms and, in particular, to consideration of the sequential binary decay mechanism. A model for this process, first discussed in a *Physical Review* paper in 1983 [3], pictures the decay as a sequence of emissions from the compound nucleus including the emission of complex fragments. Because the model is constructed using rate equations, it also enables the cooling of the system to be studied as a function of time.

The Friedman–Lynch model includes several approximations. To begin with, complex fragments are emitted only in ground states and thus do not decay further. Secondly, the emission probabilities are obtained as a modification of the Weisskopf formula so that the model does not include a treatment of angular momentum. This latter omission may be rather serious because, as we have seen in the previous chapter, angular momentum strongly influences the decay probabilities of heavy fragments. Finally, the model is based on thermodynamics although no clear identification of a heat bath is made.

The presentation of the model in [3] is somewhat involved. We therefore propose a simplified account which emphasizes the essential structure. Our presentation will not distinguish between neutrons and protons although this distinction was made in the original work.

The model is based on the Weisskopf transition rate. From equation (3.3.10) the compound nucleus decay probability per unit time and per unit emitted particle kinetic energy (ε_b) for emission of a fragment, b, from the compound system, C, to leave a residue, B, may be written (using $\hbar$ rather than h) as

$$\frac{\mathrm{d}^2 \prod_{C\to b+B}}{\mathrm{d}\varepsilon_b\,\mathrm{d}t} = \frac{g_b m_b \sigma_{bB} \rho_B(E_B^*)\varepsilon_b}{\pi^2\hbar^3\rho_C(E_C^*)}. \tag{5.3.1}$$

The notation implies that if $P_C(t)$ is the probability to find the nucleus C at time t then the rate of increase of this quantity (which is obviously negative) is $\mathrm{d}P_C/\mathrm{d}t = -P_C(t)\sum_b \mathrm{d}\prod_{C\to b+B}/\mathrm{d}t$. Friedman and Lynch make a simple improvement. The inverse cross-section, σ_{bB}, for formation of the compound nucleus in the process $b + B \to C$ is taken as

$$\sigma_{bB} = \pi R_{bB}^2\left(1 - \frac{B_C}{\varepsilon_b}\right)\theta(\varepsilon_b - B_C), \tag{5.3.2}$$

where B_C is the Coulomb barrier (0 for neutrons) associated with the pair b, B and the factor $\theta(\varepsilon_b - B_C)$ is 0 for $\varepsilon_b \le B_C$ and 1 otherwise so that fusion (or emission) below the Coulomb barrier is suppressed.

A further development of the Weisskopf formula is associated with the ratio $\rho_B(E_B^*)/\rho_C(E_C^*)$. It will be recalled that Weisskopf himself wrote this factor as $\mathrm{e}^{S_B(E_{bB}-\varepsilon_b)-S_C(E_{bB}-Q_{C\to b+B})}$ where the entropies S_B and S_C refer respectively to the residual nucleus and the compound system, and E_{bB} is the total energy in the system $b + B$, i.e., $E_{bB} = E_C^* + Q_{C\to b+B} = \varepsilon_b + E_B^*$. Friedman and Lynch begin with a similar expression. Before proceeding with their model however, it is instructive to examine the consequences of following Weisskopf by expanding the entropies using the heat bath concept. Thus, by assuming $\mathrm{d}S_B(E)/\mathrm{d}E = \mathrm{d}S_C(E)/\mathrm{d}E = 1/T$ at $E = E_{bB}$, the ratio of densities in equation (5.3.1) is written (compare with (3.3.11), (3.3.12))

$$\frac{\rho_B(E_B^*)}{\rho_C(E_C^*)} = \mathrm{e}^{-\varepsilon_b/T}\,\mathrm{e}^{Q_{C\to b+B}/T}\,\mathrm{e}^{S_B(E_{bB})-S_C(E_{bB})}. \tag{5.3.3}$$

Going a little beyond Weisskopf, we use the equidistant spacing model (section 2.5) for the entropies so that $S(E) = 2\sqrt{aE}$ (with $a = A/\varepsilon_0$, i.e., proportional to the mass number) and write the last factor in (5.3.3) as

$$\mathrm{e}^{S_B(E_{bB})-S_C(E_{bB})} = \mathrm{e}^{2\sqrt{a_B E_{bB}}-2\sqrt{a_C E_{bB}}}, \tag{5.3.4}$$

which, given that $a_B = a_C - a_b$, can be developed (by expanding the first square root) to yield

$$\begin{aligned}\mathrm{e}^{S_B(E_{bB})-S_C(E_{bB})} &= \mathrm{e}^{2\sqrt{E_{bB}}(\sqrt{a_C-a_b}-\sqrt{a_C})}\\ &\approx \mathrm{e}^{2\sqrt{E_{bB}/a_C}\left[a_C\left(1-\frac{a_b}{2a_C}-1\right)\right]} = \mathrm{e}^{-a_b T}\end{aligned} \tag{5.3.5}$$

where $T = \sqrt{E_{bB}/a_C}$ is the compound nucleus temperature not at the full excitation energy E_C^* but at the lower excitation energy $E_C^* + Q_{C\to b+B}$ (as in (5.3.3)). We thus expect the last term to reduce the emission rate for large fragments.

Friedman and Lynch do not use equation (5.3.5). Rather, they write the left-hand side of equation (5.3.3) as $\rho_B(E_B^*)/\rho_C(E_C^*) = \mathrm{e}^{\Delta S}$ where $\Delta S = S_B(E_{bB} - \varepsilon_b) - S_C(E_{bB} - Q_{C\to b+B})$ and note that the emission of a fragment with the same density as the parent nucleus is accompanied by a change of entropy

$$\Delta S = \frac{\partial S}{\partial E^*}\Delta E^* + \sigma\Delta A, \tag{5.3.6}$$

where σ is the entropy per nucleon of the parent system, E^* is the excitation energy and ΔA is the change in mass number. The calculation of ΔS is oriented towards a thermodynamical description of the system. Let us begin by denoting the excitation energies per nucleon (specific excitation energies) in B and C respectively as ε_B^* and ε_C^*. Now $E_B^* = E_C^* - \varepsilon_b + Q_{C\to b+B}$ which, in terms of specific excitation energies, is equivalent to $A_B\varepsilon_B^* = A_C\varepsilon_C^* - \varepsilon_b + Q_{C\to b+B}$. We can use this relation to write

$$\begin{aligned}\varepsilon_C^* - \varepsilon_B^* &= \frac{A_B\varepsilon_B^*}{A_C} + \frac{\varepsilon_b - Q_{C\to b+B}}{A_C} - \frac{A_C\varepsilon_B^*}{A_C} \\ &= \frac{\varepsilon_B^*(A_B - A_C)}{A_C} + \frac{\varepsilon_b - Q_{C\to b+B}}{A_C} \\ &= -\frac{A_b\varepsilon_B^*}{A_C} + \frac{\varepsilon_b - Q_{C\to b+B}}{A_C}.\end{aligned} \tag{5.3.7}$$

Friedman and Lynch now introduce an approximation by replacing the term $A_b\varepsilon_B^*$ on the right-hand side of (5.3.8) by $A_b\varepsilon^*$ where ε^* is taken to be the common excitation energy per nucleon in both the compound nucleus and the residue. The emitted fragment is thus characterized by this same excitation energy per nucleon. Then, changing the sign in (5.3.7)

$$\varepsilon_B^* - \varepsilon_C^* = \frac{A_b\varepsilon^*}{A_C} - \frac{\varepsilon_b - Q_{C\to b+B}}{A_C}. \tag{5.3.8}$$

The left-hand side of this equation is taken to be the increase in specific internal energy (which is, in fact, negative) when b is emitted from the parent nucleus, i.e., $\Delta\varepsilon^* = \varepsilon_B^* - \varepsilon_C^*$ so that the full change of entropy is written

$$\Delta S = \frac{\partial S}{\partial E^*}A_C(\varepsilon_B^* - \varepsilon_C^*) - A_b\sigma. \tag{5.3.9}$$

Noting that $\partial S/\partial E^*$ is simply the inverse of the temperature, T, we find from (5.3.8) and (5.3.9) that

$$\begin{aligned}\Delta S = \frac{A_C\Delta\varepsilon^*}{T} - A_b\sigma &= \frac{-\varepsilon_b + Q_{C\to b+B}}{T} + \frac{A_b(\varepsilon^* - T\sigma)}{T} \\ &= \frac{-\varepsilon_b + Q_{C\to b+B}}{T} + \frac{A_b f^*}{T},\end{aligned} \tag{5.3.10}$$

where, f^* (obviously defined as $f^* = \varepsilon^* - T\sigma$) is the free energy per nucleon relative to the ground state energy per nucleon, which latter quantity we can take as the negative of the binding energy per nucleon, $E_{\mathrm{g.s.}}/A = -\mathcal{B}/A(\approx -8\ \mathrm{MeV})$. Thus, $f^* = (F - E_{\mathrm{g.s.}})/A$ where F is the Helmholtz free energy.

Digressing briefly once more, we note that, for 'normal' density nuclei, a consequence of the approximate level density formula, $\rho(E^*) = \mathrm{e}^{2\sqrt{aE*}}(a = A/\varepsilon_0)$ is that the temperature variation of the *specific* internal energy of a nucleus of mass number, A, is given approximately as $U/A = -\mathcal{B}/A + T^2/\varepsilon_0$, where both $\mathcal{B}/A$, the binding energy/nucleon, and ε_0 are approximately equal to 8 MeV. As shown in the next chapter (section 6.6), one consequence is that the Helmholtz free energy per nucleon, F/A, is equal to $-(\mathcal{B}/A + T^2/\varepsilon_0)$ which implies that $f^* = -T^2/\varepsilon_0$. If this result, which we emphasize is not part of the Friedman–Lynch analysis, is used in equation (5.3.10), we recover equation (5.3.5).

Returning to the Friedman–Lynch model, we use (5.3.10) to write

$$\frac{\rho_B(E_B^*)}{\rho_C(E_C^*)} = \mathrm{e}^{-\varepsilon_b/T}\,\mathrm{e}^{Q_{C\to b+B}/T}\,\mathrm{e}^{A_b f^*(T,\rho)/T}, \tag{5.3.11}$$

in which we have indicated explicitly that f^* depends both on temperature and density.

The rate of cooling of the system is established by consideration of the conservation of the total energy. As above, we will not distinguish between protons and neutrons. Maintaining the symbols b and B to represent the emitted particle and the residue, and writing the average excitation energy of the *current* residue as $E_B^*(t)$ (at $t = 0$ before any emission takes place $E_B^* = E_C^*$), energy conservation is expressed by

$$\frac{\mathrm{d}E_B^*}{\mathrm{d}t} - \sum_b \frac{\mathrm{d}\prod_{B\to b+B'}}{\mathrm{d}t} Q_{B\to b+B'} + \sum_b \frac{\mathrm{d}E_{\mathrm{kin}}(b)}{\mathrm{d}t} = 0, \tag{5.3.12}$$

where $Q_{B\to b+B'}$ is the ground state Q-value for the emission of a fragment b from the current residue B to form a new residue B', and $\mathrm{d}E_{\mathrm{kin}}(b)/\,\mathrm{d}t$ is the average rate of production of kinetic energy which is obtained from the Weisskopf decay rate (equation (5.3.1)) as

$$\frac{\mathrm{d}E_{\mathrm{kin}}(b)}{\mathrm{d}t} = \int_{B_C}^{\varepsilon_b^{\max}} \frac{\mathrm{d}^2\prod_{B\to b+B'}}{\mathrm{d}\varepsilon_b\,\mathrm{d}t}\varepsilon_b\,\mathrm{d}\varepsilon_b, \tag{5.3.13}$$

where $\varepsilon_b^{\max}$ is the maximum possible value of ε_b for emission of b from the current residue. The rate of change of the temperature is obtained from $\mathrm{d}E_B^*/\,\mathrm{d}t$ by dividing by the specific heat $\mathrm{d}E_B^*/\,\mathrm{d}T$. For a nucleus whose temperature is approximately $\sqrt{E_B^*/a_B}$ this quantity is simply $2a_BT$. It is of course possible to calculate the average rate of change of temperature by averaging over emitted fragments (over b).

Equations (5.3.11) and (5.3.12) can be considered as the main results of the analysis. They can be exploited only if we have a precise model for f^*. There are several ways of obtaining this quantity. One method (which is closely connected to that expressed in our 'digressions') involves a generalization of the liquid drop nuclear binding energy to include temperature dependence (section 6.4). Another possibility involves the Fermi gas model which allows for both temperature and density dependence of the reduced free energy f^*.

The original Friedman–Lynch model did not include a specific dynamical mechanism for density variations. However, Friedman [19] introduced an important development by allowing for expansion (and subsequent possible contraction) of the parent system. We shall therefore briefly discuss the Fermi gas in relation to his 'expanding source' calculations. We should perhaps note that Friedman does not provide full details concerning his calculations so that the following account is not necessarily an exact representation of his model.

The mathematical description of a gas of non-interacting fermions is straightforward but somewhat lengthy. Exact calculations often require a numerical treatment. An excellent description is given in chapter 14 of the book by Greiner, Neise and Stöcker [20]. The usual form of the model (which should not be confused with the equidistant spacing model used for the construction of nuclear densities) considers M non-interacting fermions, each of mass m_N, enclosed in a volume V. The density of single particle states is

$$g(\varepsilon) = \frac{2\pi V}{h^3}(2m_N)^{3/2}\varepsilon^{1/2}. \tag{5.3.14}$$

In the limit where the temperature is small compared with the Fermi energy, an ideal gas of fermions is said to be degenerate (all single particle states up to the Fermi energy are occupied and higher states are unoccupied). Thus, at zero temperature, equation (5.3.14) may be integrated up to ε_F to yield the number of particles $A = \frac{4\pi V}{3h^3}(2m_N\varepsilon_F)^{3/2}$ (see equation (2.2.30)) so that, with $\rho = A/V$,

$$\varepsilon_F = \frac{\hbar^2(6\pi^2\rho)^{2/3}}{2m_N}. \tag{5.3.15}$$

It follows directly from (5.3.14) that, at $T = 0$ the Helmholtz free energy per particle, i.e. F/A, is simply $3\varepsilon_F/5$ (the Gibbs–Helmholtz equation shows that the zero temperature free energy is equal to the internal energy which is the average energy of the single particle states). However, for finite temperatures (see [20])

$$\frac{F(T)}{A} \approx \frac{3\varepsilon_F}{5}\left[1 - \frac{5\pi^2}{12}\left(\frac{T}{\varepsilon_F}\right)^2\right] = \frac{F(0)}{A} - cT^2, \tag{5.3.16}$$

so that

$$f^* = \frac{F(T) - F(0)}{A} = -cT^2, \tag{5.3.17}$$

where $c = \frac{\pi^2}{4\varepsilon_F}$. The density dependence of f^* is thus given by (5.3.15). We may compare the constant c in (5.3.17) with that implied by the equidistant spacing model (ESM; see before (5.3.11)) which yielded $f^* = -T^2/\varepsilon_0$. Indeed, if we recall (equation (2.5.43)) that the ESM level density parameter is $a = A/\varepsilon_0 = \pi^2 g_0/6$ and that, if spin and isospin degeneracy are not explicitly taken into account (only one nucleon per single particle level), the level density at the Fermi energy, g_0, is equal to $3A/2\varepsilon_F$, we find that $\varepsilon_0 = 4\varepsilon_F/\pi^2 = 1/c$. We may conclude that the Fermi gas model used in conjuction with Friedman–Lynch equation,

(5.3.10), is equivalent to the development of the square root in the equidistant spacing model (equation (5.3.5)).

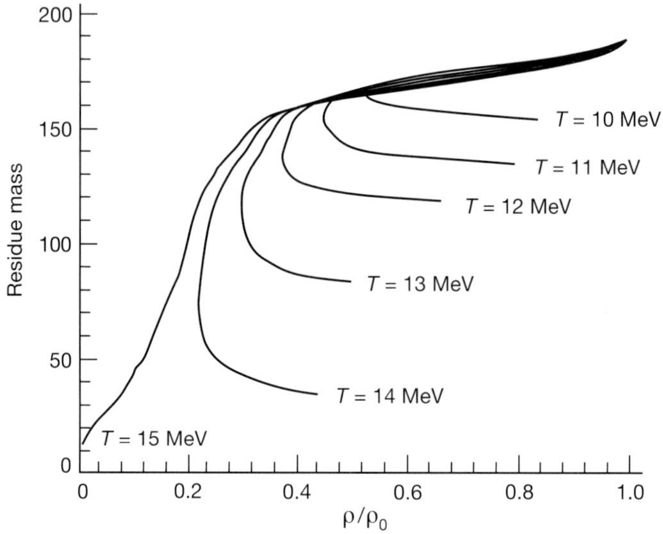

Figure 5.7. Friedman expanding source model: average value of the residue mass as a function of the density for a ^{197}Au nucleus intially at normal density, ρ_0, for temperatures varying from 10 to 15 MeV in 1 MeV steps. Except for the $T = 15$ MeV case (upper trace) where there is no residue, the initial expansion, as well as the mass loss, increases with increasing temperature and is followed by a contraction leading to a cold residue. (Reprinted from *Phys. Rev.* C, **42**, Friedman W, p 667. Copyright 1990, by the American Physical Society.)

Friedman's 'expanding source' model [19] actually follows the density of the source (residue) as a function of time. This is achieved by a rather simplified dynamical picture in which an initial collective expansion together with a restoring force is introduced into the calculation. The energy balance of the original Friedman–Lynch model (equation (5.3.12)) which determines the instantaneous temperature is thus modified by the inclusion of two additional terms. The restoring force is provided by considering the ground state energy (the negative of the binding energy) of the expanding nucleus to vary from −8 MeV/nucleon at normal density, $\rho = \rho_0$, to 0 at $\rho = 0$,

$$\frac{E_{\text{g.s.}}}{A} = 8\left[\left(1 - \frac{\rho}{\rho_0}\right)^2 - 1\right]. \tag{5.3.18}$$

Equation (5.3.18) may be thought of as a simple equation of state. Collective expansion energy is lost both due to the action of the restoring force and to the emission of fragments. The system is initialized at some fixed mass, charge, temperature, density (ρ) and expansion parameter ($\mathrm{d}\rho/\mathrm{d}t$), and the evolution of the density together with Weisskopf decay rates (which are influenced both by density and temperature) is followed at finite time steps thereby obtaining the corresponding evolution of the temperature and the mean residue mass.

An important result obtained from studies of the ^{197}Au nucleus is shown in figure 5.7. For temperatures below about 12 MeV, the system evaporates light fragments but the initial expansion is followed by a contraction so that a cold residue is formed. For higher temperatures a density of about 0.3 times normal density is reached at a temperature of about 5 MeV, whereupon the system mass drops abruptly because of the sudden emission of complex fragments. If this is actually a true picture of the decay mechanism, it seems that the process has a lot in common with the multifragmentation mechanism to be discussed in the following chapters.

5.4 The transition state approach to sequential binary decay

As mentioned in the introduction, the sequential binary decay process is normally thought of as being characterized by the emission of a complex fragment which carries both internal excitation energy and angular momentum (although this is not the case in the Friedman–Lynch model). A single step in such a process can be visualized as a straightforward extension of the evaporation mechanism which was treated successfully by the Hauser–Feshbach model. However, the complete construction of a theory which deals with angular momentum and energy dependent state densities for both the residue and the emitted fragment (rather than the single residue density which characterizes the 'standard' theory) has not been attempted. We describe, in this section, a model for emission of excited complex fragments which is based on earlier theories for fission.

The Bohr–Wheeler theory of fission considers the motion of a system along a single collective fission coordinate. It requires the existence of a maximum in the potential energy (as a function of the fission coordinate) which corresponds to a minimum in the number of microstates associated with internal excitation of the fission fragments. Statistical equililibrium is assumed to occur at the potential maximum (saddle point). If the motion along the fission coordinate from saddle to scission is adiabatic with respect to internal (single particle) degrees of freedom, the decrease in potential energy from saddle to scission appears as relative kinetic energy of the fission fragments (no energy is transferred to single particle excitations). Non-adiabatic motion, on the other hand, produces viscous 'heating' (excitation of single particle degrees of freedom) so that the relative kinetic energy of the two fragments is small. In this case, whereas the decision to fission is taken at the saddle point, the nature of the fragments may be strongly affected by the dynamics of the descent from saddle to scission. An extreme example of this point of view results in the assumption of complete statistical equilibrium at scission as in the Fong theory.

One can easily imagine that the potential energy surface for more general deformations of the original compound nucleus is rather complex. As a function of a more comprehensive set of collective coordinates (rather than just one fission coordinate), there may be several 'valleys' (including saddle points) leading to symmetric or asymmetric fission. At low excitation energies and for heavy nuclei, however, despite considerable investment, it has proved rather difficult to explain the observed mass asymmetry by following the dynamical path to fission.

Nix and Swiatecki [21] explored a liquid drop statistical model of fission in which the population of different points on the potential energy surface (defined as a function of 9 collective coordinates) at the fission saddle is determined by assuming statistical equilibrium. The dynamical evolution of any particular point on the potential energy surface can be followed to scission via a set of equations of motion for the collective coordinates (adiabatic with respect to the remaining single particle degrees of freedom) so that the statistical distribution of initial conditions at the saddle point determines the characteristics of the fission fragments. The assumption of statistical equilibrium at the saddle points implies that the influence of the pre-saddle dynamical evolution of the system which could, in principle, affect the probability of attaining a given scission configuration is not taken into account. The liquid drop statistical theory thus represents an extension of the Bohr–Wheeler model. Emission of complex fragments (asymmetric fission) is associated with the passage through a saddle configuration which *defines* the masses and charges of the emitted fragment and the residual nucleus. All available microstates corresponding to any saddle configuration are considered as equally probable.

The modification of the Bohr–Wheeler theory to include the mass asymmetry coordinate and, more generally, the influence of other collective coordinates has been further considered by Moretto [5] in the context of the adiabatic approximation. Any single collective degree of freedom q_C (with conjugate momentum p_C) involves an additional (dimensionless) factor $\mathrm{d}q_C\,\mathrm{d}p_C/h$ in the expression for the partial width. In addition, the sum of the potential and kinetic energy corresponding to these degrees of freedom must be subtracted from the total energy, E^*, in the calculation of the density of microstates.

The method can be illustrated using the mass asymmetry coordinate, q, and conjugate momentum, p. We assume these coordinates may be represented as continuous variables. This is certainly convenient for succeeding developments but also may be justified by the following simple argument. The mass asymmetry changes when a nucleon passes from one fragment to the other. During this passsage the nucleon may be found at some 'in between' point which would normally be the neck which joins the two fragments. Indeed the position of the nucleon relative to the centre of mass of the system (all other nucleons being considered as fixed) may be taken to characterize the mass asymmetry which, during the passage from one fragment to the other, can be assumed to lie in between the initial and final values. The velocity which characterizes the change of mass asymmetry is, in this case, proportional to the velocity of the nucleon.

The analogue of the Bohr–Wheeler expression for binary division is thus

$$\frac{\mathrm{d}^3 W_F}{\mathrm{d}\varepsilon_\mathrm{f}\,\mathrm{d}p\,\mathrm{d}q}\mathrm{d}\varepsilon_\mathrm{f}\,\mathrm{d}p\,\mathrm{d}q = \frac{1}{\hbar}\frac{\mathrm{d}^3\Gamma_F}{\mathrm{d}\varepsilon_\mathrm{f}\,\mathrm{d}p\,\mathrm{d}q}\mathrm{d}\varepsilon_\mathrm{f}\,\mathrm{d}p\,\mathrm{d}q$$
$$= \frac{\rho_F(E^* - B_\mathrm{f}(q) - \varepsilon_\mathrm{f} - \varepsilon)}{\rho(E^*)}\frac{\mathrm{d}\varepsilon_\mathrm{f}}{h}\frac{\mathrm{d}p\,\mathrm{d}q}{h}, \qquad (5.4.1)$$

in which ρ_F is the density of states at the 'conditional' saddle point and the function $B_\mathrm{f}(q)$ represents the potential energy at the saddle point for a given mass asymmetry. The set of saddle points viewed as a function of mass asymmetry is referred to the ridge line. The kinetic energy $\varepsilon = p^2/2M$ which characterizes motion along the ridge line is determined by the inertial parameter M and ε_f is, as in the Bohr–Wheeler theory, the kinetic energy along the fission coordinate. In some cases it may possible to separate B_f into a sum of two terms, the first of which corresponds to a particular fission barrier (e.g. for symmetric splits), and a second which depends only on the mass asymmetry.

We can get an idea of the application of the theory by supposing that the fission barrier is a quadratic function of the mass asymmetry (we may take $q = 0$ for symmetric fission). Thus, let us define $B_\mathrm{f}(q) = B_0 + \alpha q^2$ and, assuming equilibrium, treat ρ_F as a heat bath. In this case we can expand ρ_F in (5.4.1) and write

$$\frac{1}{\hbar}\frac{\mathrm{d}^3\Gamma_F(q)}{\mathrm{d}\varepsilon_\mathrm{f}\,\mathrm{d}p\,\mathrm{d}q}\mathrm{d}\varepsilon_\mathrm{f}\,\mathrm{d}p\,\mathrm{d}q = \frac{\rho_F(E^* - B_0)}{\rho(E^*)}\,\mathrm{e}^{-\beta(\alpha q^2+\varepsilon_\mathrm{f}+\varepsilon)}\frac{\mathrm{d}\varepsilon_\mathrm{f}}{h}\frac{\mathrm{d}p\,\mathrm{d}q}{h}. \qquad (5.4.2)$$

If we are interested in the decay width for a given mass asymmetry, we can integrate over ε_f and p (using the heat bath approximation to extend the integrals to infinity) thus obtaining

$$\frac{1}{\hbar}\frac{\mathrm{d}\Gamma_F(q)}{\mathrm{d}q}\mathrm{d}q = \frac{(2\pi M)^{1/2}}{h^2\beta^{3/2}}\frac{\rho_F(E^* - B_0)}{\rho(E^*)}\,\mathrm{e}^{-\beta\alpha q^2}\,\mathrm{d}q. \qquad (5.4.3)$$

Of course, our example is somewhat artificial. The Gaussian mass asymmetry distribution is obtained because we assume a quadratic form for the potential energy, $B_\mathrm{f}(q)$, along the ridge line. We may further note that the inertial parameter, M, has only a secondary effect on the mass distribution.

The above development may be thought of as implying that collective coordinates may be added as required. However, it should not be forgotten that in doing so, we are artificially increasing the number of degrees of freedom used to describe the system. Thus, in principle, the introduction of collective degrees of freedom should be accompanied by some modification of the form of the density ρ_F.

Moretto develops several illustrative examples involving various collective modes. In applications, however, only the simplest modes have been considered. Furthermore, approximations are used to simplify calculations. This is best illustrated by presenting results of a practical calculation, which has been applied to so many experimental measurements, that it can be considered to have attained the status of an independent model. The corresponding computer program was developed by Charity and is known as GEMINI [4]. The model makes a sharp distinction between decay widths for emission of light particles and those for emission of complex fragments. Widths for emission of light particles are calculated using the Hauser–Feshbach model with sharp cut-off transmission coefficients. The widths for complex fragment emission are said to derive from the transition state formalism.

A detailed derivation of the complex fragment widths starting from Moretto's work [5] is not given. The following analysis represents a possible reconstruction and involves an approximation which seems difficult to justify in the context of the generalized transition state formalism.

Let us return to equation (5.4.1) and define the width for a given mass asymmetry (q') by integration over a small region of q

$$\frac{1}{\hbar}\int_{q'}^{q'+\Delta q} \frac{\mathrm{d}^3\Gamma_F}{\mathrm{d}\varepsilon_\mathrm{f}\,\mathrm{d}p\,\mathrm{d}q}\,\mathrm{d}\varepsilon_\mathrm{f}\,\mathrm{d}p\,\mathrm{d}q = \frac{\rho_F(E^* - B_\mathrm{f}(q') - \varepsilon_\mathrm{f} - \varepsilon)}{\rho(E^*)}\frac{\mathrm{d}\varepsilon_\mathrm{f}}{h}\frac{\mathrm{d}p}{h}\Delta q. \tag{5.4.4}$$

Now expanding $\rho_F(E^* - B_\mathrm{f}(q') - \varepsilon_\mathrm{f} - \varepsilon)$ as $\rho_F(E^* - B_\mathrm{f}(q') - \varepsilon_\mathrm{f})\,\mathrm{e}^{-\beta\varepsilon}$ we again integrate over p to obtain

$$\begin{aligned}\frac{\Delta q}{\hbar}\left[\int_{-\infty}^{\infty}\frac{\mathrm{d}^3\Gamma_F(q')}{\mathrm{d}\varepsilon_\mathrm{f}\,\mathrm{d}p\,\mathrm{d}q}\,\mathrm{d}p\right]\mathrm{d}\varepsilon_\mathrm{f} &= \frac{\Delta q}{h}\frac{\rho_F(E^* - B_\mathrm{f}(q') - \varepsilon_\mathrm{f})}{\rho(E^*)}\frac{\mathrm{d}\varepsilon_\mathrm{f}}{h}\int_{-\infty}^{\infty}\mathrm{e}^{-\beta\varepsilon}\,\mathrm{d}p \\ &= \frac{\Delta q(2\pi M)^{1/2}}{h\beta^{1/2}}\frac{\rho_F(E^* - B_\mathrm{f}(q') - \varepsilon_\mathrm{f})}{\rho(E^*)}\frac{\mathrm{d}\varepsilon_\mathrm{f}}{h}.\end{aligned} \tag{5.4.5}$$

We now approximate equation (5.4.5) by removing the dimensionless factor, $\Delta q(2\pi M)^{1/2}/(h\beta^{1/2})$. In the context of Moretto's generalization of the transition state theory this may be difficult to justify, at least as far as absolute widths are concerned. The advantage is that the difficulty of having to estimate the associated inertial parameters (e.g., M for mass asymmetry) is avoided. Moretto [5] suggests that the inertial parameter, M, will vary only slowly with mass/charge asymmetry so that its influence in relative widths may be small. On the other hand, it should not be forgotten that, in the GEMINI model, the width for emission of complex particles 'competes' with that for emission of light particles and that the latter are calculated independently with the Hauser–Feshbach model.

With this approximation, and assuming we choose Δq to include all the probability for a given mass asymmetry, the width for complex fragment emission is written (Charity *et al* refer to the fission barrier as E_sad)

$$\frac{\Gamma_F(Z_1, A_1, Z_2, A_2, J)}{\hbar} = \frac{1}{h\rho(E^*)}\int_0^{E^*-E_\mathrm{sad}(J)}\rho_F(E^* - E_\mathrm{sad}(J) - \varepsilon_\mathrm{f})\,\mathrm{d}\varepsilon_\mathrm{f}, \tag{5.4.6}$$

which is, of course, simply the Bohr–Wheeler model in which the fission barrier has been replaced by the conditional barrier $E_{\rm sad}(J)$. One can remark, however, that the saddle point energy has now been made to depend explicitly on the compound nucleus angular momentum, J, as well as on the mass asymmetry. On the other hand, the only variable kinetic energy included is that corresponding to the translational (fission) degree of freedom. Thus the rotational energy is considered to be fixed for a given value of J and a given mass asymmetry as discussed in the introduction (section 5.1). Furthermore, as indicated above, the kinetic energy associated with the mass asymmetry coordinate is suppressed. It is possible to make one further development of equation (5.4.6). If we again apply the heat bath approximation by expanding the density $\rho_F(E^* - E_{\rm sad}(J) - \varepsilon_{\rm f})$ and integrating, the decay rate becomes

$$\frac{\Gamma_F(Z_1, A_1, Z_2, A_2, J)}{\hbar} = \frac{T}{h\rho(E^*)}\rho_F(E^* - E_{\rm sad}(J)) = \frac{T\rho_F(E')}{h\rho(E^*)}, \tag{5.4.7}$$

in which the excitation energy $E' = E^* - E_{\rm sad}(J)$ and the conditional saddle point temperature T is obtained from the first order Fermi gas expression for a nucleus of mass number A, i.e., $T = \sqrt{E'/a}$ with $a = A/8.5$.

The level densities, as a function of excitation energy and angular momentum in (5.4.7) are calculated using a form similar to (but not identical with) the result developed in chapter 2 i.e.,

$$\rho(E, J) = (2J + 1)\left[\frac{\hbar^2}{2I}\right]^{3/2} \frac{\sqrt{a}}{12} \frac{e^{2\sqrt{aE}}}{E^2}. \tag{5.4.8}$$

This formula is used both for the original compound nucleus (with excitation energy E^*) and for the conditional saddle point configurations (excitation energy E').

The calculation of the saddle point energies $E_{\rm sad}(J)$ is based on a rotating two touching spheroid configuration. The mass asymmetries are, of course, imposed as constraints or 'conditions'. Davies and Swierk [22] provided a general method for minimizing the potential energy at such *conditional* saddle points and showed that the two touching spheroid configuration is a reasonable approximation for the saddle point shapes (except for fission of heavy nuclei). In GEMINI these configurations are approximated as two sphere configurations, the surfaces being separated by 2 fm. In fact, the resulting saddle point energies are rescaled so that, for symmetric division, agreement is obtained with more sophisticated saddle point shapes calculated using an improved form of the liquid drop model [23] (the main improvement concerns the inclusion of a diffuse surface). The Q-values involved in the transition from the original compond system to the 'saddle' configurations are then calculated using the liquid drop model.

The scission configuration is thus represented by a rigid rotating 'dumb-bell'. The relative orbital angular momentum between the two fragments is fixed by the saddle configuration. It is anti-aligned with, and always less than, the parent angular momentum. The angular velocity, ω_{sad}, corresponding to rotation at scission is written using the ratio of moments of inertia for the initial and saddle point shapes as

$$\omega_{\text{sad}} = \frac{I_0}{I_{\text{sad}}}\omega_0, \tag{5.4.9}$$

where $\omega_0 = J/I_0$ is the angular velocity of the parent nucleus and the rotational energy of fragment 1, with moment of inertia I_1 is simply $\frac{1}{2}I_1\omega_{\text{sad}}^2$ with a similar expression for the complement. Furthermore, the rotational contribution to the relative kinetic energy of the two fragments is $\frac{1}{2}\mu R^2\omega_{\text{sad}}^2$ where R is the separation of the mass centres of the two fragments at the saddle and μ is the reduced mass. The average of the total kinetic energy of the fragments is obtained, in principle, by adding (to the rotational energy) the average energy corresponding to the probablity distribution of the kinetic energy, ε_{f}, associated with the fission coordinate,

$$\langle\varepsilon_{\text{f}}\rangle = \int_0^\infty \varepsilon_{\text{f}}\,\mathrm{e}^{-\varepsilon_{\text{f}}/T}\,\mathrm{d}\varepsilon_{\text{f}} \bigg/ \int_0^\infty \mathrm{e}^{-\varepsilon_{\text{f}}/T}\,\mathrm{d}\varepsilon_{\text{f}} = T, \tag{5.4.10}$$

and the Coulomb energy associated with the charged spheres. However, in [4], $\langle\varepsilon_{\text{f}}\rangle$ is replaced by the average energy corresponding to the evaporation formula $(2T)$. The final internal excitation energy which is to be shared between the two fragments is thus the initial excitation energy, E^*, to which we add the (negative) liquid drop Q-value corresponding to the binary division and subtract the total average kinetic energy,

$$E_{\text{I}} = E^* + Q - B_{\text{C}} - \frac{(I_1 + I_2 + \mu R^2)}{2}\omega_{\text{sad}}^2 - 2T. \tag{5.4.11}$$

In equation (5.4.11), B_{C} is the Coulomb energy associated with the saddle configuration. The term $(Q - B_{\text{C}})$ plays the role of the fission barrier for zero angular momentum. It varies strongly with the mass asymmetry and, to a lesser extent, with the charge division characterizing the two fragments

We learned in chapter 1 that the most probable distribution of the internal excitation energy occurs when the two fragments have the same temperature. In the simplified form of the Fermi gas (ESM) model for which $T_i = \sqrt{E_i/a_i}$ we find that the most probable distribution of E_{I} is such that the excitation in each fragment is proportional to its mass number (as in the first order Fong theory). By considering only this energy division, the excitation energies of the two fragments are fixed at the saddle point.

At each binary division, the excitation energy, angular momentum and kinetic energy of each of the product fragments are thus calculated. The decay process can therefore be iterated until no further emission is possible (gamma ray competition seems to have been omitted from GEMINI). The momenta of the fragments in the laboratory frame of reference can be calculated using Monte

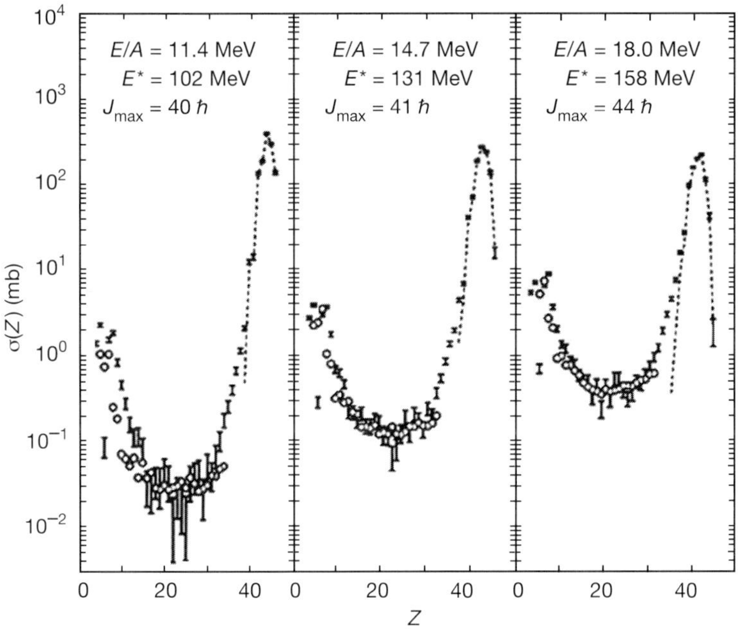

Figure 5.8. Predictions of the GEMINI code (shown with error bars determined by statistical accuracy) compared with experimental charge yields (open circles) obtained using a ^{93}Nb beam on a ^{7}Be target [4]. The yield of charges below approximately 40, which depends very strongly on the initial angular momentum distribution of the compound system, is due to the emission of complex fragments and is favoured by large angular momenta. (Reprinted from *Nucl. Phys.* A, **483**, Charity R J *et al*, p 371. Copyright 1988, with permission from Elsevier Science.)

Carlo techniques which are similar to those presented in the previous chapter for the multistep Hauser Feshbach model.

One investigation carried out by Charity and co-workers [4] involved measurements using ^{93}Nb beams at 11.4, 14.7 and 30.3 MeV/nucleon incident on targets of ^{7}Be, ^{12}C and ^{27}Al. In the experiment, it was possible to isolate a set of events corresponding to complete fusion so that the initial excitation energy could be easily estimated. The initial angular momentum distribution in the compound nucleus was assumed to vary as $2J + 1$ up to a maximum cut-off value which was considered as a parameter which could be varied to optimize the fit to the data. Despite the approximations, results of the calculations are most encouraging. Measured and predicted charge distributions for the Be target are shown in figure 5.8. Because the calculations were performed using Monte Carlo methods, the GEMINI calculations are shown as error bars which correspond to the statistical accuracy of the predictions. It is particularly interesting to note

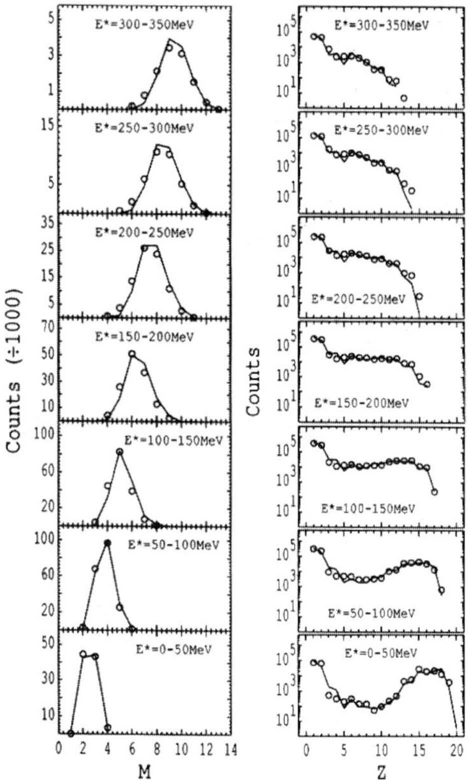

Figure 5.9. Multiplicity and charge distributions shown for several 'slices' of excitation energy of projectile-like fragments produced in the ^{40}Ca + ^{nat}Cu reaction at 35 MeV/nucleon [24]. The continuous lines are GEMINI predictions which include the effect of detector imperfections via a detector filter programme module. (Reprinted from *Phys. Rev.* C, **50**, Lleres A *et al*, p 1973. Copyright 1994, by the American Physical Society.)

that only a small span of Z-values (albeit with the largest cross-sections) can be explained by light particle evaporation acting alone.

In another study Lleres *et al* [24] examined the decay of projectile-like fragments (PLF) formed in deep inelastic collisions in the reaction ^{40}Ca + ^{nat}Cu at an incident Ca energy of 35 MeV/nucleon. The experimental situation in such studies is more complex than for studies of a single compound nucleus because it is necessary to 'reconstruct' the projectile-like parent nuclei by isolating and measuring the decay products. On the positive side, the experiment provides a set of parent nuclei with a wide range of excitation energies so that possible evolution of the decay mechanism can be studied. The mass, charge and excitation energy in any particular event is estimated by summing the kinetic energies of all light particles and heavier fragments in the PLF centre-of-mass system (obtained from the condition that the sum of PLF fragment momenta is zero), subtracting the Q-value for the process (initial nucleus $\rightarrow$ fragments) and finally correcting for unobserved neutrons. An estimation of the mean angular momentum as a function of excitation energy is also possible by examining directional correlations of light particles with heavy residues [25]. In figure 5.9 we show measured and predicted multiplicity and charge distributions for excitation energies up to about 8 MeV/nucleon. At low excitation energies the charge distributions exhibit a 'U-shape' with an important evaporation residue contribution for charges close to the parent charge, whereas at high energies this contribution moves to lower charges and is mixed with complex fragments produced by asymmetric fission. The success of the predictions over a wide range of excitation energies again provides striking evidence in support of the transition state plus Hauser–Feshbach 'GEMINI' model.

5.5 The expanding source model of Barbagallo, Richert and Wagner

The assumption used in construction of this model (for which we will use the abbreviation BRW) is that a highly excited nucleus decays by binary division or evaporation [6]. Any excited nucleus produced in such a process may further decay. Given that the decay rate may be rather rapid, one may expect to find fragments within the same spatial region a short time after the projectile–target collision. However, the model does not include the possibility of recombination of any decay products. The decay rates themselves are obtained as a natural generalization of the Weisskopf model, which contrasts strongly with that used by Friedman and Lynch in the sense that the emitted fragment and the residue are treated on as equal footing. A novel feature of the model (which also appears in Friedman's work) is that a system temperature is defined which depends not only on the populations of the various decay products but also on an externally imposed expansion of the system. The temperature decreases as the system expands. However, the word 'system' in the BRW context refers to the set of decay products each of which has a 'normal' density. The authors emphasize that the motivation for their work lies in studying the behaviour of a well-defined decay scenario. They do not advocate their approach as the unique path to understanding the decay of highly excited nuclei.

Before beginning a description of the model, we should advise the reader that we have introduced some changes in the notation of reference [6] in order to clarify the account and to achieve consistency with the description of the Weisskopf model (section 3.3). However, where possible, we have retained the BRW notation.

Let us begin with the rate equations. The time-dependent population for any nuclear species i is written as $P(i; t)$. Contrasting with the Friedman–Lynch model its evolution includes both negative and positive contributions. Starting from some initially excited parent (compound) nucleus, the populations evolve according to the rate equations

$$\frac{\mathrm{d}P(i;t)}{\mathrm{d}t} = \sum_{j>i} P(j;t)W_{j\to i} - P(i;t)\sum_{j'<i}{}' W_{i\to j'}. \tag{5.5.1}$$

The notation used in this equation implies that if $j > i$, the mass of nucleus j is greater than the mass of nucleus i. Furthermore, the prime on the second sum indicates that this sum should be carried out so as to avoid counting the same binary division twice. At the beginning of the process only the compound nucleus is present so that $P(i;0) = \delta_{i,i_C}$ where i_C denotes the compound nucleus with neutron and proton numbers N_C and Z_C, and mass A_C. At this time, only the second (loss) term of (5.5.1) is non-zero. The total numbers of neutrons and protons are of course separately conserved throughout the process.

Let us begin by considering the emission of light particles. The specific transition rate $W_{j\to i}$ is given by energy integration of the Weisskopf transition rate $\frac{\mathrm{d}^2\prod_{j\to i}}{\mathrm{d}\varepsilon_{j-i}\,\mathrm{d}t}$ which is the probability per unit time and per unit kinetic energy that the fragment $(j-i)$ is emitted from the parent j to leave the residue i. Modifications to the basic Weisskopf transition rate include the more realistic inverse (fusion) cross-section used by Friedman and Lynch. Thus σ_{bB} is given by equation (5.3.2). A further modification is introduced in the phase space factor by replacing the emitted particle mass, m_b, by the reduced mass μ_{bB}. This modification, which would appear if the phase space density of states is calculated in the centre-of-mass system, seems to be not quite complete as no corresponding modification is made to the kinetic energy, ε_b, which is the sum of the energies of the emitted fragment and residual nucleus. Be that as it may, using the notation of section 3.3, the specific rate of emission of a ground state light particle per unit energy is written

$$\frac{\mathrm{d}^2\prod_{C\to b+B}}{\mathrm{d}\varepsilon_b\,\mathrm{d}t} = \frac{g_b\mu_{bB}\sigma_{bB}\rho_b(E_B^*)\varepsilon_b}{\pi^2\hbar^3\rho_C(E_C^*)}. \tag{5.5.2}$$

The BRW model introduces an improvement on (5.5.2) by explicitly including angular momentum magnetic degeneracy factors, not only for the emitted fragment (g_b in equation (5.5.2)), but also for the parent nucleus and the residual nucleus so that, in their work, the specific rate of emission of a light particle (which corresponds, in the BRW notation, to the transformation of the parent nucleus, j, to the residue, i) integrated over the light particle energies is

$$W_{j\to i} = \frac{\mathrm{d}\prod_{j\to i}}{\mathrm{d}t} = \frac{g_{j-i}\mu_{j-i,i}\sigma_{j-i,i}}{\pi^2\hbar^3 g_j\rho_j(E_j^*)}\int_{E_C(j-i)}^{\varepsilon_{\max}} g_i\rho_i(E_i^*)\varepsilon_{j-i}\,\mathrm{d}\varepsilon_{j-i}, \tag{5.5.3}$$

where the lower limit of integration represents the Coulomb barrier for the emission of the light particle. In equation (5.5.3), the residual nucleus excitation energy is expressed in terms of ε_{j-i} as $E_i^* = E_j^* + Q_{j\to j-i,i} - \varepsilon_{j-i}$ so that this equation represents a straightforward convolution. The upper limit, $\varepsilon_{\max}$, which is the maximum kinetic energy (with respect to the residue) available for the light particle, is the sum of the parent excitation energy and the emission Q-value, i.e., $\varepsilon_{\max} = E_j^* + Q_{j\to j-i,i}$. This quantity depends explicitly on the excitation energy of the parent and thus introduces time dependence into the energy integrated transition probabilities. If the light particle has energy $\varepsilon_{\max}$, the residue, i, is produced in its ground state.

The extension to the case where the emitted nucleus, $(j-i)$, is a fragment with its own excited states rather than a (ground state) light particle should now be obvious. It is sufficient to replace the density $\rho_i(E_i^*)$ by the convolution

$$\int \rho_i(E_i^*)\rho_{j-i}(E_{j-i}^*)\,\mathrm{d}E_i^*$$

where the sum of the excitation energies $E_\mathrm{I} = E_i^* + E_{j-i}^* = E_j^* + Q_{j\to j-i,i} - \varepsilon_{j-i}$. The equation corresponding to (5.5.3) now implies a triple convolution. The rate corresponding to a specific relative kinetic energy, ε_{j-i} (the generalization of equation (5.5.2)) is

$$\frac{\mathrm{d}^2\prod_{j\to i}}{\mathrm{d}\varepsilon_{j-i}\,\mathrm{d}t} = \frac{g_i g_{j-i}}{g_j}\frac{\mu_{j-i,\mathrm{i}}\,\sigma_{j-i,\mathrm{i}}\,\varepsilon_{j-i}}{\pi^2\hbar^3\rho_j(E_j^*)}\left[\int_0^{E_\mathrm{I}} \rho_i(E_i^*)\rho_{j-i}(E_\mathrm{I}-E_i^*)\,\mathrm{d}E_i^*\right]. \tag{5.5.4}$$

In the case of light particle emission, fixing the kinetic energy relative to the residue is sufficient to fix the excitation energy of the residue. The same procedure applied to the case where both fragments may be excited fixes only the sum of the internal excitation energies so that the computational problem becomes rather complicated. Barbagallo, Richert and Wagner therefore adopt approximate methods (a second order expansion about the maximum of the integrand) to evaluate the convolution in equation (5.5.4). Particle and complex fragment multiplicities are entirely determined as a function of time by the rate equations.

Like the Weisskopf model, the BRW model does not explicitly include a treatment of angular momentum. The level density of any complex fragment is written as a function of energy using the approximate Fermi gas form

$$\rho(E_k^*) = \frac{0.2\,\mathrm{e}^{2\sqrt{a_k E_k}}}{A_k^{5/3}}. \tag{5.5.5}$$

However, an additional factor $f(J_k) = (2J_k+1)\,\mathrm{e}^{-J_k(J_k+1)/2\sigma_k^2}$ which depends on angular momentum is used to estimate the mean values of the spin degeneracy factors g_k. Thus, $g_k = 2\langle J_k\rangle + 1$ where

$$\langle J_k\rangle = \frac{\int_0^\infty J_k f(J_k)\,\mathrm{d}J_k}{\int_0^\infty f(J_k)\,\mathrm{d}J_k}. \tag{5.5.6}$$

As in equation (2.5.54), the width of the spin distribution is taken to depend on the excitation energy. Thus

$$\sigma_k^2 = 0.0888\sqrt{a_k E_k^*}\,A_k^{2/3}, \tag{5.5.7}$$

which, if we take the temperature $T = \sqrt{E_k^*/a_k}$, the level density parameter $a_k = A_k/8.0$ and the nuclear radius parameter $r_0 = 1.077$, is equivalent to $0.009569 \times T \times M_k R_k^2$ (i.e., is consistent with equation (2.5.56)). Of course the g_k factors are only approximate values because little is known about the real angular momentum distributions of the fragments.

A further interesting step taken by the authors of reference [6] concerns the cooling and expansion of the system. The notation is somewhat unconventional so that careful explanation is required. The cooling is studied by defining a global 'Fermi gas' temperature.

$$T(t) = \left[\frac{E_{\text{tot}} - \sum_i B(i)P(i;t) - E_{\text{kin}}(t) - E_{\text{coul}}(t)}{a_C}\right]^{1/2}, \tag{5.5.8}$$

in which a_C is the level density parameter ($a_C = A_C/8$) and E_{tot} is defined (unconventionally) as the *sum* of the parent nucleus excitation energy and a quantity $B(i_C)$ (B_A in [6]), which is actually the negative of the compound nucleus binding energy (recall that the binding energy is normally a positive quantity and that the total mass-energy of the nucleus (N, Z) in its ground state is $931.494A + Z \times 7.289 + N \times 8.071 - \mathcal{B}(N, Z)$, see (2.2.22) and (2.2.26)). Thus, E_{tot} is the total energy of the system referred to the free nucleon state. The second term, in which $B(i)$ is again the negative of the binding energy of fragment i, corrects E_{tot} for the presence of evaporated fragments. The two terms taken together would be expected to represent a *generalization* of the maximum summed excitation energy after emission of a single fragment, i.e., $E_j^* + Q_{j\to j-i,i}$, and would be expected to be equal to this quantity if only one fragment had been emitted (i.e., only one non-zero P_i). However, the Coulomb part of the binding energy is not included in $B(i)$ but in the term E_{coul} so that $B(i)$ should be thought of as a modified (negative) binding energy.

The kinetic energy term is obtained by defining the average *rate of creation* of kinetic energy in a single emission process at time t'. We write (in a slightly different notation from that used in reference [6]),

$$\frac{\mathrm{d}\langle\varepsilon_{j-i}(t')\rangle}{\mathrm{d}t'} = \int_{E_C(j-i)}^{\varepsilon_{\max}} \frac{\mathrm{d}^2 \prod_{j\to i}}{\mathrm{d}\varepsilon_{j-i}\,\mathrm{d}t}\varepsilon_{j-i}\,\mathrm{d}\varepsilon_{j-i}, \tag{5.5.9}$$

which is essentially the same as the corresponding quantity in the Friedman–Lynch model. The average of the total kinetic energy for any emission process at time t may now be constructed as the integral of the production rate of kinetic energy from $t' = 0$ to t, the integrand being weighted by the probability of existence of the parent nucleus j. The total kinetic energy is then given by the sum over all possible binary divisions (again avoiding double counting).

$$E_{\text{kin}}(t) = \sum_j \sum_{i<j}' \int_0^t P(j;t')\frac{\mathrm{d}\langle\varepsilon_{j-i}(t')\rangle}{\mathrm{d}t'}\,\mathrm{d}t'. \tag{5.5.10}$$

Finally the Coulomb energy is calculated by imposing a specific model for the expansion dynamics. A mean expansion velocity is defined from the kinetic energies of the fragments by writing

$$v_{\mathrm{exp}}(t) = \left[\frac{\sum_j \sum'_{i<j} P(j;t) 2E_{\mathrm{kin}}(j-i;t)/M_{j-i}}{\sum_j P(j;t)} \right]^{1/2}, \tag{5.5.11}$$

where M_{j-i} and $E_{\mathrm{kin}}(j-i;t)$ are respectively the mass and the average kinetic energy of the fragment $(j-i)$ at time t. The 'size' of the system is defined using this velocity as

$$R(t) = R(t=0) + \int_0^t v_{\mathrm{exp}}(t')\,\mathrm{d}t', \tag{5.5.12}$$

where $R(t=0)$ is the radius of the parent nucleus, and the distance separating any two fragments is approximated by

$$R_{ij}(t) = \frac{R(t)}{A^{1/3}}(A_i^{1/3} + A_j^{1/3}), \tag{5.5.13}$$

where A is the parent mass number and A_k the mass number of the fragment k. The Coulomb energy is then given by the sum of the self Coulomb energies of the fragments and the sum of the interaction energies. Thus,

$$E_{\mathrm{coul}}(t) = \sum_{j=1}^{A} P(j;t)E_{\mathrm{coul}}(j) + \frac{\mathrm{e}^2}{2}\sum_{j,i}\frac{P(i;t)Z_i P(j;t)Z_j}{R_{ij}(t)}, \tag{5.5.14}$$

so that by substituting (5.5.10) and (5.5.14) in equation (5.5.8) the evolution of the temperature is obtained as a function of time.

By solving the rate equations numerically it is possible to study the 'chemistry' of the system (i.e., the production of fragments) as it expands and cools. A particular study was made of the nucleus ^{40}Ar. At low initial excitation energies (for initial temperatures $T_0 < 3$ MeV) the process is dominated by light particle evaporation. However, at $T_0 = 6$ MeV, intermediate mass fragments are formed in a rather short time interval (of the order of 10^{-22} s). These fragments are of course excited and may decay further (by evaporation of light particles) on a much longer timescale. However, to within evaporative corrections they may be considered to survive during the expansion process. This result is most interesting because it goes some way towards a unification of the fast binary sequential decay mechanism and the multifragmentation mechanism which pictures the fragments as being formed in some limited spatial region in a time which is again of the order of 10^{-22} s. It is also in qualitative agreement with the result obtained by Friedman.

Figure 5.10 shows some features of the temporal evolution of the decay process. It will be observed that the Coulomb and kinetic energies change little beyond 10^{-20} s and that the energy corresponding to Coulomb interactions

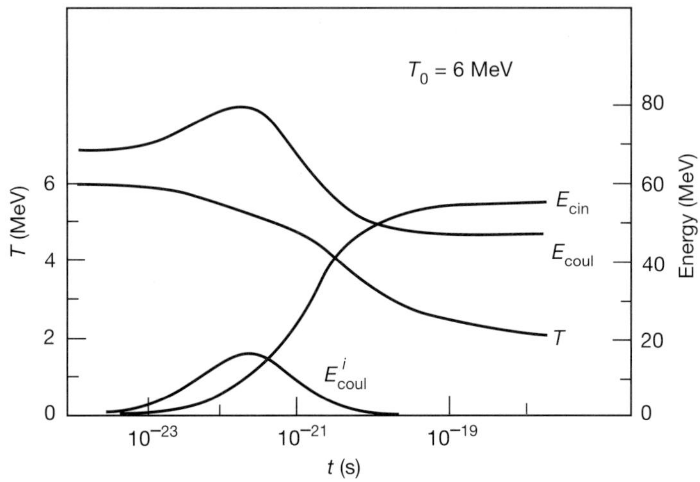

Figure 5.10. Evolution of temperature and various components of the total energy in the Barbagallo, Richert and Wagner model [6] for an ^{40}Ar nucleus with an initial temperature of 6 MeV. E_{cin} is the kinetic energy of the fragments, E_{coul}, the total Coulomb energy and E^i_{coul}, the inter-fragment Coulomb interaction energy. Despite the sequential character of the decay mechanism, E^i_{coul} falls to zero after only about 10^{-20} s indicating that the complex fragments are produced in a rather small spatio-temporal region. (Reprinted from *Z. Phys.* A, **324**, Barbagallo C, Richert J and Wagner P, p 97. Copyright 1986, Springer-Verlag GmbH & Co.)

between fragments falls to zero beyond this point. All 'chemical' activity has ceased beyond 10^{-18} s.

It will be evident from the presentations of this and the preceding two sections that there is no standard model for sequential binary decay. The transition state model (GEMINI) possesses the major advantage of including a treatment (albeit approximate) of angular momentum. This is probably the reason why it has been widely adopted by experimentalists. Both GEMINI and the BRW model treat the emitted fragment and the residual nucleus on an equal footing. The Friedman model does not include this feature. However, it allows for the study of some thermodynamical properties of the expansion of the parent system which may be the compound nucleus or any member of the chain of subsequent residues.

5.6 Characterization of decay mechanisms

We have, so far, encountered two distinct decay mechanisms which are referred to as sequential (multistep) evaporation and sequential binary division. We have also mentioned the multifragmentation process which we defined as the prompt disintegration of the excited compound nucleus into light particles and complex fragments. It would certainly be satisfying to think that all such mechanisms could be treated in one global statistical theory. Whether the nucleus decays by multifragmentation or, say, by evaporation would, in such a theory, be simply a question of evaluating the number of 'decaying' microstates corresponding to any particular decay mechanism.

The reason why this strategy has, in general, not been adopted is that in many experimental situations one particular decay mechanism is much more probable than the others. Thus, for example, at excitation energies below about 2 MeV/nucleon, sequential evaporation is clearly the dominant mechanism. Under these circumstances, it would not be very efficient to calculate the multifragmentation width (which may even be zero if corresponding Q-values exceed the excitation energy) at each evaporation step. This 'economical' approach is less obviously applicable at higher energies where sequential evaporation, sequential binary decay and multifragmentation may all compete. Nevertheless, experimentalists have usually resorted to comparing specific models (such as GEMINI) with data in the hope that the observations can be found to be strongly compatible with one particular mechanism while rejecting the others.

Support for a predominant decay mechanism can sometimes be found by analysing characteristics of event sets in the hope of attributing 'generic' characteristics to a particular decay mechanism. Furthermore, supposing that such characteristics exist, they may be used as 'selection criteria' to isolate a subset of events corresponding to the desired mechanism. That is the topic to be discussed in this section. As a simple illustration we may consider the so-called charge (mass) Dalitz plot (figure 5.11) which is constructed using only the three highest charges (or masses) in any event. Let Z_1, Z_2 and Z_3 be the three highest charges and d_1, d_2 and d_3 be the distance of a point from the three labelled sides of an equilateral triangle. Then by defining, for each event,

$$d_i = \frac{Z_i}{Z_1 + Z_2 + Z_3}, \tag{5.6.1}$$

($i = 1$ to 3) we obtain a plot which may be of considerable use in identifying the mechanism. For example if, as is usually the case in sequential evaporation, events contain just one large fragment and several evaporated particles, events will be found concentrated in one corner of the triangle whereas for binary decay followed by evaporation (which produces two large fragments), events would be expected to lie close to the mid-point of one side. On the other hand, for events in which three or more large fragments are produced we would expect to find a 'cloud' of points near the centre of the triangle. Because the labelling of the sides is arbitrary it is usual to symmetrize the plot by randomly assigning the labels one to three to the three largest fragments in any event in the set.

The Dalitz plot just presented involves only observed charges and is, therefore, referred to as a 'static' representation of a set of events. Dynamical representations which involve, in some way, the velocities of the fragments can also be constructed. These representations are, in general, more useful for discriminating

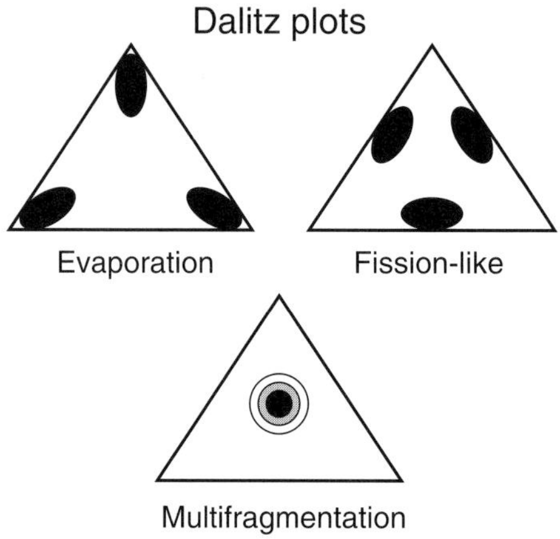

Figure 5.11. Idealized picture of the symmetrized Dalitz plot for the three largest charges in a partition. Such plots may be used to select specific types of events by 'gating' on selected regions of the triangle.

between decay mechanisms. As a first example, we consider an extension of the analysis of section 4.6. The aim of this analysis was to derive a precise value of the average, over events, of the square of the momentum of the heaviest fragment (evaporation residue). In sequential binary decay it often occurs that there is just one binary division followed by a series of sequential evaporations. In this situation the qualitative arguments developed in section 4.6 concerning negative recoil correlations still apply to the heaviest fragment provided that such a fragment is produced by evaporation following the initial binary division. In general, however, it is more useful to construct the *spectrum* of p_R^2 values (in the spirit of section 4.6 the heaviest fragment is a residue which is why we use the subscript 'R'). In the 'p_R^2' spectrum, sequential binary division events are concentrated at small values when compared with events generated in a multifragmentation process. Examples of spectra are shown in figure 5.12 [26]. The main reason for the difference is that, in a prompt multifragmentation process, momentum conservation implies that fragments are 'pushed away' from the heaviest fragment by Coulomb forces roughly in the same direction. The correlation between the momenta of any two of the lighter fragments would therefore be expected to be positive whereas, according to the analysis of section 4.6, it is strictly negative for pure sequential emission. The multifragmentation process thus induces large recoil momenta in the heaviest fragment.

Another method used to identify the decay mechanism is referred to as the coplanarity–sphericity plot [27]. For an event with multiplicity N, this analysis begins with the construction of the momentum tensor in a Cartesian reference

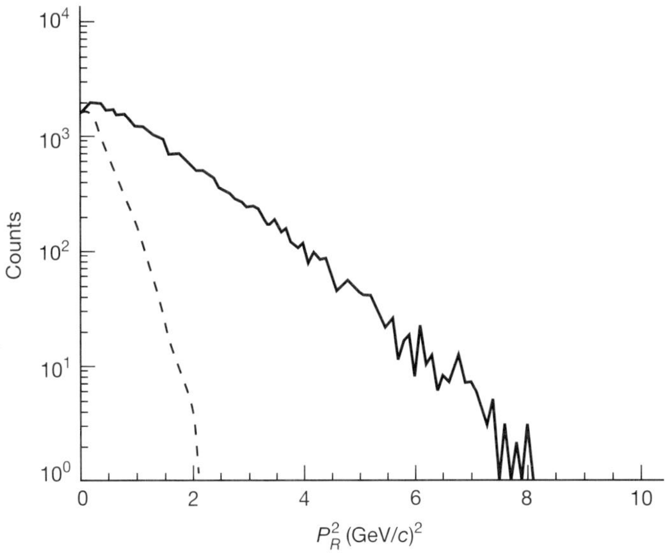

Figure 5.12. Spectrum of values of the square of the momentum of the heaviest fragment in events produced in central collisions of $^{32}S + ^{27}Al$ at 35 MeV/nucleon. The experimental result (thick trace) [26] is incompatible with the binary sequential decay prediction (broken curve).

frame fixed in the laboratory system of coordinates,

$$Q_{i,j} = \sum_{n=1}^{N} \gamma_n p_i^{(n)} p_j^{(n)}, \tag{5.6.2}$$

in which $p_i^{(n)}$ is the ith Cartesian component of the momentum of the nth fragment in the event, and γ_n is a weighting factor assigned to the nth fragment. The matrix Q with components $Q_{i,j}$ is obviously symmetric and is thus specified by six independent quantities. Three of these can be taken as the axes of an ellipsoid and the remaining three as the Euler angles which define the orientation of the ellipsoid in the original laboratory Cartesian frame. In the rotated frame, Q is diagonal.

The choice of γ_n determines the physical character of the tensor. Thus if we set $\gamma_n = 1/2m_n$, where m_n is the mass of the nth fragment, $Q_{i,j}$ is referred to as a kinetic energy flow tensor [28,29], whereas if $\gamma_n = 1/|p_n|$, we are dealing with momentum flow [30].

In order to gain some appreciation of the significance of $Q_{i,j}$, let us confine our study to the kinetic energy tensor. Consider a rather artificial event composed of six fragments. Divide the fragments into three pairs and let the first pair have oppositely directed momenta p_1 and p_{-1} along the axis labelled '1', the second pair have oppositely directed momenta along axis 2, and the third pair have oppositely directed momenta along axis 3 (in any such construction, the total momentum in the centre-of-mass system must vanish). One quickly finds that off-diagonal components of the kinetic energy tensor vanish and that the diagonal components are equal to the summed energy associated with the corresponding axis. Thus,

$$Q_{1,1} = \frac{p_1^2}{2m_1} + \frac{p_{-1}^2}{2m_2} \tag{5.6.3}$$

is the total energy associated with axis 1 with similar results for $Q_{2,2}$ and $Q_{3,3}$. We can renormalize the components of Q by dividing by the total energy. This latter quantity is just the trace of the matrix Q which is independent of the choice of coordinate system. If the three energies are identical and we use the symbol λ to refer to renormalized energies (which are now dimensionless quantities), we find $\lambda_1 = \lambda_2 = \lambda_3 = 1/3$. These three quantities are taken as a measure of the axes of the ellipsoid. In this particular case, our idealized event would be characterized as 'spherical'. If we reduce the energy associated with the axis 2 (e.g., by increasing the masses of particles 3 and 4) λ_2 will be reduced and λ_1 and λ_3 increased so that the event looks more 'disc-shaped'.

More generally, the representation of the matrix $Q_{i,j}$ in the rotated frame in which Q is diagonal produces three energies ($Q_{1,1}$, $Q_{2,2}$, $Q_{3,3}$) which can be renormalized to provide the dimensionless ellipsoid axes λ_1, λ_2, λ_3 ($\lambda_1 + \lambda_2 + \lambda_3 = 1$) and arranged so that $\lambda_1 \geq \lambda_2 \geq \lambda_3$. The renormalization implies that we are concerned with the form of the ellipsoid rather than its overall size. The form is thus defined by two parameters constructed from the set (λ_1, λ_2, λ_3). We define the sphericity,

$$s = \tfrac{3}{2}(1 - \lambda_1) = \tfrac{3}{2}(\lambda_3 + \lambda_2), \tag{5.6.4}$$

and the coplanarity

$$c = \frac{\sqrt{3}}{2}(\lambda_2 - \lambda_3). \tag{5.6.5}$$

Spherical event shapes in momentum space are characterized by $s = 1, c = 0$ ($\lambda_1 = \lambda_2 = \lambda_3 = 1/3$). If s is close to 0 (which implies c is also close to zero) the event is cigar shaped, whereas if c is close to $\sqrt{3}/4$ and s is near 3/4 the event is disc-shaped ($\lambda_1 = \lambda_2 = 0.5$, $\lambda_3 = 0$). One would expect 'spherical' events in multifragmentation involving many fragments. However, if only a few fragments are involved, momentum conservation confines the fragments to a plane so that a disc-shaped event is observed. Unfortunately, a disc-shaped event is also expected for sequential evaporation in which fragments are preferentially emitted in a plane perpendicular to the compound nucleus angular momentum, so that it is wise to confine the search for distinctions between multifragmentation and sequential binary decay to events with high multiplicity. For back to back emission characterizing fission-like events, a cigar-shape is expected with most of the energy concentrated along the major axis of the ellipsoid. Incidentally, it may have occurred to the reader that the ellipsoid axes may be represented on a Dalitz plot in exactly the same way as the three heaviest charges. This presentation is preferred by some authors [30].

A more detailed characterization of event shapes in momentum space was proposed by Fox and Wolfram [31]. These authors characterize the angular correlation between fragment directions in terms of Legendre polynomials. The second order moment which is, apparently, the most useful for decay of highly excited nuclei is written

$$H(2) = \frac{1}{H(0)} \sum_{i=1}^{N} \sum_{j=1}^{N} |\boldsymbol{p}_i|\,|\boldsymbol{p}_j| \frac{3\cos^2\theta_{ij} - 1}{2}, \qquad (5.6.6)$$

where $\boldsymbol{p}_i$ is the vector momentum of fragment i and

$$H(0) = \sum_{i=1}^{N} \sum_{j=1}^{N} |\boldsymbol{p}_i|\,|\boldsymbol{p}_j|.$$

An examination of (5.6.6) shows that, in events for which $H(2)$ is small, particle directions tend to be uncorrelated (the integral of $3x^2 - 1$ for $x = \cos\theta_{ij}$ distributed uniformly in the range -1 to $+1$ is 0). Strong correlations (e.g., cigar-shaped events with $\cos^2\theta_{ij}$ close to 1) have $H(2)$ values close to 1. The sum over i and j may, in some cases, be restricted to include only complex fragments. An example of the use of (5.6.6) is provided by the work of Mario *et al* [32] and is shown in figure 5.13. The figure shows sphericity–coplanarity plots for subsets of data in the reaction Xe + Sn at 50 MeV/nucleon. The different subsets were selected using the second Fox–Wolfram moment. With increasing $H(2)$ the subsets show the (not unexpected) evolution from a spherical shape, through disc shapes to forms which are characterized by small values of both sphericity and coplanarity and are thus cigar shaped.

Yet another global variable which can be used, in principle, to distinguish between sequential binary decay and multifragmentation is the so called two fragment velocity correlation function. This is defined as

$$1 + R(\boldsymbol{v}_1, \boldsymbol{v}_2) = C \frac{\sum_{\text{events}} \sum_{\text{pairs}} Y_{12}(\boldsymbol{v}_1, \boldsymbol{v}_2)}{\sum_{\text{events}} \sum_{\text{pairs}} Y_{12}^{\text{mix}}(\boldsymbol{v}_1, \boldsymbol{v}_2)}, \qquad (5.6.7)$$

where $Y_{12}(\boldsymbol{v}_1, \boldsymbol{v}_2)$ is the yield of coincidences between a pair of fragments with vector velocities $\boldsymbol{v}_1$ and $\boldsymbol{v}_2$. The denominator is constructed by taking fragments 1 and 2 from different events whereas the sum in the numerator is over pairs of fragments in the same event. The normalization constant, C, is fixed by requiring that the number of pairs is the same in numerator and denominator. The theory of pair correlations is quite complex (the reader is referred to the work of Gong *et al* [33]). However, when the only interaction taken into account is the Coulomb repulsion between the fragments then, as shown by Kim *et al* [34], it is useful

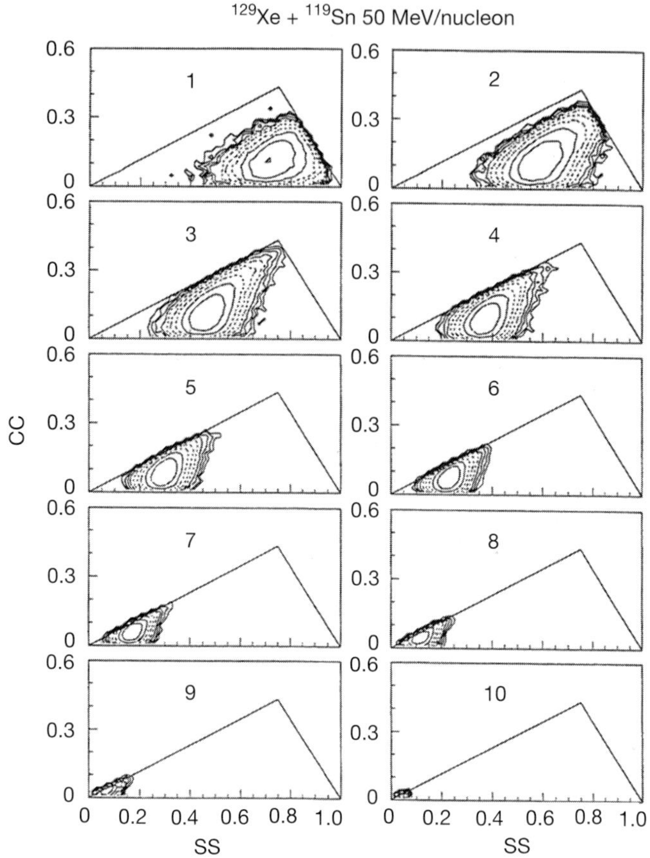

Figure 5.13. Coplanarity–sphericity plots [32] in the reaction ^{129}Xe (50 MeV/nucleon) + ^{119}Sn reaction. The events have been divided into ten bins, each bin corresponding to increasing values of the Fox–Wolfram moment $H(2)$. The figure shows that the two methods of selecting events are strongly correlated. Small $H(2)$ values correspond to spherical event shapes whereas large values indicate cigar shapes.

to display the correlation simply as a function of the so-called reduced velocity. This quantity is defined for any two fragments with charges Z_1 and Z_2 and vector velocities $\boldsymbol{v}_1$ and $\boldsymbol{v}_2$ as

$$v_{\text{red}} = \frac{|\boldsymbol{v}_1 - \boldsymbol{v}_2|}{\sqrt{Z_1 + Z_2}}. \tag{5.6.8}$$

With this definition $R(\boldsymbol{v}_1, \boldsymbol{v}_2)$ can be written $R(v_{\text{red}})$. An event with N-fragments furnishes $N(N-1)/2$ independent values of v_{red} but as with the Fox–Wolfram $H(n)$ variable, light particles are often excluded from the analysis [34,35]. Values of $R(v_{\text{red}}) > 0$ indicate that the particular value of v_{red} is favoured in the event set,

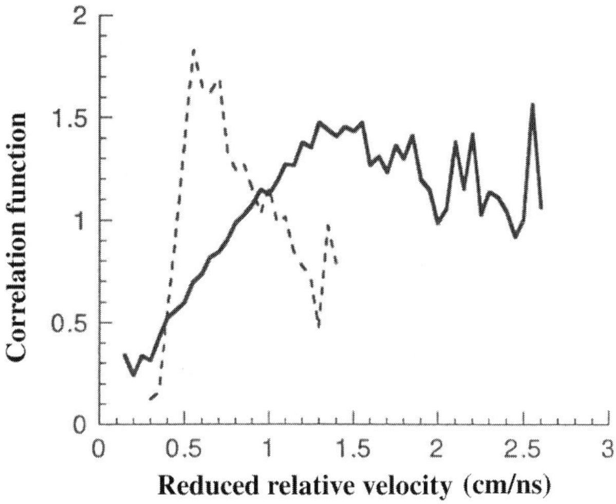

Figure 5.14. Velocity correlation function versus v_{red} for events with three complex fragments observed in the $^{32}S + ^{27}Al$ at 35 MeV/nucleon [36]. As in figure 5.12 the experimental result (thick trace) is incompatible with the sequential binary division mechanism which exhibits a well-defined 'Coulomb peak' (broken trace).

whereas $R(v_{red}) < 0$ indicates an anti-correlation, i.e., a depreciation of fragment pairs with the given value of v_{red}. It is easy to give a rough justification for the use of (5.6.8). Thus, if we consider two touching fragments with masses M_1 and M_2, and charges Z_1 and Z_2, the relative velocity corresponding to the Coulomb repulsion is

$$v_{12} = \sqrt{\frac{2E_{12}}{\mu}} = \sqrt{\frac{2Z_1 Z_2 e^2 (M_1 + M_2)}{R_{12} M_1 M_2}} \approx \sqrt{\frac{2Z_1 Z_2 e^2 (A_1 + A_2)}{m_N R_{12} A_1 A_2}}, \quad (5.6.9)$$

where m_N is the nucleon mass, R_{12} is the distance between the mass centres of the two fragments, E_{12} is the Coulomb barrier energy and $\mu = M_1 M_2/(M_1 + M_2)$ is the reduced mass. If we assume $A = 2Z$ for both fragments we can write

$$\frac{v_{12}}{\sqrt{Z_1 + Z_2}} = \sqrt{\frac{e^2}{m_N R_{12}}} = v_{red}, \quad (5.6.10)$$

so that if R_{12} varies only slowly with the selected pair of fragments we may expect a Coulomb 'bump' in the correlation function plotted as a function of v_{red}. This correlation is observed especially in sequential binary processes. In multifragmentation, variations in R_{12} tend to wash out the characteristic structure. This can be seen in an example which is taken from the work of Chabane *et al* [36] and shown in figure 5.14. Incidentally, (5.6.10) suggests that some further improvement may be obtained by writing the fragment radii as $r_0 A^{1/3}$ so that, with $A = 2Z$, $R_{12} = 2^{1/3} r_0 (Z_1^{1/3} + Z_2^{1/3})$. Then

$$v'_{\text{red}} = v_{12} \sqrt{\frac{Z_1^{1/3} + Z_2^{1/3}}{Z_1 + Z_2}} = \sqrt{\frac{e^2}{2^{1/3} m_N r_0}}. \tag{5.6.11}$$

Before leaving this section it is as well to state that our list of correlations is not exhaustive. Other correlations which are mainly used to distinguish between peripheral (large impact parameter) and central (small impact parameter) heavy ion collisions may be found in the literature. The interested reader is referred to the works of Peter *et al* [17] and Pawlowski *et al* [37] which give quite detailed accounts of, or references to, the various techniques.

References

[1] Geller R 1996 *Electron Cyclotron Resonance Ion Sources and ECR Plasmas* (Bristol: Institute of Physics Publishing)

[2] Désesquelles P 1995 *Ann. Phys., Paris* **20** 1; Désesquelles P, Bondorf J P, Mishustin I and Botvina A S 1996 *Nucl. Phys.* A **604** 183

[3] Friedman W A and Lynch W G 1983 *Phys. Rev.* C **28** 16

[4] Charity R J *et al* 1988 *Nucl. Phys.* A **483** 371; Charity R J *et al* 1988 *Nucl. Phys.* A **476** 516

[5] Moretto L G 1975 *Nucl. Phys.* A **247** 211

[6] Barbagallo C, Richert J and Wagner P 1986 *Z. Phys.* A **324** 97

[7] Péghaire A *et al* 1990 *Nucl. Instr. and Meths. in Phys. Res.* A **295** 365

[8] Drain D *et al* 1989 *Nucl. Instr. and Meths. in Phys. Res.* A **281** 528

[9] De Souza R T *et al* 1990 *Nucl. Instr. and Meths. in Phys. Res.* A **295** 365

[10] Stracener D W, Sarantites D G, Sobotka L G, Elson J, Hood J T, Majka J, Abenante V, Chbihi A and Hensley D C 1990 *Nucl. Instr. and Meths. in Phys. Res.* A **294** 485

[11] Pouthas J *et al* 1995 *Nucl. Instr. and Meths. in Phys. Res.* A **357** 418

[12] Lynen U *et al* 1989 *Report GSI-02-89* unpublished

[13] Gobbi A *et al* 1993 *Nucl. Instr. and Meths. in Phys. Res.* A **324** 156

[14] Kwiatkowski K *et al* 1995 *Nucl. Instr. and Meths. in Phys. Res.* A **360** 571

[15] Van der Werf S Y 1978 *Nucl. Instrum. Methods* **153** 221

[16] Abramowitz M and Stegun I R 1970 *Handbook of Mathematical Functions* (New York: Dover) Ch. 24

[17] Peter J *et al* 1990 *Nucl. Phys.* A **519** 611

[18] Brandan M E, Cole A J, Désesquelles P, Giorni A, Heuer D, Lleres A, Menchaca Rocha A and Michaelian K 1993 *Nucl. Instr. and Meths. in Phys. Res.* A **334** 461

[19] Friedman W 1988 *Phys. Rev. Lett.* **60** 2125; Friedman W 1990 *Phys. Rev.* C **42** 667

[20] Greiner W, Neise L and Stöcker H, 1994 *Thermodynamics and Statistical Mechanics* (Berlin: Springer)

[21] Nix J R and Swiatecki W J 1965 *Nucl. Phys.* **71** 1

[22] Davies K and Swierk A J 1985 *Phys. Rev.* C **31** 915

[23] Swierk A J 1986 *Phys. Rev.* **33** 2039

[24] Lleres A, Cole A J, Désesquelles P, Giorni A, Heuer D, Viano J B, Chambon B, Cheynis B, Drain D and Pastor C 1994 *Phys. Rev.* C **50** 1973

[25] Lleres A *et al* 1993 *Phys. Rev.* C **48** 2573

[26] Heuer D *et al* 1994 *Phys. Rev.* C **50** 1943

[27] Cugnon J and L'Hote D 1983 *Nucl. Phys.* A **397** 519

[28] Cebra D A, Howden S, Karn J, Nadasen A, Ogilvie C A, Vander Molen A, Westfall G D, Wilson W K, Winfield J S and Norbeck E 1990 *Phys. Rev. Lett.* **64** 2246

[29] Fai G and Randrup J 1983 *Nucl. Phys.* A **404** 551

[30] Desesquelles P, Cole A J, Giorni A, Heuer D, Lleres A, Viano J B, Chambon B, Cheynis B, Drain D, and Pastor C 1993 *Phys. Rev.* C **48** 1828

[31] Fox G C and Wolfram S 1978 *Phys. Rev. Lett.* **41** 1581

[32] Marie N 1995 *PhD Thesis*, University of Caen, France

[33] Gong W G, Bauer W, Gelbke C K and Pratt S A 1991 *Phys. Rev.* C **43** 781

[34] Kim Y D, de Souza R T, Gelbke C K and Gong W G 1992 *Phys. Rev.* C **45** 387

[35] Kämpfer B *et al* 1993 *Phys. Rev.* C **48** R955

[36] Chabane A 1996 *PhD Thesis*, University J. Fourier, Grenoble, France

[37] Pawlowski P *et al* 1996 *Phys. Rev.* C **54** R10

Chapter 6

Statistical models for multifragmentation

We must even be prepared for the collision
to lead to the explosion of the whole nucleus.

Niels Bohr

6.1 Introduction

This is the first of two chapters devoted to current statistical theories of the multifragmentation decay mechanism.

At excitation energies above about 4 MeV/nucleon an excited nucleus decays by disintegrating into a number of complex fragments and light particles. If most of the complex fragments are produced simultaneously, this process is referred to as prompt multifragmentation or, following our convention, simply multifragmentation. The primary complex fragments may of course be excited and decay on a much longer timescale by evaporation of light particles or even complex fragments. Multifragmentation thus contrasts strongly with evaporation which, via a (presumed) succession of states of temporal equililibrium, produces several light particles together with a more massive evaporation residue. The contrast with the sequential binary decay mechanism is less marked since this latter mechanism may also lead to events containing several complex fragments.

First indications that multifragmentation might be an important mode of decay of highly excited nuclei appeared in publications in the late 1970s and early 1980s [1,2]. Subsequently, more detailed theoretical analyses were carried out, usually based, in some way or another, on the nuclear equation of state. The most well-known equation of state relates the pressure, P, and the temperature, T, of a gas in equilibrium, to the particle density, N/V. Thus, Boyle's Law, $PV = Nk_{\rm B}T$ ($k_{\rm B}$ is Boltzmann's constant) is an equation of state for an ideal gas (measuring temperature in energy units we would write $PV = NT$ (see section 1.7)).

The qualitative similarity between the nucleon–nucleon interaction and the interaction between atoms of a monatomic gas (long range attraction and short range repulsion) led to speculation that the liquid gas phase transition, which may be observed for a real fluid (and, indeed, in infinite nuclear matter [3]), may have a counterpart in finite nuclei and that multifragmentation corresponded, in fact, to this phase transition. Unfortunately, while theoretical works certainly stimulated interest in this subject, they did not succeed in establishing a definitive consensus concerning the multifragmentation mechanism. Examples of the liquid–gas phase transition approach can be found in the papers by Siemens [4] and by Jaqaman, Mekjian and Zamick [5]. These works generally do not consider the fact that multifragmentation is mainly observed as a result of heavy ion collisions. Therefore, they do not explicitly take into account the particular conditions imposed by the entrance channel.

The dynamical constraint was taken into account in a paper by Bertsch and Siemens [6] who suggested that an initially excited and compressed system produced in a head-on collision of two heavy ions would tend to oscillate adiabatically with alternate cycles of expansion and contraction. The initial compression is due to the projectile–target impact. Expansion is accompanied by the conversion of internal excitation energy into collective kinetic energy and, in the compression cycle, the inverse process takes place. The particular form of the equation of state employed by Bertsch and Siemens identifies a region of the pressure versus density (P–ρ) plot which corresponds to the mechanically unstable 'spinodal' region ($\mathrm{d}P/\mathrm{d}\rho \leq 0$). Their analysis suggested that a nuclear system which attains this region would not oscillate but would undergo breakup with copious production of nucleons and alpha particles. In this picture, multifragmentation occurs essentially due to a dynamical instability on the first expansion cycle and is thus a fast process. Consequently, the thermodynamical description of multifragmentation as a liquid–gas phase transition, such as that envisaged in references [4] and [5] is, according to Bertsch and Siemens, 'not relevant'.

Support for the spinodal break-up picture has appeared in many works. A recent example can be found in the paper by Peilert *et al* [7] involving quantum molecular dynamical simulations of the decay of nuclei for which the excitation energy is in the form either of thermal heating with no compression, or of compression with no thermal heating. The calculations show that compression, which may drive the system into the zone of spinodal instability, is an essential ingredient for multifragmentation. The Bertsch–Siemens result also seems to have motivated recent calculations by Colonna, Chomaz and collaborators [8]. It appears that yields of primary fragments produced via spinodal break-up may be seriously modified by subsequent fragment–fragment interactions. However, the authors conclude that significant deviations from statistical model predictions should be observable.

Since the early 1980s, interest in multifragmentation, including further work on identification of the liquid–gas phase transition, has grown rapidly and, as discussed in the previous chapter, has been strongly stimulated by the increase in available beam energies and, especially since about 1985, by the construction of multidetectors (multiple detector arrays). The current status of the subject of multifragmentation is exposed in reviews by Konopka and Stöcker [9] and by Pochodzalla [10] (writing in 1996 and 1997). According to the former authors, no definitive experimental proof of the liquid–gas phase change in finite nuclei has yet been found.

It is also fair to say that, at the present time (1998), no compelling experimental evidence in support of spinodal, as opposed to statistical multifragmentation, has been published. There is thus continuing interest in statistical models of multifragmentation which are relevant only for the description of the ultimate stage of the reaction but which, by variation of initial conditions (excitation energy or temperature, nucleon density), may exhibit behaviour which is characteristic of a phase transition. Generally speaking, these models are highly successful in reproducing results of experiments and thus may be used to extract the densities and excitation energies (or temperatures) which best characterize experimental data sets. We shall be meeting with several examples in the course of this and the next chapter.

This first 'multifragmentation' chapter is concerned with the more conventional models. Specifically, after some preliminaries exposed in sections 6.2 and 6.3, we shall be examining models which picture a multifragmentation 'configuration' as a set of complex fragments and light particles initially localized in a small 'freeze-out' spatial region. We shall describe successively the Copenhagen liquid drop multifragmentation model and a related canonical approximation (section 6.4) as well as the Berlin Metropolis sampling method (sections 6.5 and 6.6). In section 6.7 we propose an analysis of finite size corrections to the freeze-out volume, and in section 6.8 we discuss an adaptation of the Bohr–Wheeler transition state method proposed by workers in Berkeley.

The following chapter will be concerned with multifragmentation viewed as a phase transition. We shall investigate the analogy between multifragmentation and percolation and also discuss attempts to measure simultaneously the excitation energy and temperature of an excited nucleus (the caloric curve). Finally we will present recent work concerned with the vaporization limit.

6.2 Events, partitions and macrostates

We must first define, in the context of nuclear collisions, what we mean by an event or macrostate. Let the sum of projectile and target mass numbers be A_C and the sum of charge numbers be Z_C. Now suppose that the reaction products are detected in a perfect (100% efficient) multidetector so that the sum of the mass (charge) numbers of the detected fragments is equal to $A_C(Z_C)$. Following any given projectile–target collision, let the number of detected fragments with mass number A and charge number Z be $n_{A,Z}$. Then a specification of any macrostate can be made using the vector $\boldsymbol{n} = \{n_{1,0}, n_{1,1} \ldots n_{A,Z} \ldots n_{A_C,Z_C}\}$ which is usually referred to as a partition. This definition is somewhat restrictive. A measured event is not only characterized by the partition but also by the vector momenta (or velocities) of each and every fragment contained in $\boldsymbol{n}$. This latter specification is conventionally considered to be the most detailed definition of a macrostate which is useful in the statistical approach (γ rays are not usually taken into account). Theories for multifragmentation may be constrained to predict only partition probabilities or may include prediction of dynamical characteristics. In the following discussion we shall concentrate on macrostates defined as partitions. In real experiments this definition is usually further restricted so that macrostates coincide with charge partitions rather than those of mass and charge. This is mainly due to difficulties with complex fragment mass identification. We may further note that most multidetectors are insensitive to neutrons.

In many cases (although not always) the theory is set up as a Monte Carlo event generator. Statistical model event generators do not attempt to follow the full dynamical evolution of each individual projectile–target collision. Rather, the model is used to generate only the final stage of the process (at which point equilibrium is assumed). The set of fragments corresponding to a given partition (macrostate) is considered to be distributed randomly in some small spatial region. The fragments may interact via Coulomb and nuclear forces and are usually doted with some initial velocities. Thus, if required, the subsequent dynamical evolution of the system can be simulated by numerical integration of the equations of motion. This procedure yields the asymptotic vector momentum of each fragment.

Any complex fragment, produced as a member of a partition in the initial multifragmentation process, may be excited and, thereby, be subject to further decay which is thought to take place on a timescale which is significantly longer than the multifragmentation timescale. It follows that the observed partitions which contain only 'cold' nuclei are in fact a transformed version of the originals so that any model calculation may have need of 'evaporation corrections' before being compared with experiment. These corrections may either be introduced statically or, if the dynamical evolution is followed explicitly in each event, as stochastic modifications of the equations of motion.

Another difficulty (which is a real nuisance) arises because of pre-equilibrium emission which leads to a distribution of parent nucleus masses, charges, momenta and excitation energies. This necessitates the incorporation of the effects of pre-equilibrium emission in model predictions. The influence of detector imperfections must also be included via a 'filter' programme unit.

If we limit our ambition to the study of partitions of a parent nucleus with charge number , Z_C, and mass number, A_C, we may define a (discrete) probability distribution $P(\boldsymbol{n})$ with $\sum_n P(\boldsymbol{n}) = 1$. The measured set of events is then considered as a sample of this distribution. Thus if N_t is the total number of events (or trials) and $\phi(\boldsymbol{n})$ is the number of times the partition $\boldsymbol{n}$ was observed (the frequency), then we may consider the quantities

$$\pi(\boldsymbol{n}) = \frac{\phi(\boldsymbol{n})}{\sum_n \phi(\boldsymbol{n})} = \frac{\phi(\boldsymbol{n})}{N_t}, \tag{6.2.1}$$

as experimental estimations of $P(\boldsymbol{n})$. The $P(\boldsymbol{n})$ values themselves contain all possible information concerning the partitions. Thus, for example, the average number of fragments of charge Z, which is referred to as the inclusive charge distribution, is given by

$$\langle n_Z \rangle = \sum_{A=1}^{A_C} \sum_{\boldsymbol{n}} n_{A,Z}(\boldsymbol{n}) P(\boldsymbol{n}) = \sum_{\boldsymbol{n}} n_Z(\boldsymbol{n}) P(\boldsymbol{n}), \tag{6.2.2}$$

and the probability distribution of the total multiplicity ($N = \sum n_{A,Z}(\boldsymbol{n})$) is

$$P(N) = \sum_{\boldsymbol{n}} P(\boldsymbol{n})\, \delta\Big(\sum n_{A,Z}(\boldsymbol{n}) - N\Big). \tag{6.2.3}$$

We may further deduce useful sum rules. For example, in any partition, $\boldsymbol{n}$,

$$\sum_{A=1}^{A_C} \sum_{Z=1}^{Z_C} Z\, n_{A,Z}(\boldsymbol{n}) = \sum_{Z=1}^{Z_C} Z n_Z(\boldsymbol{n}) = Z_C \tag{6.2.4}$$

(obviously a similar relation holds for mass numbers). Furthermore, by multiplying both sides of equation (6.2.4) by $P(\boldsymbol{n})$ and summing over $\boldsymbol{n}$ (using equation (6.2.2)) we obtain the same relation for the mean values, i.e., $\sum_{Z=1}^{Z_C} Z \langle n_Z \rangle = Z_C$.

The detailed implementation of multifragmentation statistical models implies an estimation of the probabilities $P(\boldsymbol{n})$ (the normalized statistical weight or fraction of microstates) to be assigned to each possible macrostate. Thus, in principle, it is necessary to generate all partitions of a given parent mass and charge one by one (with no repetitions) in order to construct sums such as (6.2.2) and (6.2.3). This programme can rarely be carried out rigorously. The difficulty is due simply to the sheer number of macrostates. We have already seen that the decay of heavy nuclei at low excitation energies can be modelled as a competition between neutron emission and the various mass and charge partitions associated with fission. The number of such partitions is of the order of 10^3. In multifragmentation, partitions of all multiplicities (not just two) may be observed. If we consider only charge partitions of a total charge, Z_C, their number may be estimated using the Hardy–Ramanujan formula [11] in which the leading term is

$$N_{\text{part}}(Z_C) \approx \frac{\exp\Big(\pi\sqrt{\frac{2Z_C}{3}}\Big)}{4Z_C\sqrt{3}}. \tag{6.2.5}$$

For $Z_C = 100$, we find $N_{\text{part}} \approx 2.10^8$. The problem is, of course, considerably aggravated by the need to include the mass as well as the charge of fragments even if, as is practically unavoidable, only known (i.e., stable or relatively long-lived) nuclei are considered as fragments. It is thus extremely time consuming to calculate all statistical weights so that equation (6.2.1) cannot be straightforwardly implemented. As an alternative, one may attempt to make a Monte Carlo calculation in which partitions are generated randomly (allowing repetitions) such that the frequency, $\phi_{\mathrm{M}}(\boldsymbol{n})$, which is the number of times the partition, $\boldsymbol{n}$, is generated, is proportional to the statistical model probability $P_{\mathrm{M}}(\boldsymbol{n})$. Then, theoretical estimates of mean values can be made using the analogue of equation (6.2.1). This procedure may not seem obvious but can in fact be carried out with reasonable accuracy (see section 6.5).

It is important to realize that numerical procedures used to generate partitions in Monte Carlo simulations may imply a built in statistical bias. For example, if we generate partitions of mass number A using the one-dimensional bond percolation model presented in section 1.4 (breaking bonds in a linear chain) partitions are generated with the statistical weight which is approximately $\frac{N!}{n_1!n_2!...n_A!}$ ('approximately' because of the constraint of total mass). Thus, considerable care is required if theories are implemented using Monte Carlo event generators.

6.3 Framework for multifragmentation simulations

This section is intended to provide a framework for understanding the approximations involved in the Copenhagen [12] and Berlin [13] multifragmentation models. The particular presentation is based on the Bohr–Wheeler (BW) theory for neutron emission. It is most important to have a good grasp of the physical picture behind this theory and we therefore begin by revisiting the presentation of chapter 3, albeit with a somewhat different emphasis. An important aspect of the BW theory concerns the specification of states of motion of an emitted neutron relative to the residual nucleus. Microstates corresponding to neutrons near the surface of the nucleus are assumed to be those of a classical gas (nuclear forces are introduced only via the Q-value). That is why we refer to 'evaporation' of neutrons. In practice, of course, this can only be a rough approximation so that the success of the evaporation model of neutron emission widths and energy spectra is a little surprising.

The framework to be presented includes further approximations and, for this reason, should probably not be applied to light particle emission (we have the Hauser–Feshbach theory so this is no great loss). It should be considered simply as providing a qualitative link between evaporation and multifragmentation which serves to highlight problems in the formulation of a theory for the latter process. The BW theory is based on the notion of transition states. However, the states involved in the following description of the decay process are *spatially* identical with scission configurations. We will therefore refer to them as *scission* states.

We had best begin by providing a definition of these states. The probability of the scission ($C \to B + b$) in a time Δt from a compound nucleus (C.N.), C, with excitation energy E_C^*, is

$$W_S \Delta t = \frac{\text{Number of C.N. states which decay to } b + B \text{ in } \Delta t}{\text{Total number of C.N. states at } E_C^*}. \tag{6.3.1}$$

We define a set of scission states, with density ρ_{bB}, as those states of the compound nucleus for which b lies in a thin spherical shell of width ΔR at the surface of B with any *arbitrarily directed* velocity, v_b. Decay takes place when the particle b leaves the 'scission' shell moving outwards from the parent nucleus at some angle θ to the outwardly directed normal. The quantity Δt in equation (6.3.1) is to be understood as a small time interval in the sense that $v_b \Delta t \leq \Delta R$ for all possible values of v_b. The numerator on the right of equation (6.3.1) thus represents a subset of the scission states. We can therefore write (replacing numbers of states by densities and multiplying by $1 = \rho_{bB}/\rho_{bB}$)

$$W_S \Delta t = \frac{\rho_{bB}}{\rho_C(E_C^*)} \left\langle \frac{\rho_{bB}(\theta, v_b, \Delta t)}{\rho_{bB}} \right\rangle, \tag{6.3.2}$$

or, with an obvious definition of the scission probability, $P_S(\Delta t)$,

$$W_S \Delta t = \frac{\rho_{bB}}{\rho_C(E_C^*)} P_S(\Delta t). \tag{6.3.3}$$

In equation (6.3.2), $\rho_{bB}(\theta, v_b, \Delta t)$ designates the subset of scission states (with the velocity, v_b, directed at an angle θ) which actually do decay in the time interval, Δt, and the average is over velocities and angles. We may also write down a similar relation by considering states of ρ_{bB} for which the relative velocity, v_b, of the pair (b, B) is fixed. In this case, the equation corresponding to (6.3.3) is

$$W_S(v_b) \Delta t = \frac{\rho_{bB}}{\rho_C(E_C^*)} P_S(v_b, \Delta t), \tag{6.3.4}$$

where $P_S(v_b, \Delta t)$ is now simply an average over (solid) angles. For particles with velocities directed along the outwardly pointing normal to the surface of B, $P_S(\theta = 0, v_b, \Delta t)$ is obviously $v_b \Delta t/\Delta R$. Averaging over solid angles reduces this probability by a factor of two (i.e., eliminates inwardly directed trajectories) and by a second factor of two corresponding to the average of $v_b \cos\theta$ over outward directions. Thus $P_S(v_b, \Delta t) = v_b \Delta t/4\Delta R$ (see equation (3.3.17)).

Elimination of inwardly directed trajectories is a non-trivial feature because it implies that b will not escape if it has to pass through the body of the emitting nucleus. Indeed, the notion of non-obstruction is an important aspect of the theory. Consider the case of an energetic neutron somewhere in the body of an excited nucleus. If 'left to itself' this neutron may have a finite probability to escape through the surface of the nucleus within the interval Δt. However, it is highly likely that during the course of its flight it will experience one or more collisions which will prevent its escape. For such states $P_S = 0$. On the other hand, a neutron which is already at the surface of the nucleus with an outwardly directed velocity is unlikely to collide with another nucleon in the course of its flight. That is why scission states are best chosen to correspond to particles located in the surface of the emitter.

At this point, it is as well to emphasize that, in principle, the introduction of scission states is a matter of convenience and that their definition is largely a matter of choice. This is obvious from (6.3.1) and (6.3.3). Any set of states of $b+B$ for which one can calculate P_S (even if $P_S \geq 1$) will do. The above definition was made because it is easy to write down the density of states for a single particle, b, with a randomly directed velocity enclosed in a volume ΔV (approximately $4\pi R^2 \Delta R$). The reader may care to reflect on the 'Weisskopf' choice. This is quite different (but naturally yields an identical result) because V (rather than ΔV) is taken to be a large volume which surrounds the emitting nucleus. Only a small fraction of the states in this volume evolve into compound nuclear states under time reversal. Furthermore, as pointed out in section 3.3, the notion of the inverse reaction itself arises because of the choice made by Weisskopf. It serves only to identify the fraction of phase space (b, B) states which actually result from decay of the compound nucleus.

Just as we averaged over angles in equation (6.3.4), for fixed v_b, so we can average over velocities. The average value of the velocity, v_b, in the scission probability may be calculated by considering the states of B to act as a heat bath for the states of relative motion. In this case, in close correspondence with evaporation theory (section 3.3), we quickly find that $\langle v_b \rangle = \sqrt{8T'/m_b \pi}$ where $T' = \mathrm{d}\ln[\rho_B(E^*)]/\mathrm{d}E^*$ at $E^* = E_B^*$ so that we can rewrite (6.3.3) as

$$W_S = \frac{\rho_{bB}(E_{bB})}{\rho_C(E_C^*)} \frac{1}{\Delta R} \sqrt{\frac{T'}{2\pi m_b}} = \frac{\rho_{bB}(E_{bB})}{\rho_C(E_C^*)} \frac{1}{\tau_S}, \tag{6.3.5}$$

in which $E_{bB} = E_B^* + \varepsilon_b$ is the total energy in the (bB) system and τ_S is a characteristic scission time (a measure of the time for the light particle to escape from the scission shell). We now make three approximations. First, we equate the scission state temperature, T', which depends for example on the Q-value and thus, in principle, on the decay channel, with the compound nucleus temperature, $T = \mathrm{d}\ln[\rho_C(E^*)]/\mathrm{d}E^*$ at $E^* = E_C^*$, so that the temperatures for all scission channels are identical. Secondly, we assume that the scission region has the same thickness ΔR for all emitted particles. Finally, we simply ignore changes induced by differences in masses. We thus suppose that differences in ρ_{bB} for the various competing channels are much more important than differences due to the $\sqrt{m_b}$ factor. With these assumptions τ_S is constant and the decay rates are simply proportional to the corresponding densities of scission states.

The density of scission states, ρ_{bB}, is a convolution involving internal excitation (E_B^*) of the residual nucleus, B, and the density, $\rho_b(\varepsilon_b)$, of the phase space states corresponding to the relative motion of the pair (b, B) with b constrained to lie at the surface of the compound nucleus. Given the assumed constancy of τ_S, the probability of decay per unit time is therefore simply proportional to this convolution, i.e.,

$$W_S = \frac{1}{\tau_S \rho_C(E_C^*)} \int_0^{E_{bB}} \rho_B(E_{bB} - \varepsilon_b)\rho_b(\varepsilon_b)\,\mathrm{d}\varepsilon_b. \tag{6.3.6}$$

While the specific decay rate (equivalently the mean lifetime of any given nucleus) may be of interest in evaluating the internal consistency of the model, we may, for prediction purposes, be content with the branching ratio for any particular two-body exit channel which we label by designating the emitted particle $\nu = n, p, d, t, \alpha \ldots$ etc (rather than 'bB'). We can simply write the branching ratio for emission of any light particle as

$$P(\nu) = \frac{\rho_\nu(E_\nu)}{\sum_{\nu'} \rho_{\nu'}(E_{\nu'})} \tag{6.3.7}$$

where the denominator represents the sum over all accessible channels. The energy, E_ν (previously E_{bB}), is, of course, the energy available for distribution between internal excitation of the residue and kinetic energy of the emitted particle.

A major step is now to extend equation (6.3.7) to any type of decay channel including multifragment decay. We emphasize that this supposes that the characteristic scission time associated with multifragmentation is the same for all multifragment channels. This may be a better assumption for multifragmentation than for light particle emission since multifragmentation may be thought of as due to an expansion of the parent nucleus characterized by a unique expansion velocity and a more or less constant inertial parameter (the equivalent of m_b in equation (6.3.6)). We thus replace (6.3.7) by its analogue for multifragmentation

$$P(\boldsymbol{n}) = \frac{\rho_{\boldsymbol{n}}(E_{\boldsymbol{n}})}{\sum_{\boldsymbol{n}'} \rho_{\boldsymbol{n}'}(E_{\boldsymbol{n}'})} \tag{6.3.8}$$

where we have used the vector symbol $\boldsymbol{n}$ as in the previous section to refer to a partition of the parent mass and charge.

The theory (maybe one should say 'hypothesis') represented by equation (6.3.8) requires us to evaluate the density of scission microstates corresponding to each and every partition. For neutron emission we recall that this was a relatively easy job because scission states were constructed to correspond to a neutron lying just outside the range of nuclear forces of the residual nucleus in a spherical shell of width ΔR and with random momentum components. Thus, if we fix the position and momentum of the centre of mass (equation (1.2.32)) and let ε_{n} represent the neutron energy *relative* to the residue we can write

$$\rho_{\mathrm{n}}(E_{\mathrm{n}}) = \int_0^{E_{\mathrm{n}}} \rho_B(E_{\mathrm{n}} - \varepsilon_{\mathrm{n}}) 2\pi \frac{(4\pi R^2 \Delta R)(2\mu_{\mathrm{n}})^{3/2} (c_{\mathrm{n}} \varepsilon_{\mathrm{n}})^{1/2}}{h^3} \, \mathrm{d}\varepsilon_{\mathrm{n}}, \tag{6.3.9}$$

where μ_{n} is the reduced mass, $E_{\mathrm{n}} = E_C^* + Q_{\mathrm{n}}$ and $\rho_B(E_{\mathrm{n}} - \varepsilon_{\mathrm{n}})$ is the density of states corresponding to internal excitation of the residual nucleus at energy $E_B^* = E_C^* + Q_{\mathrm{n}} - \varepsilon_{\mathrm{n}}$. The factor c_{n} which is the ratio of residue mass to total mass transforms the total kinetic energy, ε_{n}, into the corresponding centre-of-mass energy of the neutron and the residue. As discussed in chapter 1 (see equation (1.2.31)), fixing the centre-of-mass position and momentum is equivalent to reducing the dimension of phase space by h^3. That is why, in equation (6.3.9), a single factor of h^3 appears in the denominator.

In the Weisskopf theory the factor πR^2 was associated with an inverse cross-section. However, one may take a somewhat different point of view. Thus, as shown in figure 6.1, it is possible to identify the scission state volume $\Delta V = 4\pi R^2 \Delta R$ with a 'freeze-out' volume for a massive residue and a small emitted particle by insisting that the vector which characterizes the separation between the centres of mass of the two fragments is confined to ΔV. The origin of this vector should normally be the centre of mass of the parent nucleus but for small emitted particles may be taken as the centre of mass of the residue as indicated in the figure. Inasmuch as it refers to a scission spatial configuration, the word 'freeze-out', when used as an adjective, can be thought of as synonymous with 'scission state'.

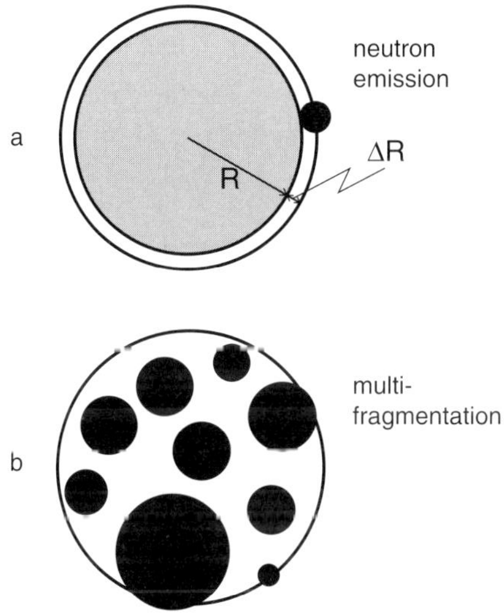

Figure 6.1. (*a*) Neutron emission from a heavy parent nucleus: the neutron is confined to the surface of the parent nucleus, i.e., to the volume element $\Delta V = 4\pi R^2 \Delta R$ which plays the role of freeze-out volume. More precisely (especially for light systems) one would require the centre of mass of the parent-neutron system to be fixed and the relative position coordinate to lie within ΔV. (*b*) Multifragmentation: the fragments may be disposed randomly inside the freeze-out volume provided they do not overlap. In principle, one should fix the position of the centre of mass and the associated momentum.

For emission of a proton, the radial position may be considered to lie in a small region ΔR near the top of the Coulomb barrier with potential energy B_{p} so that we might expect the density of scission states $\rho_{\mathrm{p}}(E_{\mathrm{p}})$ to be of the same form as the density for neutron emission except that $E_{\mathrm{p}} = E_C^* + Q_{\mathrm{p}} - B_{\mathrm{p}}$. A similar formula may be applied to the emission of all light particles. However, we need to generalize (6.3.9) to the multifragment configuration. Since, in this case, we cannot distinguish between emitted fragments and residues we have to suppose (at least as a first guess) that a typical configuration corresponds to a random distribution of fragments within some small spatial region. We may further simplify the model by not insisting on fixing the position and momentum of the centre of mass. Using equation (1.3.10) we thus write (do not confuse 'n' which refers to a neutron with '***n***' which represents a multifragmentation partition)

$$\rho_{\boldsymbol{n}}(E_{\boldsymbol{n}}) = \int_0^{E_{\boldsymbol{n}}} \rho_{\mathrm{I}}(E_{\boldsymbol{n}} - \varepsilon_{\boldsymbol{n}}) \left[\frac{V_{\boldsymbol{n}}^N}{\prod_{A,Z} n_{A,Z}!} \times \left(\frac{\prod_{A,Z} (2\pi M_{A,Z})^{3n_{A,Z}/2}}{h^{3N}} \right) \frac{\varepsilon_{\boldsymbol{n}}^{3N/2-1}}{\Gamma(3N/2)} \right] \mathrm{d}\varepsilon_{\boldsymbol{n}}, \qquad (6.3.10)$$

in which $E_{\boldsymbol{n}} = E_C^* + Q_{\boldsymbol{n}} - B_{\boldsymbol{n}}$ and $n_{A,Z}$ is the number of fragments with mass number A (mass $M_{A,Z}$) and charge number Z. The factor in square brackets is the density of microstates corresponding to position and momentum coordinates for N-fragments with summed kinetic energy, $\varepsilon_{\boldsymbol{n}}$, enclosed in a volume $V_{\boldsymbol{n}}$ with no restrictions on the centre-of-mass position and momentum (equation (1.3.10)). The product $\prod_{A,Z} n_{A,Z}!$ in the denominator is inserted to avoid overcounting of microstates in partitions which contain identical nuclei. The freeze-out volume, $V_{\boldsymbol{n}}$, is the spatial region which encloses the centres of mass of the fragments at scission. It may depend on the partition $\boldsymbol{n}$ but may also be considered to be constant. Our supposition that the centre of mass of each fragment may occupy any point in $V_{\boldsymbol{n}}$ implies that all the volume is available to each fragment. This is obviously a significant approximation (compare with the neutron case) for two reasons. The first is that it implies inclusion of spatial configurations in which two or more fragments overlap (see section 6.7). The second is that fixing the position of the centre of mass reduces the power of $V_{\boldsymbol{n}}$ by one so we might expect to see $V_{\boldsymbol{n}}^{N-1}$ rather than $V_{\boldsymbol{n}}^N$. The latter difference is important for estimation of absolute numbers of microstates. It has no consequence for branching ratios providing $V_{\boldsymbol{n}}$ is constant.

Inclusion of the potential energy corresponding to the interfragment Coulomb interactions as a quantity depending only on the partition is obviously a further approximation. The 'barrier', $B_{\boldsymbol{n}}$, should be considered as an average value, the average being carried out for a given partition over all spatial configurations (this is, in fact, quite a reasonable approximation in the sense that the standard deviation of the mean is usually rather small).

Equation (6.3.10) again represents a convolution of two densities. However, multifragment (scission) states generally contain several excited fragments so that the density of states for internal excitation represented by $\rho_{\mathrm{I}}(E_{\boldsymbol{n}} - \varepsilon_{\boldsymbol{n}}) \equiv \rho_{\mathrm{I}}(E_C^* + Q_{\boldsymbol{n}} - B_{\boldsymbol{n}} - \varepsilon_{\boldsymbol{n}})$ must, itself, be a convolution of the densities corresponding to internal excitation of all members of the partition. Thus

$$\rho_{\mathrm{I}}(E_n - \varepsilon_n) = \int_0^{E_n-\varepsilon_n} \rho_1(E_1^*) \int_0^{E_n-\varepsilon_n} \rho_2(E_2^*) \ldots \int_0^{E_n-\varepsilon_n} \rho_j(E_j^*) \ldots \times \delta\left[\sum E_j^* - (E_n - \varepsilon_n)\right] \mathrm{d}E_1^* \, \mathrm{d}E_2^* \ldots \mathrm{d}E_j^* \ldots \quad (6.3.11)$$

We have seen, in chapter 1, how such integrals can be approximated by locating the maximum of the integrand. We may also remark that microstate densities of fragments (particles) which have no excited states can be represented by delta functions in the integrand.

Whereas we could make a simple picture leading to the absolute decay rate for the separation of a neutron and a residual nucleus, we have, for the moment, no such picture for the multifragment scission configuration. There are two related difficulties. The first is that freeze-out volume, V_n, is, *a priori*, unknown. If we assume that this quantity is constant, we should probably choose a value which is sufficiently large so that fragments have little chance of overlapping. Otherwise equation (6.3.10) is inconsistent. Having done this, a second difficulty appears. The phase space contribution is obtained assuming the fragments have randomly oriented ('inwardly and outwardly directed') velocities. If we now imagine a large freeze-out volume and we *reverse the direction of time* it must be the case that the fragments will coalesce to form the original compound system. If they do not do so we have the 'Weisskopf scenario' where only a fraction of the assumed phase space leads to compound nucleus formation under time reversal. The equivalent condition is guaranteed for neutron scission states because inwardly directed trajectories are immediately 'buried' in the residual nucleus. For multifragmentation, the correction must be estimated numerically. The use of branching ratios (6.3.8) is equivalent to the assumption that the correction is partition independent so that it can be ignored.

Introducing angular momentum is not a simple matter although, in principle, it is easy to constrain the scission configuration to have a fixed angular momentum. It is also relatively straightforward to correct the phase space density to take account of the additional constraint. However, one must further specify the moment of inertia of the multifragment configuration and thus decide if the configuration rotates as a rigid body or if the rotation of the fragments themselves (around internal axes) is decoupled from the rotation of the set of mass centres. Most theories do not include a detailed treatment of angular momentum.

Branching ratios can be calculated for any type of designated macrostate. Thus, as explained in section 6.2, rather than $P(\boldsymbol{n})$, we could in principle calculate $P(\boldsymbol{n}, \boldsymbol{p}(\boldsymbol{n}))$ where the $3N$-dimensional vector $\boldsymbol{p}(\boldsymbol{n})$ specifies the three Cartesian momentum components of all N-fragments of the partition $\boldsymbol{n}$ (as indicated earlier this latter calculation is a many body problem and must be solved numerically). At a more macroscopic level, we might be interested simply in the multiplicity distribution, $P(N)$. However, as mentioned above, for the purposes of explaining model approximations, the partition degree of complexity is usually a useful level at which to define macrostates.

Several groups have proposed models based on the freeze-out 'scenario'. Although we shall describe in detail only a few of these models, it is appropriate at this point, to cite, as further examples, the works of Mekjian [14] who seems to have pioneered the statistical approach to multifragmentation, of Fai and Randrup [15] and of Randrup and Koonin [16]. We should also state that the development described above provides one possible framework for calculations of multifragmentation yields using the techniques of equilibrium statistical mechanics. For possible alternatives (involving rather different approximations) we recommend the reader to the papers of Chung [17], and of Cerutti *et al* [18]. In the first of these works, fragments are formed by distributing nucleons randomly in the freeze-out volume and assembling fragments using an agglomeration mechanism based on the proximity of nucleons in position space (a variant requires also proximity in momentum space). Such models are referred to as coalescence models. In the second approach the freeze-out configurations are simulated using a site-bond percolation model. Nucleons are distributed randomly on a simple cubic lattice on which the number of available sites exceeds the number of nucleons. Fragments are defined as unconnected sets of interconnected occupied sites. The relation between coalescence and percolation models has been discussed by Das Gupta, Gale and Haglin [19].

In the following sections we shall see how the pure statistical mechanical framework (counting microstates as in equation (6.3.10)) is implemented in practical simulations. In each of the descriptions we shall assume that decay takes place from an excited parent nucleus with precisely defined parent mass number A_C (mass M_C), charge number, Z_C, and excitation energy, E_C^*. In practice, due to pre-equilibrium emission and/or to specific features of the reaction mechanism, the models often need to be operated for a range of values of A_C, Z_C and E_C^*.

6.4 The Copenhagen heated liquid drop model

The Copenhagen statistical multifragmentation model [12] (often referred to as SMM) begins from equation (6.3.10) by expanding the density of states corresponding to internal excitation of fragments and integrating over ε_n. Thus, the density of states corresponding to internal excitation is assumed to act as a heat bath ($\rho_\mathrm{I}(E_n - \varepsilon_n) = \rho_\mathrm{I}(E_n)\,\mathrm{e}^{-\varepsilon_n/T}$). Assuming a partition independent freeze-out volume $V_n = V_F$, we can use (6.3.10) to write

$$\begin{aligned}\rho_n(E_n) &= \rho_{\mathrm{I}}(E_n)\left[\prod_{A,Z}\frac{(2\pi M_{A,Z})^{3n_{A,Z}/2}}{n_{A,Z}!}\right]\frac{V_F^N}{h^{3N}}\int_0^\infty \mathrm{e}^{-\varepsilon_n/T}\frac{\varepsilon_n^{3N/2-1}}{\Gamma(3N/2)}\,\mathrm{d}\varepsilon_n \\ &= \rho_{\mathrm{I}}(E_n)\left[\prod_{A,Z}\frac{(M_{A,Z})^{3n_{A,Z}/2}}{n_{A,Z}!}\right]\left[\frac{V_F^N}{h^{3N}}(2\pi T)^{3N/2}\right] \\ &= \rho_{\mathrm{I}}(E_n)\left[\prod_{A,Z}\frac{A^{3n_{A,Z}/2}}{n_{A,Z}!}\right]\left[\frac{V_F}{\lambda_{\mathrm{T}}^3}\right]^N, \end{aligned} \tag{6.4.1}$$

in which $M_{A,Z} \approx A \times m_N$ and the thermal wavelength, λ_{T}, is given in the terms of the nucleon mass, m_N, as

$$\lambda_{\mathrm{T}} = \sqrt{\frac{2\pi\hbar^2}{m_N T}}, \tag{6.4.2}$$

(with reference to equation (6.3.5), λ_{T} is the de Broglie wavelength of a nucleon with the 'thermal' velocity $\sqrt{T/2\pi m_N}$). In fact, the authors of reference [12] allow for some multiplicity dependence of the freeze-out volume but this is not essential for understanding the main structure of the model.

In equation (6.4.1), neither the centre-of-mass momentum nor the position of the centre of mass is fixed, so that these quantities would be expected to be dispersed around average values.

The density of states for internal excitation, $\rho_{\mathrm{I}}(E_n)$, is represented in terms of the internal entropy S_{I} as

$$\rho_{\mathrm{I}}(E_n)\Delta E_n = \mathrm{e}^{S_{\mathrm{I}}}. \tag{6.4.3}$$

Here the model uses a further approximation which is to represent the convolution integral (6.3.11) (with summed energy E_n rather than $E_n - \varepsilon_n$) by the maximum of the integrand. In chapter 1 (section 1.3) we saw that this is equivalent to assuming that each fragment is characterized by the same temperature in which case the integral can be represented as a simple product. Approximating $\rho_{\mathrm{I}}\Delta E_n$ by a product of terms for equal temperature fragments is equivalent to expressing the total entropy as a sum over all members of the partition. Thus

$$S_{\mathrm{I}} = \sum_{A,Z} n_{A,Z} S_{A,Z}. \tag{6.4.4}$$

The nature of this step should be clearly understood. The temperature T which characterizes the thermal wavelength λ_{T} is indeed given by $1/T = \mathrm{d}\ln[\rho_{\mathrm{I}}(E^*)]/\mathrm{d}E^*$ at $E^* = E_n$ and the expression (6.4.1) is a (good) approximation to the full microcanonical sum. However, replacing the convolution integral which represents the density of states corresponding to internal excitation by the maximum of the multidimensional integrand is a new approximation. The convolution cannot be rigorously reduced to a product because there is no 'external' heat bath. Nevertheless, equation (6.4.1) with (6.4.3) still represents an approximate expression for the microcanonical density of states. One consequence of this strategy is that the characteristic temperature would be expected to depend on the partition.

The entropy, S_{I}, is obtained in the Copenhagen model using a temperature-dependent liquid drop model in which the entropy for any given fragment is expressed as the derivative of the free energy, $F_{A,Z}$, (at constant volume).

$$S_{A,Z} = -\frac{\partial F_{A,Z}}{\partial T}. \tag{6.4.5}$$

To calculate the sum in (6.4.4) we therefore need to specify the free energy for each fragment as a function of the temperature. This is achieved using a liquid drop model for the internal energy. It will be recalled that the internal energy, which in the present context is written $U(A, Z, T)$, is the average energy of a nucleus associated with a heat bath system characterized by the temperature, T. The energy is usually specified with respect to the mass (energy) of Z free protons and $A - Z$ free neutrons. The internal energy at $T = 0$ (ground state) is thus the negative of the binding energy defined in section 2.2. and, at finite temperature, is expressed using the liquid drop model (compare with equation (2.2.28)) as

$$\begin{aligned} U(A, Z, T) = {} & \left(W_0 + \frac{T^2}{\varepsilon_0}\right)A + \left(\beta(T) - T\frac{\mathrm{d}\beta(T)}{\mathrm{d}T}\right)A^{2/3} \\ & + \frac{3}{5}\frac{Z^2\mathrm{e}^2}{R_{A,Z}} + \gamma\frac{(A-2Z)^2}{A}. \end{aligned} \tag{6.4.6}$$

The first (volume) term in equation (6.4.6) is referred to as the 'bulk' contribution ($W_0 = -16$ MeV). In the equidistant spacing Fermi gas model presented in chapter 2, the excitation energy is given by $E^* = aT^2$ where $a = A/\varepsilon_0$. We explained in section 2.5 that this relation also applies to real nuclei provided that a surface diffuseness correction to the level density parameter, a, is made. This results in $\varepsilon_0 \approx 8.5$ MeV. In the present case, the surface is treated separately (although no correction for the diffuseness is included) so that the Copenhagen model uses the value obtained from studies of infinite nuclear matter ($\varepsilon_0 = 16$ MeV).

The free energy corresponding to the surface tension is written as $\beta(T)$ (the notation could be improved) so that using the Gibbs–Helmholtz equation ($U(T) = F - T\partial F/\partial T$) the surface contribution to the internal energy is $\beta(T) - T\partial\beta(T)/\partial T$ (again at constant volume). The asymmetry coefficient is denoted 'γ' rather than 'a_A' and the pairing energy is ignored.

Defining $U_B(A, Z, T) = \left(W_0 + \frac{T^2}{\varepsilon_0}\right)A$, the bulk (volume) contribution to the free energy, F_B, can be obtained by writing the Gibbs–Helmholtz equation as an inhomogeneous differential equation

$$\frac{\partial F_B}{\partial T} + \frac{-F_B}{T} + \left(\frac{AW_0}{T} + \frac{AT}{\varepsilon_0}\right) = 0. \tag{6.4.7}$$

The homogeneous equation has the solution $F_B(T) = cT$ and a particular solution to the inhomogeneous equation is thus (see any text on linear first order differential equations)

$$-T\int\left(\frac{AW_0}{T} + \frac{AT}{\varepsilon_0}\right)\frac{\mathrm{d}T}{T} = -T\left(\frac{-AW_0}{T} + \frac{AT}{\varepsilon_0}\right), \tag{6.4.8}$$

so that the full solution is

$$F_B(A, Z, T) = cT + AW_0 - \frac{AT^2}{\varepsilon_0}. \tag{6.4.9}$$

The entropy $-\partial F_B/\partial T$ (at constant volume) is immediately obtained from (6.4.9) as

$$S_B(A, Z, T) = -\frac{\partial F_B}{\partial T} = -c + \frac{2AT}{\varepsilon_0}, \tag{6.4.10}$$

from which the constant c is eliminated by setting $S_B(A, Z, T = 0) = 0$ (one may remark that the same result is obtained by substituting $T = \sqrt{E^*/a}$ in the level density $\mathrm{e}^{2\sqrt{aE^*}}$).

The surface contribution to the entropy is perhaps the most significant feature of the model. In a real liquid drop, the surface tension (the surface free energy per unit area) vanishes at the critical temperature of the liquid–gas phase transition. A parameterization of the free energy which includes this effect while maintaining the calculated behaviour at lower temperatures [20] leads to

$$\beta(T) = \beta_0\left[\frac{(T_c^2 - T^2)}{(T_c^2 + T^2)}\right]^{5/4}, \tag{6.4.11}$$

where $T_c(\approx 12$ MeV) is the critical temperature and β_0 is the surface energy coefficient at $T = 0$ (here taken as 18 MeV). It follows (from (6.4.6)) that the surface contribution to the entropy is

$$S_S(A, Z, T) = -\frac{\mathrm{d}\beta(T)}{\mathrm{d}T}A^{2/3}. \tag{6.4.12}$$

The asymmetry contribution to the liquid drop free energy is assumed to be temperature independent so that it does not contribute to the entropy. A possible contribution from the pairing energy (which would be expected to be temperature dependent) is ignored. Finally, the Coulomb energy, which is the sum of the internal energy of the fragments together with the interaction energy, is also assumed to be temperature independent (this does not seem unreasonable) so that the total liquid drop entropy of a fragment is obtained by adding the bulk and surface contributions. We find (equations (6.4.10)–(6.4.12))

$$S(A, Z, T) = \left[\frac{2}{\varepsilon_0} + \frac{5\beta_0 T_c^2}{A^{1/3}(T_c^2 + T^2)^2}\left[\frac{(T_c^2 - T^2)}{(T_c^2 + T^2)}\right]^{1/4}\right] AT. \qquad (6.4.13)$$

Very light nuclei require special attention. Nuclei smaller than ^{4}He are considered to be in their ground states. Consequently, they make no contribution to the liquid drop entropy. With this condition, equations (6.4.1), (6.4.4) and (6.4.13) are sufficient to provide an approximate estimation of the density of microstates to be associated with any given partition.

There is another interesting feature which is put to good use in the model. An average value of the total energy of the multifragment freeze-out configuration can be estimated. This involves calculating a 'liquid drop' Q-value using equation (6.4.6). The original excitation energy of the parent nucleus is given as

$$E_C^* = -Q_{\boldsymbol{n}} + B_{\boldsymbol{n}} + E_{\mathrm{I}} + \varepsilon_{\boldsymbol{n}}, \qquad (6.4.14)$$

where $B_{\boldsymbol{n}}$ is the energy corresponding to inter-fragment Coulomb interactions (averaged over spatial configurations), and the average value of $\varepsilon_{\boldsymbol{n}}$ is $3NT/2$. The right-hand side of (6.4.14) obviously depends on the temperature, T, so that, by adjusting T for each partition, it is possible to ensure that, on average, each and every partition is produced from an excited parent (compound) nucleus with a fixed excitation energy. Of course, this means that the temperature will fluctuate about an average value. However, in practice it turns out that these fluctuations are not too important so that the overall coherence of the model is not seriously affected. One consequence of this procedure is that the temperature and therefore the statistical weight, $\rho_{\boldsymbol{n}}(E_{\boldsymbol{n}})$, is influenced by the Coulomb interaction energy.

The Coulomb term in equation (6.4.14) requires precision. To a first approximation the Coulomb energy of a given partition, $\boldsymbol{n}$, can be written as the energy of a uniformly charged sphere of radius R_F, i.e., $\frac{3}{5}\frac{Z_C^2 e^2}{R_F}$. However, it is necessary to correct this expression because the total charge, Z_C, is 'clustered' inside the fragments. If the total charge were to be distributed uniformly over the freeze-out volume, a fragment with charge number, Z, and mass number, A, would occupy a volume, AV_F/A_C. Assimilating this volume to a spherical cell, we find that the radius of the cell is

$$R_{\text{cell}} = \left[\frac{3V_F A}{4\pi A_C}\right]^{1/3}, \qquad (6.4.15)$$

and that the self Coulomb energy of the charged cell is $\frac{3}{5}\frac{Z^2 e^2}{R_{\text{cell}}}$. The Coulomb energy of the fragment considered as the collapse of a cell charge into a sphere with radius $R_{A,Z}$ is obviously $\frac{3}{5}\frac{Z^2 e^2}{R_{A,Z}}$. The electrostatic potential outside the cell radius is not affected by the collapse, so other contributions to the total Coulomb energy are unchanged. The formation of a single fragment thus leads to an increase in the Coulomb energy of

$$\Delta E_{A,Z} = \frac{3}{5}\frac{Z^2 e^2}{R_{A,Z}}\left[1 - \frac{R_{A,Z}}{R_{\text{cell}}}\right] = \frac{3}{5}\frac{Z^2 e^2}{R_{A,Z}}\left[1 - \left(\frac{\rho_F}{\rho_0}\right)^{1/3}\right], \tag{6.4.16}$$

where the second expression on the right arises from the consideration that if the freeze-out density is $\rho_F = (3V_F/4\pi)^{1/3}$ and if the fragments are formed with the 'normal' nuclear density, ρ_0, then $R_{A,Z}/R_{\text{cell}} = (\rho_F/\rho_0)^{1/3}$.

An approximation to the total Coulomb energy of the freeze-out configuration is thus obtained by supposing that (6.4.16) holds for a single fragment even when the remaining charge is not uniformly distributed but is itself contained in other fragments. In this case the total *interaction* Coulomb energy of a multifragment configuration is the total Coulomb energy minus the self Coulomb energy of the fragments

$$\begin{aligned} B_n &= \left(\frac{3}{5}\frac{Z_C^2 e^2}{R_F} + \sum_{A,Z} n_{A,Z}\Delta E_{A,Z}\right) - \sum_{A,Z} n_{A,Z}\left(\frac{3}{5}\frac{Z^2 e^2}{R_{A,Z}}\right) \\ &= -\frac{3}{5}\frac{Z_C^2 e^2}{R_F} - \sum_{A,Z} n_{A,Z}\left[\frac{3}{5}\frac{Z^2 e^2}{R_{A,Z}}\left(\frac{\rho_F}{\rho_0}\right)^{1/3}\right] \\ &= \left(\frac{\rho_F}{\rho_0}\right)^{1/3}\left[\frac{3}{5}\frac{Z_C^2 e^2}{R_C} - \sum_{A,Z} n_{A,Z}\left(\frac{3}{5}\frac{Z^2 e^2}{R_{A,Z}}\right)\right], \end{aligned} \tag{6.4.17}$$

where R_C is the radius of the compound nucleus at normal density, ρ_0. For future reference, we note in passing that the term in square brackets in the last line of (6.4.17) is equal to the Coulomb contribution to the liquid drop multifragmentation Q-value.

Knowledge of the statistical weight, $\rho_n(E_n)$, to be assigned to any given partition, may be used in a Monte Carlo event generation procedure. The model implies that fragments and light particles in the partition are initially disposed at random positions with their mass centres constrained to lie inside the freeze out volume. The fragment velocities are random and the mean kinetic energy of each fragment is $3T/2$. Furthermore, each fragment experiences the Coulomb repulsion of all other members of the partition under consideration. Starting from any particular configuration, the subsequent trajectories of the fragments can thus be calculated by integrating the corresponding set of coupled differential equations (there are standard methods for doing this). This is most useful for comparison with experiment because detection efficiency and evaporation corrections can be directly included in the simulations.

There are two distinct ways of carrying out Monte Carlo simulations. In both methods, the Monte Carlo event generation procedure requires selection of a large number of partitions. In the work reported in [12] the multiplicity distribution plays an important role. Successive events are generated by sampling the unbiased multiplicity distribution, $P_{A_C}(N)$, which represents the number of ways the compound nucleus (i.e., the integer, A_C) can be divided into exactly N parts. Having selected N, a particular partition is generated by making N–1 random cuts on a line of A_C points. Charges are then assigned to each fragment by following (as far as possible) the liquid drop predictions for the most stable isotopes. Finally, the statistical weight for the selected partition is calculated using equation (6.4.1).

The set of events (macrostates) generated in this way is not strictly random (i.e., selected with uniform probability) but hopefully associated with a significant fraction of the statistical weight. If this is true then, for example, the average value of an observable, O, which depends only on the partition, may be estimated in terms of the weights, ρ_n, as

$$\langle O \rangle = \frac{\sum_n O(\boldsymbol{n})\rho_n}{\sum_n \rho_n}. \tag{6.4.18}$$

In most cases, it is difficult to be sure that the generated event set is indeed associated with a substantial fraction of the statistical weight. The problem is common to most simulations of this type. The only way to improve confidence in the results is by running the simulation program for larger and larger numbers of events and verifying the convergence of interesting observables (i.e., verifying that $\langle O \rangle$ in (6.4.18) does not significantly change when the number of events is increased by, say, a factor of ten).

An alternative method which is used in most modern simulations involves sampling the space of microstates in such a way that the frequencies of generated macrostates are not uniform but actually reflect the statistical weights. This will be explained in more detail in section 6.5. For the moment, however, let us just say that, in this case, mean values are estimated using an equation of the form of (6.4.18).

An example of the capacity of the SMM code to fit experimental data is shown in figure 6.2. The data were obtained using a 600 MeV/nucleon Au beam incident on C, Al, Cu and Pb targets using the ALADIN forward spectrometer at GSI in Darmstadt [21]. The role of the target is secondary in the experiment. It serves to produce highly excited projectile-like nuclei which subsequently decay by multifragmentation. Thus, the data presented in the figure, which concerns specific characteristics of the charge distribution, are similar for all targets. The input for the calculation is the correlated distribution of initial mass and excitation energy of projectile-like fragments. Both quantities depend on the impact parameter characterizing the collision. One would expect that, the smaller the impact parameter, the greater the excitation energy and the loss of mass from the projectile. The correlation required to fit the data confirms this expectation. The accuracy of the predictions is rather striking.

Before leaving this section we wish to consider a canonical model [22] which is similar to but not identical with that of reference [12]. Let us begin again with (6.4.1) and write again that the density of states corresponding to internal

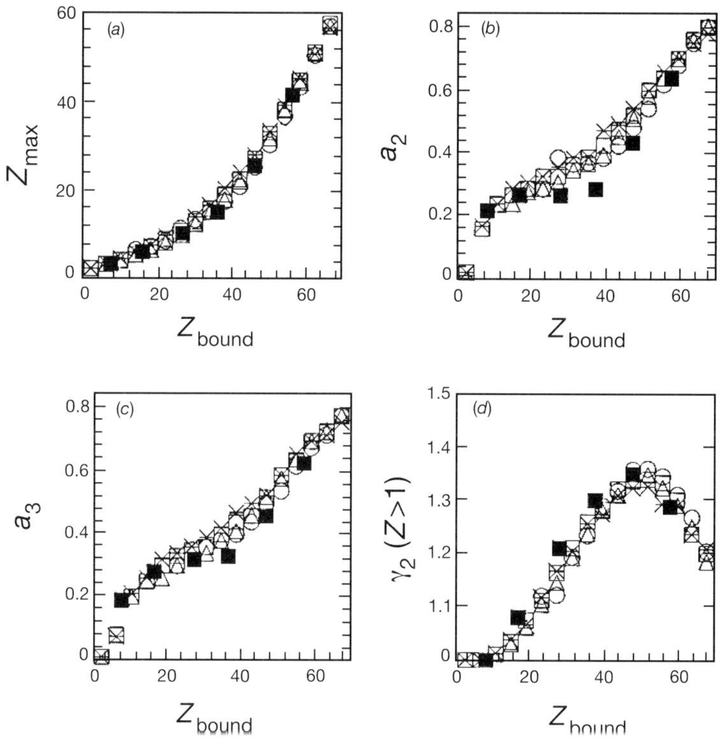

Figure 6.2. Data obtained using the ALADIN apparatus with a 600 MeV/nucleon Au beam incident on C, Al, Cu and Pb targets [21]. As can be seen, the data, which corresponds to projectile fragmentation, are very similar for all four targets. The quantity Z_{bound} is the summed charge of complex fragments (light particles excluded), Z_1(or Z_{max}), Z_2 and Z_3 are the largest, second largest and third largest charges in any event and $\langle Z\rangle = (Z_1 + Z_2 + Z_3)/3$. (*a*) Z_{max}, maximum charge in the event. (*b*) the asymmetry $a_2 = (Z_1 - Z_2)/(Z_1 + Z_2)$. (*c*) $a_3 = \sqrt{(Z_1 - \langle Z\rangle)^2 + (Z_2 - \langle Z\rangle)^2 + (Z_3 - \langle Z\rangle)^2}\,/\sqrt{6}\langle Z\rangle$. (*d*) The γ_2 moment (see section 7.2). The predictions (black squares) were made using the Copenhagen SMM code. (Reprinted from *Nucl. Phys.* A, **561**, Barz H W *et al*, p 466. Copyright 1993, with permission from Elsevier Science.)

excitation of the fragments with total energy $E_n = E_C^* + Q_n - B_n$ is given by (see (6.3.11))

$$\rho_{\mathrm{I}}(E_n) = \int_0^{E_n} \rho_{\mathrm{I}}(E_1^*) \int_0^{E_n} \rho_2(E_2^*) \ldots \int_0^{E_n} \rho_j(E_j^*) \ldots \times \delta\left[\sum E_j^* - E_n\right] \mathrm{d}E_1^*\,\mathrm{d}E_2^* \ldots \mathrm{d}E_j^* \ldots, \tag{6.4.19}$$

where $\rho_j(E_j^*)$ is the microstate density of the jth *complex* fragment with mass number A_j at an excitation energy E_j^*. As in the Copenhagen model, we suppose that the integrand in (6.4.19) is strongly peaked around its maximum value (which occurs when all fragments have the same energy per nucleon, i.e., the same temperature) so that the multiple integral is numerically (if not dimensionally) well approximated by its maximum value. However, in contrast with the Copenhagen model, we do not introduce a critical temperature and simply take the microstate densities of complex fragments to be given by

$$\rho_j(E_j^*) = \mathrm{e}^{2\sqrt{a_j E_j^*}}, \tag{6.4.20}$$

with $a_j = A_j/\varepsilon_0$ ($\varepsilon_0 \approx 8.5$ MeV), so that the maximum value approximation to the multidimensional integral gives (recall that the maximum occurs when, for all j, $\sqrt{E_j^*/a_j} = T$)

$$\rho_{\mathrm{I}}(E_n) \approx \mathrm{e}^{2\sqrt{aE_n}}, \tag{6.4.21}$$

with

$$a = \sum_j a_j = \frac{1}{\varepsilon_0}\sum_j A_j \approx A_C/\varepsilon_0 = a_C, \tag{6.4.22}$$

where A_j is the mass number of the jth complex fragment in the partititon.

The important approximation is, of course, that (6.4.21) is supposed to hold even for partitions which contain light particles whose level densities are delta functions. In practice, the sum $\sum_j a_j$ in (6.4.22) would be expected not only to be less than a_C but to show some dependence on the partition. This variation would, in turn, induce some additional partition dependence of the temperature especially for high multiplicity partitions which contain a significant number of light particles.

In spite of the approximate nature of (6.4.22) we adopt (6.4.21) and expand the square root in the exponential so that with $1/T = \mathrm{d}(2\sqrt{a_C E^*})/\mathrm{d}E^*$ at $E^* = E_C^*$ and $\Delta E = E_C^* - E_n$

$$\rho_{\mathrm{I}}(E_n) = \mathrm{e}^{2\sqrt{a_C(E_C^*-\Delta E)}} \approx \mathrm{e}^{2\sqrt{a_C E_C^*}}\,\mathrm{e}^{-\Delta E/T}. \tag{6.4.23}$$

With $\Delta E = B_n - Q_n$ we can now write the full density of states (6.4.1) as

$$\begin{aligned}\rho_n(E_n) &= \mathrm{e}^{2\sqrt{a_C E_n}}\left[\prod_{A,Z}\frac{A^{3n_{A,Z}/2}}{n_{A,Z}!}\right]\left[\frac{V_F}{\lambda_{\mathrm{T}}^3}\right]^N \\ &= \mathrm{e}^{2\sqrt{a_C E_C^*}}\left[\prod_{A,Z}\frac{A^{3n_{A,Z}/2}}{n_{A,Z}!}\right]\left[\frac{V_F}{\lambda_{\mathrm{T}}^3}\right]^N \mathrm{e}^{-(B_n-Q_n)/T}. \end{aligned}\tag{6.4.24}$$

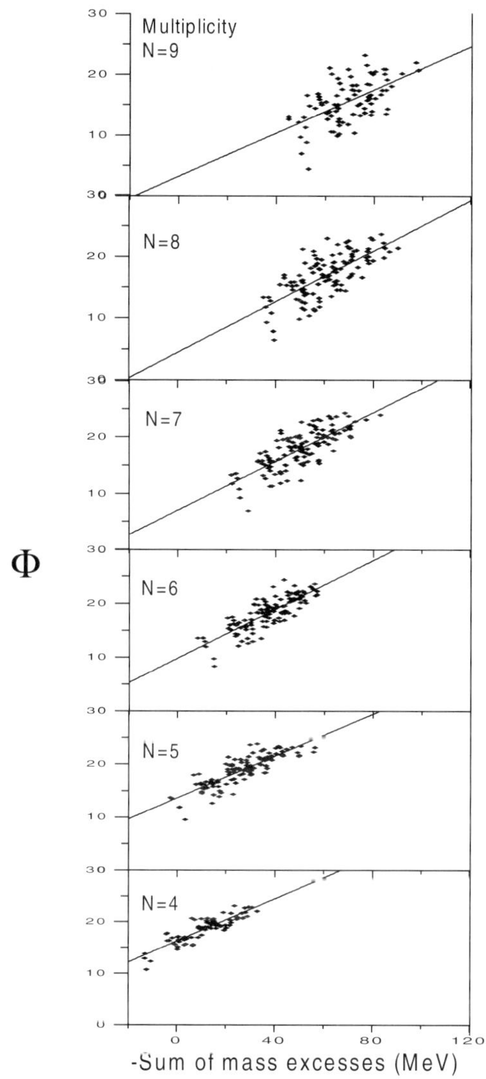

Figure 6.3. The function $\Phi(\boldsymbol{n})$ defined in equation (6.4.27) plotted against the negative of the sum of mass excesses (to within a constant, the Q-value) for partitions of fixed multiplicity, N, observed in central collisions of $^{32}S + ^{27}Al$ at 35 MeV/nucleon [22]. The slope of the straight lines fitted to the data indicates a temperature of 5.5 MeV.

As in the Copenhagen model, it is important to emphasize that while equation (6.4.24) has the appearance of a canonical weight, it in fact represents an approximation to the full microcanonical sum for a given partition of the original isolated

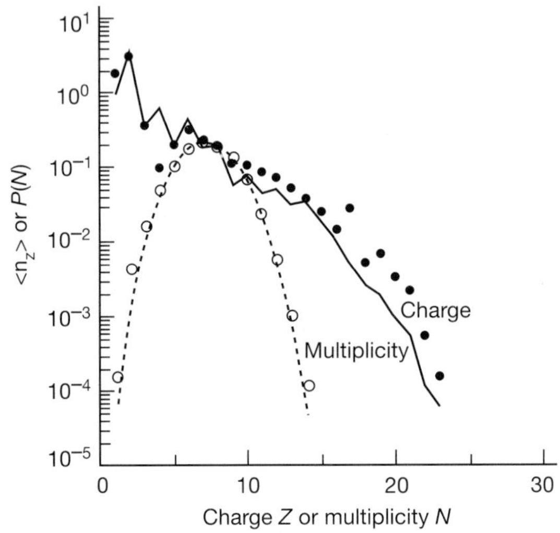

Figure 6.4. 'Boltzmann' prediction for charge (continuous line) and multiplicity (broken curve) distributions compared with the measured values (extracted from the partition probabilities) in the $^{32}S + ^{27}Al$ at 35 MeV/nucleon [22].

system. Note that $B_{\boldsymbol{n}}$ which we take to be the average, over fragment positions, of the Coulomb interaction energy of the partition $\boldsymbol{n}$, appears explicitly in the statistical weight.

Suppose now that we have a set of measured events. We estimate the partition probabilities from the observed frequencies and then separate the probabilities according to multiplicity. If the partition frequencies correspond to the weights given by (6.4.24), then for a given multiplicity we can write the logarithm of the partition probability as

$$\ln[P_{\boldsymbol{n}}] = 2\sqrt{a_C E_C^*} + N \ln[V_F] + \frac{3}{2}\sum_{A,Z} n_{A,Z} \ln[A] - \sum_{A,Z} \ln[n_{A,Z}!] \\ - 3N \ln[\lambda_{\mathrm{T}}] - \frac{B_{\boldsymbol{n}}}{T} + \frac{Q_{\boldsymbol{n}}}{T} - \ln[\Omega], \tag{6.4.25}$$

where $\ln[\Omega]$ is a normalization constant (in fact Ω is, approximately, the microcanonical sum of states). Let us concentrate for a moment on the Coulomb energy. By examination of equation (6.4.17), we find that $B_{\boldsymbol{n}}$ would be expected to depend, to first order, on the multiplicity rather than on the details of the partition. We may thus attempt to expand the term $B_{\boldsymbol{n}}$ as a polynomial in N. Investigation of this problem based either on numerical simulation [22] or directly using equation (6.4.17) shows that the first order expansion is indeed adequate over a wide range of multiplicity. Thus we write

$$B_{\boldsymbol{n}} = a_C + b_C N, \tag{6.4.26}$$

and use (6.4.25) (grouping constants in round brackets) to define

$$\begin{aligned}\Phi(\boldsymbol{n}) &= \ln[P_{\boldsymbol{n}}] - \frac{3}{2}\sum_{A,Z} n_{A,Z}\ln[A] + \sum_{A,Z}\ln[n_{A,Z}!] \\ &= \left(2\sqrt{a_C E_C^*} - \ln[\Omega] - \frac{a_C}{T}\right) \\ &\quad + N\left[\ln[V_F] - 3\ln[\lambda_T] - \frac{b_C}{T}\right] + \frac{Q_{\boldsymbol{n}}}{T}.\end{aligned} \tag{6.4.27}$$

Using the first line of equation (6.4.27), $\Phi(\boldsymbol{n})$ can be directly computed for any partition whose probability has been estimated from the experiment. For events with a fixed multiplicity, the right-hand side of equation (6.4.27) may be considered to vary linearly as the sum of the mass excesses of the fragments. Thus, writing $Q_{\boldsymbol{n}}$ as the difference between the mass excess of the parent nucleus and the sum of mass excesses of the fragments ($\Delta(\text{parent}) - \sum \Delta(\text{fragments})$) we obtain

$$\begin{aligned}\Phi(\boldsymbol{n}) = &\left(2\sqrt{a_C E_C^*} - \ln[\Omega] - \frac{a_C}{T} + \frac{\Delta(\text{parent})}{T}\right) \\ &+ N\left[\ln[V_F] - 3\ln[\lambda_T] - \frac{b_C}{T}\right] + \frac{-\sum_{A,Z} n_{A,Z}\Delta(A,Z)}{T}.\end{aligned} \tag{6.4.28}$$

Plotting $\Phi(\boldsymbol{n})$ versus $-\sum_{A,Z} n_{A,Z}\Delta(A,Z)$ for fixed N, we find the slope is the inverse of the temperature (which may depend on the multiplicity).

Equation (6.4.28) represents one of the simplest methods of analysing partition probabilities because it involves a simple straight line fit to the data. This procedure has actually been carried out [22] using measured partition probabilities for multifragmentation in the $^{32}S + ^{27}Al$ reaction at 37.5 MeV/nucleon (^{32}S incident energy). Of course the problem is rather more complicated than may be suggested by equation (6.4.28). Thus, in the analysis of the experiment (which was carried out using the AMPHORA multidetector), it was necessary to take into account the fact that charge, rather than charge and mass, partitions were measured. Corrections for detection efficiency, pre-equilibrium emission and post-multifragmentation evaporation were also considered.

Results of the analysis are shown in figures 6.3 and 6.4. In figure 6.3, the quantity $\Phi(\boldsymbol{n})$ is plotted versus the negative of the sum of mass excesses of the fragments in any given partition. The increased dispersion observed with increasing multiplicity is probably due to the progressive breakdown of the approximation embodied in equation (6.4.22) (too many light particles). The extracted temperature, which turns out to be almost independent of multiplicity, is 5.5 MeV.

A most useful feature of the 'temperature plots' shown in figure 6.3 is that the intercepts of these plots may themselves be plotted against the multiplicity N.

Thus, if we define $\alpha(N)$ as

$$\alpha(N) = \left(2\sqrt{aE_C^*} - \ln[\Omega] - \frac{a_C}{T} + \frac{\Delta(\text{parent})}{T}\right) + N\left[\ln[V_F] - 3\ln[\lambda_\mathrm{T}] - \frac{b_C}{T}\right], \tag{6.4.29}$$

we see that, if the temperature is known (especially if it is independent of N), the slope $\ln[V_F] - 3\ln[\lambda_\mathrm{T}] - b_C/T$ may be used to determine the freeze-out volume (in practice, this calculation must be performed numerically because b_C depends weakly on V_F). The analysis of the AMPHORA data suggests that the freeze-out volume corresponds to a nucleus which has expanded to about 2.7 times its normal size.

Given the extracted values of the freeze-out volume and the temperature, it is possible to calculate all partition probabilities. The corresponding predictions for the charge and multiplicity distributions (equations (6.2.2) and (6.2.3)) are compared with the experimental data in figure 6.4. The experimental distributions are well reproduced by the simple 'Boltzmann' theory in both cases.

6.5 The Metropolis algorithm

In many applications it is useful to dispose of a simulation which generates a set of 'theoretical' events which may be (numerically) corrected for imperfections of the detection system (and possibly for evaporation) and thus compared directly with the measured event set. One way of setting up such a simulation involves the Metropolis algorithm [23]. Its success is due to the fact that only relative weights rather than absolute probabilities need be known. For this reason the technique is sometimes referred to as importance sampling.

It would be possible to introduce importance sampling in a purely formal manner. However, given the preoccupation of this book (in particular the application of the method to multifragmentation), it is probably more useful to employ the terminology of statistical mechanics as applied to partition probabilities. Consider, therefore, a set of macrostates defined by the vectors $\boldsymbol{n}$. As a concrete example we could consider the set of charge partitions of a parent nucleus with charge Z_C and define $\boldsymbol{n} = \{n_1, n_2 \ldots n_{Z_C}\}$ where n_k is the number of fragments of charge k and $\sum_k kn_k = Z_C$ so that a given vector $\boldsymbol{n}$ refers to one particular charge partition. A physical measurement of a (large) number of events can be considered as an estimation of the partition probabilities. The measured values are to be compared with the theoretical multivariate probability distribution $P_\mathrm{M}(\boldsymbol{n})$ arising from some model of the physical process.

The simulation should generate a set of events such that the frequency distribution $\phi_\mathrm{M}(\boldsymbol{n})$ (the number of times that the partition, $\boldsymbol{n}$, was generated) corresponds to the model probability distribution $P_\mathrm{M}(\boldsymbol{n})$, i.e.,

$$\mathrm{Lim}N_{t\to\infty}\frac{\phi_\mathrm{M}(\boldsymbol{n})}{N_t} = P_\mathrm{M}(\boldsymbol{n}), \tag{6.5.1}$$

where, of course, the number of trials (events) $N_t = \sum_n \phi_M(\boldsymbol{n})$. The average value of any observable $O(\boldsymbol{n})$ is then estimated, for any value of N_t, as

$$\langle O \rangle = \frac{\sum_n O(\boldsymbol{n})\phi_M(\boldsymbol{n})}{\sum_n \phi_M(\boldsymbol{n})}. \tag{6.5.2}$$

Any arbitrary frequency distribution, $\phi(\boldsymbol{n})$, can be considered as the components of a 'frequency' vector developed on the basis set $\boldsymbol{n}$. Two distinct frequency vectors with components $\phi(\boldsymbol{n})$ and $\phi'(\boldsymbol{n}')$ are connected by a stochastic transition matrix $T(\boldsymbol{n}' \leftarrow \boldsymbol{n})$ such that

$$\phi'(\boldsymbol{n}') = \sum_{n} T(\boldsymbol{n}' \leftarrow \boldsymbol{n})\phi(\boldsymbol{n}). \tag{6.5.3}$$

To understand equation (6.5.3), it is sufficient to note that an operator which changes any vector $\boldsymbol{n}$ into a set of vectors $\boldsymbol{n}'$ with probabilities $T(\boldsymbol{n}' \leftarrow \boldsymbol{n})$ produces a change in the frequency distribution. (You either transform the system vector and keep the basis vectors fixed or transform the basis and keep the system fixed. $T(\boldsymbol{n}' \leftarrow \boldsymbol{n})$ represents a transformation of the basis.) We also require $T(\boldsymbol{n}' \leftarrow \boldsymbol{n}) > 0$, i.e., that the transition matrix, T, 'connects' every pair of basis vectors of the space.

The conservation of probability implies of course that $\sum_{n'} T(\boldsymbol{n}' \leftarrow \boldsymbol{n}) = 1$ (provided that the $\boldsymbol{n}' = \boldsymbol{n}$ term is included in the sum). If this seems obscure, imagine that $\phi(\boldsymbol{n})$ is non-zero for just one partition, $\boldsymbol{n}^*$, and 0 for all others. The action of $T(\boldsymbol{n}' \leftarrow \boldsymbol{n}^*)$ is to distribute the events contained in $\phi(\boldsymbol{n}^*)$ over all macrostates, $\boldsymbol{n}'$. Conservation of events then implies $\sum_{n'} T(\boldsymbol{n}' \leftarrow \boldsymbol{n}^*) = 1$ and it is easy to see that the same relation holds for any initial state, $\boldsymbol{n}$, with associated frequency $\phi(\boldsymbol{n})$. Obviously, there is no corresponding rule for $\sum_n T(\boldsymbol{n}' \leftarrow \boldsymbol{n})$.

Dividing both sides of equation (6.5.3) by N_t gives (at least in the limit of large N_t)

$$P'(\boldsymbol{n}') = \sum_{n} T(\boldsymbol{n}' \leftarrow \boldsymbol{n})P(\boldsymbol{n}), \tag{6.5.4}$$

where $P'(\boldsymbol{n}')$ and $P(\boldsymbol{n})$ are two distinct probability distributions. Equation (6.5.4) may be interpreted by saying that we *apply* the transformation T to the probability distribution P.

Given the properties of T, one can show (the derivation is not trivial: see Cox and Miller [24]) that, by repeated application of the transformation matrix, we arrive at a probability distribution $P_E(\boldsymbol{n})$ which does not change. We will refer to $P_E(\boldsymbol{n})$ as the equilibrium distribution. It represents a fixed point of T and can be shown to be unique. The stationary property may be ensured if, at equilibrium, $P_E(\boldsymbol{n})$ satisfies the law of detailed balance which states that for distinct states $\boldsymbol{n}'$ and $\boldsymbol{n}$,

$$T(\boldsymbol{n}' \leftarrow \boldsymbol{n})P_E(\boldsymbol{n}) = T(\boldsymbol{n} \leftarrow \boldsymbol{n}')P_E(\boldsymbol{n}'). \tag{6.5.5}$$

This is because, in a single 'update' the change in the occupation probability of the state $\boldsymbol{n}'$ may be obtained as the sum of a loss term which represents transitions 'out' of the state $\boldsymbol{n}'$ to all other states and a gain term (transitions from all other states into $\boldsymbol{n}'$)

$$\Delta P(\boldsymbol{n}') = -\sum_{\boldsymbol{n}\neq\boldsymbol{n}'} T(\boldsymbol{n} \leftarrow \boldsymbol{n}')P(\boldsymbol{n}') + \sum_{\boldsymbol{n}\neq\boldsymbol{n}'} T(\boldsymbol{n}' \leftarrow \boldsymbol{n})P(\boldsymbol{n}). \tag{6.5.6}$$

If detailed balance is satisfied $T(\boldsymbol{n} \leftarrow \boldsymbol{n}') = T(\boldsymbol{n}' \leftarrow \boldsymbol{n})\frac{P(\boldsymbol{n})}{P(\boldsymbol{n}')}$ so that

$$\Delta P(\boldsymbol{n}') = -\sum_{\boldsymbol{n}} T(\boldsymbol{n}' \leftarrow \boldsymbol{n})\frac{P(\boldsymbol{n})}{P(\boldsymbol{n}')}P(\boldsymbol{n}') + \sum_{\boldsymbol{n}} T(\boldsymbol{n}' \leftarrow \boldsymbol{n})P(\boldsymbol{n}) = 0. \tag{6.5.7}$$

Let us see what equations (6.5.4) and (6.5.5) imply in practice. We may begin with some arbitrary set of N_t events which can be characterized by the frequency distribution $\phi_0(\boldsymbol{n})$. Each event corresponds to a particular partition, $\boldsymbol{n}$, and indeed, may be considered to constitute a special frequency distribution with unit frequency at $\boldsymbol{n}$ and zeros elsewhere. We apply T, successively, to each of these special distributions (*each of the N_t events*), and sum the results to calculate the new frequency distribution (we could obtain the same result by operating on the frequency distribution itself but we choose to go the 'long way round'). We now reiterate the operation making successive transformations of the set of events until, according to some criterion which we may have imposed, the frequency distribution is considered to be stationary.

The actual choice of T obviously determines the equilibrium distribution and depends on the physics of the problem. Metropolis *et al* proposed to construct the transition matrix using the model probability distribution. Thus, for distinct macrostates $\boldsymbol{n}$ and $\boldsymbol{n}'$,

$$\begin{aligned} T(\boldsymbol{n}' \leftarrow \boldsymbol{n}) &= 1 && \text{if} \quad P_{\mathrm{M}}(\boldsymbol{n}') > P_{\mathrm{M}}(\boldsymbol{n}), \\ T(\boldsymbol{n}' \leftarrow \boldsymbol{n}) &= \frac{P_{\mathrm{M}}(\boldsymbol{n}')}{P_{\mathrm{M}}(\boldsymbol{n})} && \text{if} \quad P_{\mathrm{M}}(\boldsymbol{n}') \leq P_{\mathrm{M}}(\boldsymbol{n}). \end{aligned} \tag{6.5.8}$$

We see immediately that this choice implies that the equilibrium distribution is in fact the model distribution. Thus, for two arbitrarily chosen macrostates $\boldsymbol{n}$ and $\boldsymbol{n}'$, if $P_{\mathrm{M}}(\boldsymbol{n}') > P_{\mathrm{M}}(\boldsymbol{n})$ then $T(\boldsymbol{n}' \leftarrow \boldsymbol{n})P_E(\boldsymbol{n}) = P_E(\boldsymbol{n})$ and $T(\boldsymbol{n} \leftarrow \boldsymbol{n}')P_E(\boldsymbol{n}') = P_E(\boldsymbol{n}')[P_{\mathrm{M}}(\boldsymbol{n})/P_{\mathrm{M}}(\boldsymbol{n}')]$ so that if $P_E(\boldsymbol{n}) = P_{\mathrm{M}}(\boldsymbol{n})$, the detailed balance condition $T(\boldsymbol{n}' \leftarrow \boldsymbol{n})\ P_E(\boldsymbol{n}) = T(\boldsymbol{n} \leftarrow \boldsymbol{n}')\ P_E(\boldsymbol{n}')$ (both equal to $P_E(\boldsymbol{n})$) leading to (6.5.7) is satisfied. In addition, the initial set of events which has been successively transformed to form the equilibrium set may be considered to represent a sample corresponding to $P_{\mathrm{M}}(\boldsymbol{n})$ so that the frequency distribution is expected to obey (6.5.1). The Metropolis algorithm thus succeeds in generating a sample of a known probability distribution using only ratios of probabilities. For example, using known macrostate weights, $\rho_{\mathrm{M}}(\boldsymbol{n})$, we could rewrite equation (6.5.8) as

$$\begin{aligned} T(\boldsymbol{n}' \leftarrow \boldsymbol{n}) &= 1 && \text{if} \quad \rho_{\mathrm{M}}(\boldsymbol{n}') > \rho_{\mathrm{M}}(\boldsymbol{n}), \\ T(\boldsymbol{n}' \leftarrow \boldsymbol{n}) &= \frac{\rho_{\mathrm{M}}(\boldsymbol{n}')}{\rho_{\mathrm{M}}(\boldsymbol{n})} && \text{if} \quad \rho_{\mathrm{M}}(\boldsymbol{n}') \leq \rho_{\mathrm{M}}(\boldsymbol{n}). \end{aligned} \tag{6.5.9}$$

Although the Metropolis alogorithm is rightly considered as a sophisticated technique to be applied in Monte Carlo based simulations, the part of the procedure explicitly involving Monte Carlo methods is very simple. Thus, if a transition is to be made with a probability $T(\boldsymbol{n}' \leftarrow \boldsymbol{n})$ and refused with probability $1 - T(\boldsymbol{n}' \leftarrow \boldsymbol{n})$, it is sufficient to generate a uniform random deviate, R, between zero and one and determine whether $R \leq T(\boldsymbol{n}' \leftarrow \boldsymbol{n})$ (acceptation) or whether $R > T(\boldsymbol{n}' \leftarrow \boldsymbol{n})$ (refusal). In fact, we are dealing with the most elementary application of algorithm 4.4 involving a one-dimensional sum which contains only two terms.

In the above description, we assumed that we possessed some trial frequency distribution (equivalently, a set of events) which may be used to 'seed' the iteration process. The number of iterations required to generate the equilibrium distribution from this initial distribution obviously depends in some way on the seed distribution. Thus, the actual choice of the seed distribution and the number of iterations required may vary. One possible scheme is to seed the system with a single macrostate and generate events one by one. From say the Nth event, we make some small change in the corresponding macrostate and propose the new macrostate as a new event. This event is accepted or rejected according to the transition probability. If accepted, the new macrostate represents the $N + 1^{\text{th}}$ event and is, in its turn, transformed. If rejected, one registers the old macrostate as the $N + 1^{\text{th}}$ event (the initial state is transformed into itself) and a new attempt is made to transform this macrostate. In this way, a train of events is generated and as N becomes very large the most recent subset of N_t events ($N_t < N$) can be taken as the equilibrium sample. It is probably clear from this discussion that some experience with the algorithm is necessary in any given problem in order to decide just when to begin accepting events.

One final comment: the method would be expected to work efficiently if the probability distribution includes a single region of highly probable macrostates. On the contrary, in a situation involving a large number of (almost) equally probable macrostates, the method may not work well because a significant part of the probability may not be sampled at all. For example, if we recall the 10^8 partitions for $Z_{\text{p}} = 100$ (see after (6.2.5)) we may conclude that, if all partitions are equally probable, a generated set of say 10^7 events would 'see' at most 10% of the total macrostate probability. A problem may also occur if the probability is concentrated in several 'narrow' peaks. In this case, it is difficult to ensure that all of the regions of significant probability are discovered, explored, and thereby represented in the event chain.

6.6 The Berlin Metropolis sampling multifragmentation model

The Berlin model (MMMC for microcanonical Metropolis Monte Carlo) developed by Gross and collaborators [13] is based on a multifragmentation 'scenario' which resembles that used in the Copenhagen model. Fragments are confined to a region of space referred to as the freeze-out volume. A partition is composed of neutrons and charged fragments (including, of course, light charged particles). Neutrons are singled out for special treatment as we shall see below.

In fact two models have been investigated, a canonical model in which the freeze-out volume is considered to be associated with a ficticious heat bath system and a microcanonical model which, over the last few years, seems to have constituted the principal preoccupation of the Berlin school. Some aspects of the model will not be presented in detail. However, the present account should provide a basic understanding of the principles involved.

We shall begin with the canonical model because, at least as far as the practical operation is involved, it is not very different from the microcanonical version. We shall then discuss the extensions necessary for the microcanonical option. Both models use Metropolis importance sampling.

In the canonical version of the Berlin model the total energy of a mass-charge partition, $\boldsymbol{n}$, confined in the freeze-out volume is expressed as a sum over charged fragments and neutrons. There are N_C charged fragments and N_ν neutrons. The total neutron mass is thus $M_\mathrm{n} = N_\nu m_\mathrm{n}$ where m_n is the neutron mass. The centre-of-mass momentum is set to zero.

$$\sum_{j=1}^{N_C} \boldsymbol{p}_j + \sum_{i=1}^{N_\nu} \boldsymbol{p}'_i = 0, \qquad (6.6.1)$$

where the index j refers to charged fragments and $\boldsymbol{p}'_i$ is the laboratory momentum of a neutron. The momentum, $\boldsymbol{P}$, associated with the motion of the centre of mass of the N_ν neutrons is obviously $\boldsymbol{P} = -\sum_{j=1}^{N_C} \boldsymbol{p}_j$.

The detailed description of the dynamics of the neutrons in the centre-of-mass system is a little complicated. The relevant masses and momenta are associated with 'ficticious' particles. For example, in the particular case of three particles with masses m_1, m_2 and m_3, we could proceed by constructing the reduced mass of the first two (we could call this μ_1) and its associated momentum, $\boldsymbol{p}_1$, and then the reduced mass (and momentum) of the ficticious particle (mass μ_1) and the third particle (mass m_3) i.e., $\mu_2 = \mu_1 m_3/(\mu_1 + m_3)$. For four particles, one could construct the reduced masses and momenta of two pairs and then the inter-pair reduced mass and momentum. For neutrons the primary masses are, of course, all equal. Although this construction process is somewhat fastidious (and not unique), there are some global features which are independent of the precise mode of construction. Thus, for example, the total energy for N_ν neutrons in the centre of mass is simply $\sum_{i=1}^{N_\nu - 1} p_i^2/2\mu_i$ where μ_i is the reduced mass associated with one of the set of $N_\nu - 1$ momenta, $\boldsymbol{p}_i$. The total energy of a system of charged fragments and neutrons can thus be written (substituting $-\sum_{j=1}^{N_C} \boldsymbol{p}_j$ for $\boldsymbol{P}$) as

$$E_{\boldsymbol{n}} = E_\mathrm{C} + \sum_{j=1}^{N_C} \left[\frac{p_j^2}{2m_j} + E_j^* - \mathcal{B}_j \right] + \sum_{i=1}^{N_\nu - 1} \frac{p_i^2}{2\mu_i} + \frac{\left(\sum_{j=1}^{N_C} \boldsymbol{p}_j \right)^2}{2M_\mathrm{n}}. \qquad (6.6.2)$$

In equation (6.6.2), the energy is measured from the state composed of free neutrons and protons which, of course, is common to all partitions. Note that the Coulomb energy, E_C (the mutual Coulomb repulsion between charged fragments), is included explicitly in $E_{\boldsymbol{n}}$. The three terms in square brackets represent, respectively, the charged fragment kinetic energy, the internal excitation energy and the binding energy of the fragment labelled j. The sum of the last two terms, which we designate by $E'_{\rm n}$, represents the kinetic energy of the neutrons in the laboratory system.

Equation (6.6.2) can be written so as to emphasize the unobserved neutrons and the equally unobserved internal excitation energy of the complex fragments. If $E_{\rm I}$ is the total internal excitation energy and $E_{\rm n}$ the energy of the neutrons in the neutron centre-of-mass system (do not confuse $E_{\rm n}$ with the total energy, $E_{\boldsymbol{n}}$), we can write

$$E_{\boldsymbol{n}} = E'_{\boldsymbol{n}} + E_{\rm I} + E_{\rm n} = E'_{\boldsymbol{n}} + \sum_{j=1}^{N_C} E_j^* + \sum_{i=1}^{N_\nu - 1} \frac{p_i^2}{2\mu_i}, \tag{6.6.3}$$

where, by definition,

$$E'_{\boldsymbol{n}} = E_{\rm C} + \sum_{j=1}^{N_C} \left[\frac{p_j^2}{2m_j} - \mathcal{B}_j \right] + \frac{\left[\sum_{j=1}^{N_C} \boldsymbol{p}_j \right]^2}{2M_{\rm n}}, \tag{6.6.4}$$

depends only on the identity and the spatial and momentum coordinates of the charged fragments.

The canonical weight corresponding to (6.6.3) can be written directly as

$$W_{\boldsymbol{n}}(E'_{\boldsymbol{n}}, E_{\rm I}, E_{\rm n})\, {\rm d}E'_{\boldsymbol{n}}\, {\rm d}E_{\rm I}\, {\rm d}E_{\rm n} = {\rm e}^{-E_{\boldsymbol{n}}/T} \rho'_{\boldsymbol{n}}(E'_{\boldsymbol{n}}) \rho_{\rm I}(E_{\rm I}) \rho_{\rm n}(E_{\rm n})\, {\rm d}E'_{\boldsymbol{n}}\, {\rm d}E_{\rm I}\, {\rm d}E_{\rm n}, \tag{6.6.5}$$

where $\rho_{\rm I}(E_{\rm I})$ is the density of microstates corresponding to internal excitation of charged fragments and $\rho_{\rm n}(E_{\rm n})$ is the density corresponding to the $6(N_\nu - 1)$ neutron degrees of freedom. The objective in writing (6.6.5) is obviously to integrate over unobserved quantities.We can define a partially integrated weight

$$W_{\boldsymbol{n}}(E'_{\boldsymbol{n}})\, {\rm d}E'_{\boldsymbol{n}} = {\rm e}^{-E'_{\boldsymbol{n}}/T} \rho'_{\boldsymbol{n}}(E'_{\boldsymbol{n}})\, {\rm d}E'_{\boldsymbol{n}} \int_0^\infty \rho_{\rm I}(E_{\rm I})\, {\rm e}^{-E_{\rm I}/T}\, {\rm d}E_{\rm I} \times \int_0^\infty \rho_{\rm n}(E_{\rm n})\, {\rm e}^{-E_{\rm n}/T}\, {\rm d}E_{\rm n}. \tag{6.6.6}$$

The integral over internal excitation energies of complex fragments is straightforward. In the canonical approach the implicit convolution turns into a product (see section 1.4)

$$W_{\rm I} = \prod_{j=1}^{N_C} \left[\int_0^{\delta_j} \rho_j(E_j^*)\, {\rm e}^{-E_j^*/T}\, {\rm d}E_j^* \right], \tag{6.6.7}$$

which by separating out the ground states ($E_j^* = 0$) and supposing that excited states lie above 0.1 MeV, can be written

$$W_{\rm I} = \prod_{j=1}^{N_C}\left[1 + \int_{0.1}^{\delta_j} \rho_j(E_j^*)\,{\rm e}^{-E_j^*/T}\,{\rm d}E_j^*\right]. \qquad (6.6.8)$$

The upper limit, δ_j (rather than ∞) occurs because the Berlin model allows each complex fragment to be excited only up to the particle emission threshold, δ_j. This is an important feature because it implies that complex fragments, once formed, decay only by gamma ray emission so that particle evaporation corrections do not need to be included. While useful for computational purposes, this assumption seems rather drastic. The sum of states in equation (6.6.8) can be calculated explicitly provided the low lying states of all complex fragments are available or can be modelled by a suitably parametrized density function.

The integral over neutron degrees of freedom is a little more complicated because the total centre-of-mass momentum, $\boldsymbol{P}$, but not the position, is constrained. Separating this coordinate, the microcanonical density of states (number of states per unit energy and per unit vector momentum) for a set of N_ν identical particles (total mass $M_{\rm n} = N_\nu m_{\rm n}$) with total energy $E'_{\rm n}$ and total momentum, $\boldsymbol{P} = \boldsymbol{P}_{\rm n}$, restricted to a volume $V_{\rm n}$ is (see equation (1.2.31))

$$\rho_{\rm n}(E'_{\rm n}) = \left[\frac{V_{\rm n}}{h^3}\right]^{N_\nu} \int {\rm d}^3 p_1\,{\rm d}^3 p_2 \ldots {\rm d}^3 p_{N_\nu-1}\,\delta\left(\sum_{i=1}^{N_\nu-1}\frac{p_i^2}{2\mu_i} + \frac{P^2}{2M_{\rm n}} - E'_{\rm n}\right)$$
$$\times\,\delta(\boldsymbol{P}-\boldsymbol{P}_{\rm n})\,{\rm d}^3 P$$
$$= \left[\frac{V_{\rm n}}{h^3}\right]^{N_\nu} \int {\rm d}^3 p_1\,{\rm d}^3 p_2 \ldots {\rm d}^3 p_{N_\nu-1}\,\delta\left[\sum_{i=1}^{N_\nu-1}\frac{p_i^2}{2\mu_i} - \left(E'_{\rm n} - \frac{P_{\rm n}^2}{2M_{\rm n}}\right)\right], \qquad (6.6.9)$$

which, with $E_{\rm n} = E'_{\rm n} - \frac{P_{\rm n}^2}{2M_{\rm n}}$, and the centre-of-mass energies associated with the reduced masses μ_i written as ε_i, becomes (note that fixing the neutron centre of mass momentum implies that the density depends on $E_{\rm n}$ rather than on $E'_{\rm n}$)

$$\rho_{\rm n}(E_{\rm n}) = \left[\frac{V_{\rm n}}{h^3}\right]^{N_\nu} (2\pi)^{N_\nu-1}\left[\prod_{i=1}^{N_\nu-1}(2\mu_i)^{3/2}\right]$$
$$\times \int_0^{E_{\rm n}} \varepsilon_1^{1/2} \int_0^{E_{\rm n}} \varepsilon_2^{1/2} \ldots \int_0^{E_{\rm n}} \varepsilon_{N_\nu-1}^{1/2}$$
$$\times\,\delta\left[\sum_{i=1}^{N_\nu-1}\varepsilon_i - E_{\rm n}\right]{\rm d}\varepsilon_1\,{\rm d}\varepsilon_2 \ldots {\rm d}\varepsilon_{N_\nu-1}, \qquad (6.6.10)$$

where the integration over each energy has been indicated explicitly. The Laplace transform of the multiple integral in (6.6.10) is given by the convolution theorem (section 1.3)

$$\int_0^\infty \left[\int_0^{E_n} \varepsilon_1^{1/2} \int_0^{E_n} \varepsilon_2^{1/2} \ldots \int_0^{E_n} \varepsilon_{N_\nu-1}^{1/2} \times \delta\left[\sum_{i=1}^{N_\nu-1} \varepsilon_i - E_n \right] d\varepsilon_1\, d\varepsilon_2 \ldots d\varepsilon_{N_\nu-1} \right] e^{-E_n/T}\, dE_n$$
$$= \prod_{i=1}^{N_\nu-1} \left[\int_0^\infty \varepsilon_i^{1/2} e^{-\varepsilon_i/T}\, d\varepsilon_i \right] = \left(\frac{\sqrt{\pi}}{2}\right)^{N_\nu-1} T^{3(N_\nu-1)/2}, \quad (6.6.11)$$

so that with the identity (see Messiah [25] chapter IX or prove it by induction starting with two particles)

$$\prod_{i=1}^{N_\nu} m_i = M_n \prod_{i=1}^{N_\nu-1} \mu_i, \quad (6.6.12)$$

the transform of (6.6.10) is thus

$$\int_0^\infty \rho_n(E_n)\, e^{-E_n/T}\, dE_n$$
$$= \left[\frac{V_n}{h^3}\right]^{N_\nu} (2\pi)^{N_\nu-1} \left[\prod_{i=1}^{N_\nu-1} (2\mu_i)^{3/2}\right] \left(\frac{\sqrt{\pi}}{2}\right)^{N_\nu-1} T^{3(N_\nu-1)/2}$$
$$= \left[\frac{V_n}{h^3}\right]^{N_\nu} \frac{(2\pi m_n T)^{3N_\nu/2}}{(2\pi N_\nu m_n T)3/2} \quad (6.6.13)$$

Two differences in the present case are that the spin degrees of freedom lead to a (quantum mechanical) correction of $(2s_n + 1)^{N_\nu} = 2^{N_\nu}$ and that we are dealing with N_ν identical particles so that the partition sum contains a factor of $N_\nu!$ in the denominator. With these modifications, the statistical weight associated with neutron degrees of freedom becomes

$$W_n = \frac{2^{N_\nu}}{N_\nu!} \int_0^\infty \rho_n(E_n)\, e^{-E_n/T}\, dE_n = \frac{1}{N_\nu!} \frac{\left[2V_n\left(\frac{m_n T}{2\pi\hbar^2}\right)^{3/2}\right]^{N_\nu}}{(2\pi M_n T)^{3/2}}. \quad (6.6.14)$$

Equation (6.6.6) is now written

$$W_n(E_n')\, dE_n' = e^{-E_n'/T} \rho_n'(E_n')\, dE_n' W_I W_n, \quad (6.6.15)$$

where $\rho_n'(E_n')$ is the density of states corresponding to spatial configurations of charged particles within the freeze-out volume which are compatible with $N_\nu \geq 0$

in the sense that total mass is fixed. In fact, if no further integration is performed, $\rho_n'(E_n')$ should be a uniform distribution (microstates are equally probable). Thus, one may imagine a Metropolis simulation in which a charged particle partition and spatial configuration is randomly selected (thus fixing E_{C}, N_ν and $\sum_j \mathcal{B}_j$), following which a set of charged particle momenta is chosen at random. The momentum of the centre of mass of the neutrons which make up the full mass is fixed by equation (6.6.1). Such a configuration is a candidate for an event and could be accepted or rejected using the Metropolis transition probability (6.5.8) obtained from the canonical model probability distribution (6.6.15) which we can write as

$$W_n = \mathrm{e}^{-E_{\mathrm{C}}/T}\,\mathrm{e}^{\sum_j \mathcal{B}_j/T}\prod_j \mathrm{e}^{-p_j^2/2M_jT}\,\mathrm{e}^{-\left(\sum \boldsymbol{p}_j\right)^2/2M_{\mathrm{n}}T}\,W_{\mathrm{I}}W_{\mathrm{n}}. \tag{6.6.16}$$

The difficulty is that this strategy would probably give rise to a most inefficient programme. It will be recalled that a chain of events is established in one version of the Metropolis method by presenting successive ‘candidates’ and evaluating the transition probability for a transition from the current event to the candidate. In practice, it is necessary to guide the choice of event candidates in such a way that the probability of acceptance is, on avererage, not too close to the limits zero or one. Thus, for example, in the Berlin canonical model, the charged particle momenta in a candidate event are not chosen at random but are sampled from the distribution

$$W(p_1, p_2 \ldots p_{N_C}) = \frac{1}{S}\left[\mathrm{e}^{-\left(\sum \boldsymbol{p}_j\right)^2/2M_nT}\prod_j \mathrm{e}^{-p_j^2/2m_jT}\right], \tag{6.6.17}$$

where S is a normalization constant.

Let us now turn to the microcanonical model and say straightaway that, as far as its operation is concerned, it does not differ substantially from the canonical model. Of course, the Boltzmann weighting is absent so that all microstates for which the total energy is equal to a fixed value are equally probable. However, as in the canonical model, we define macrostates by assembling subsets of the full set of degrees of freedom (the use of the word ‘microcanonical’ may, therefore, seem a little inappropriate but is generally accepted). Once again, unobserved neutron degrees of freedom are separated. The corresponding density of states $\rho_{\mathrm{n}}(E_{\mathrm{n}})$ is now a function of the sum of the neutron energies in their centre of mass rather than a function of the temperature T. The model also groups the internal excitation degrees of freedom by writing a density of internal states (which of course is a convolution product) as $\rho_{\mathrm{I}}(E_{\mathrm{I}})$. The absence of a heat bath has two consequences. The first is that the simplifications (essentially the factorization) characteristic of the canonical distribution is no longer available. The second is that total energy must be conserved so that the simple product of canonical weights $W_{\mathrm{I}}W_{\mathrm{n}}$ has to be abandoned in favour of a factor $\rho_{\mathrm{I}}(E_{\mathrm{I}})\rho_{\mathrm{n}}(E_{\mathrm{n}})\,\mathrm{d}E_{\mathrm{I}}\,\mathrm{d}E_{\mathrm{n}}$.

The techniques employed in reference [13] (Zhang *et al*) depend on the number of neutrons in the multifragmentation partition. Their detailed exposition is somewhat lengthy so that we are obliged to refer the interested reader to the original work. We can, however, write down the microcanonical equivalent of equation (6.6.15) i.e., the microcanonical weight which depends on the total energy E_n.

$$W_n(E_n)\,\mathrm{d}E_n'\,\mathrm{d}E_\mathrm{I}\,\mathrm{d}E_\mathrm{n} = \rho_n'(E_n')\rho_\mathrm{I}(E_\mathrm{I})\rho_\mathrm{n}(E_\mathrm{n}) \times \delta(E_n' + E_\mathrm{I} + E_\mathrm{n} - E_n)\,\mathrm{d}E_n'\,\mathrm{d}E_\mathrm{I}\,\mathrm{d}E_\mathrm{n}. \quad (6.6.18)$$

The density $\rho_\mathrm{n}(E_\mathrm{n})$ corresponding to neutrons in their centre-of-mass system can be directly estimated from (6.6.14) by performing the inverse Laplace transform ($\beta = 1/T$)

$$\rho_\mathrm{n}(E_\mathrm{n}) = \frac{1}{2\pi i}\int_{c-i\infty}^{c+i\infty} W_\mathrm{n}(\beta)\,\mathrm{e}^{\beta E_\mathrm{n}}\,\mathrm{d}\beta. \quad (6.6.19)$$

In equation (6.6.14) the β dependent factor is $1/\beta^{3(N_\nu-1)/2}$. The inverse transform of this factor is

$$\frac{1}{2\pi i}\int_{c-i\infty}^{c+i\infty} \frac{1}{\beta^{\frac{3(N_\nu-1)}{2}}}\,\mathrm{e}^{\beta E_\mathrm{n}}\,\mathrm{d}\beta = \frac{E_\mathrm{n}^{\frac{3N_\nu}{2}-\frac{5}{2}}}{\Gamma\left(\frac{3(N_\nu-1)}{2}\right)}. \quad (6.6.20)$$

Then from (6.6.14) (redefining ρ_n so as to include the factor $2^{N_\nu}/N_\nu!$) we obtain

$$\rho_\mathrm{n}(E_\mathrm{n}) = \frac{1}{N_\nu!}\,\frac{\left[2V_\mathrm{n}\left(\frac{m_\mathrm{n}}{2\pi\hbar^2}\right)^{3/2}\right]^{N_\nu}}{(2\pi N_\nu m_\mathrm{n})^{3/2}}\,\frac{E_\mathrm{n}^{\frac{3N_\nu}{2}-\frac{5}{2}}}{\Gamma\left(\frac{3(N_\nu-1)}{2}\right)}. \quad (6.6.21)$$

The degrees of freedom corresponding to internal excitation energy of charged fragments may also be grouped to contribute a single energy dependent factor. In this case, we need to evaluate the familiar convolution integral

$$\rho_\mathrm{I}(E_\mathrm{I}) = \int_0^{\delta_1} \rho_1(E_1^*) \int_0^{\delta_2} \rho_2(E_2^*) \ldots \int_0^{\delta_j} \rho_j(E_j^*) \ldots \times \delta\left[\sum E_j^* - E_\mathrm{I}\right] \mathrm{d}E_1^*\,\mathrm{d}E_2^* \ldots \mathrm{d}E_j^* \ldots. \quad (6.6.22)$$

However, the cut-offs δ_j imply that this is not a simple affair. Furthermore, because of these cut-offs, the total microstate density does not exhibit a smooth increase as a function of the total energy. Such behaviour is often interpreted as evidence of a phase change (see chapter 7).

Equation (6.6.18) could be used as the basis for a Metropolis simulation. However, as stated above, the MMMC code employs several time-saving (transformation) techniques. Evidently, the choice of candidates for successive events is complicated by the need to conserve energy. In practice, in order to try and ensure that the final set of events represents a substantial fraction of the total statistical weight, the code is run by generating some 10^6 events before the generation of the final event set (typically 10^7 events).

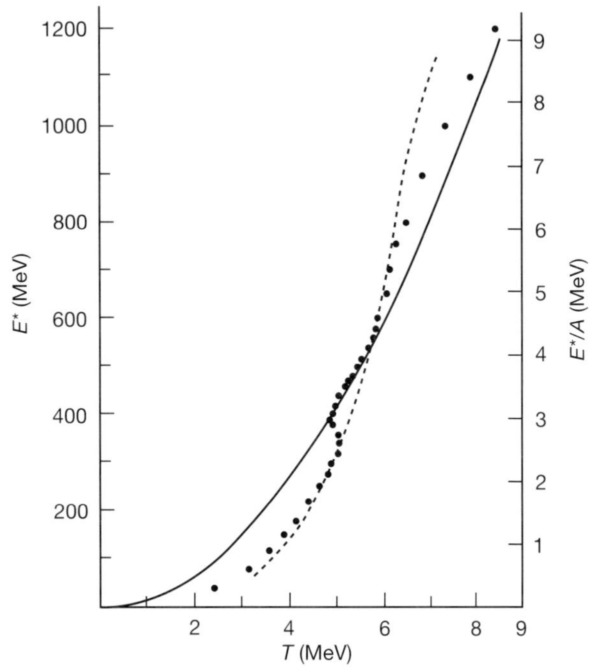

Figure 6.5. Excitation energy versus temperature obtained in the microcanonical simulation (points) and canonical simulation (broken curve) with the MMMC code [13]. The solid curve represents $E^* = aT^2$ with $a = A/8$. The temperature was 'measured' in the microcanonical simulation by assuming the average kinetic energy of neutrons to be $3N_\nu T/2$. The kink observed in the E^*–T plot may be considered as an indication of a phase transition.

Let us recapitulate. The Berlin model relies heavily on the fact that, due to the non-observation of certain well-defined subsets of degrees of freedom, one may integrate over these variables and form a macrostate probability distribution which can be used to define transition probabilities in a Metropolis Monte Carlo simulation code. The macrostate probability distribution is somewhat easier to simulate in the canonical version of the model. However, the principle is the same for both canonical and microcanonical versions of the model.

One important result obtained using the microcanonical simulation is shown in figure 6.5 [13]. If the microcanonical density of states, discarding the neutron degrees of freedom, is considered as a heat bath, the mean neutron kinetic energy can be taken as a measure of the temperature so that $\langle E_n \rangle = 3N_\nu T/2$. With this definition of temperature, it seems that there is some evidence for a phase change in the ^{131}Xe system insofar as the temperature, T, stays constant (at about 5 MeV) for excitation energies of the order of 3–4 MeV/nucleon. This interesting observation can be considered as an incitation for experimentalists to look for a similar effect in real, rather than simulated, data.

6.7 Corrections for finite size of fragments

A particular difficulty which occurs with multifragmentation models which seek to explicitly evaluate the microcanonical entropy (or the canonical partition sum) concerns corrections to the spatial component in the density of microstates which arise because of the finite size of the fragments. We have seen in chapter 1 how this problem is formulated for a real gas. The present pre-occupation is a little different because we wish only to estimate the 'loss of volume' which occurs when fragments are considered as hard spheres which are not allowed to overlap [26]. In the following, we assume a constant freeze-out volume V_F. However, at the end of this section it will be obvious that a multiplicity dependence of the freeze-out volume could be easily included in the analysis.

Let us suppose N-fragments to be confined such that their mass centres are randomly disposed within a spherical 'freeze-out' volume with radius R_F. In the limit of point-like non-interacting fragments (ideal gas approximation), the spatial contribution to the density of microstates is, as we have seen previously,

$$\int \mathrm{d}\boldsymbol{r}_1 \int \mathrm{d}\boldsymbol{r}_2 \ldots \int \mathrm{d}\boldsymbol{r}_N = \int_0^{R_F} 4\pi r_1^2 \,\mathrm{d}r_1 \int_0^{R_F} 4\pi r_2^2 \,\mathrm{d}r_2 \ldots \int_0^{R_F} 4\pi r_N^2 \,\mathrm{d}r_N = V_F^N, \tag{6.7.1}$$

so that the correction to the ideal gas approximation can be expressed as

$$\Lambda(R_F) = \chi(R_F) V_F^N, \tag{6.7.2}$$

where the factor $\chi (\chi \leq 1)$ formally expresses the suppression of spatial configurations which contain interpenetrating (overlapping) fragments.

One possible method of obtaining χ involves using simple Monte Carlo methods. The technique consists in making a large number, N_{MC}, of attempts to dispose the N-fragments at random positions in the freeze-out volume. If N_R trials have the property that, for all i, j,

$$\sqrt{(x_i - x_j)^2 + (y_i - y_j)^2 + (z_i - z_j)^2} \geq (R_i + R_j), \tag{6.7.3}$$

an estimate of the probability of overlap is given as $\chi = N_R/N_{\mathrm{MC}}$ so that spatial contribution to the phase space density is estimated as

$$\Lambda_{\mathrm{MC}}(R_F) = V_F^N \frac{N_R}{N_{\mathrm{MC}}}. \tag{6.7.4}$$

As explained in section 4.2 (see equation (4.2.4)), if N_R is small, the relative error in χ (σ_χ/χ) is closely approximated by $\sqrt{1/N_R}$ and may thus be rather large. A difficulty thus arises when the freeze-out volume, V_F, is sufficiently small so that the calculation time spent to obtain a large number of 'successes' becomes prohibitively long.

In the analytical approach, the probability that a configuration is retained is the probability that all distinct fragment pairs do not overlap. Therefore, the problem is solved if we can calculate the overlap probability for any given pair of fragments in the set. Following the treatment of a real gas, we can approximate χ as the product of two body terms. In this approach, the influence of fragments other than the specific pair considered is ignored, so that one might expect the approximation to work well in situations where simultaneous overlap of more than two fragments can be neglected.

Let us position the centre of the freeze-out volume at the origin of our coordinate system (see figure 6.6) and place the fragment labelled i at the radial coordinate r_i. Assuming $2R_F \geq (R_i + R_j)$ for all i, j we now place a second fragment, j, within the volume. To avoid overlap with fragment i, we need to exclude the centre of mass of fragment j from the volume

$$V_{ij} = \frac{4\pi}{3}(R_i + R_j)^3, \tag{6.7.5}$$

so that if r_i and r_j are such that the volume V_{ij} is *entirely enclosed* within V_F, the

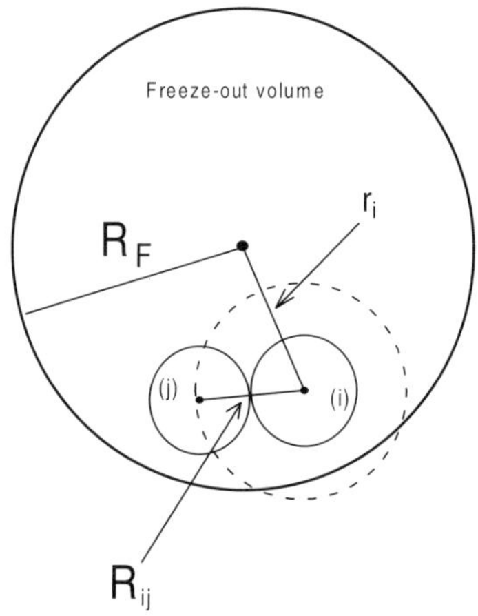

Figure 6.6. Two body overlaps in the freeze-out volume. If the ith fragment is placed at r_i the mass centre of the jth fragment must lie outside the sphere, represented by the broken circle, in order not to overlap. In the figure, part of this volume lies outside the freeze-out volume.

probability that fragment j will overlap with fragment i is given by

$$\Phi_{ij}(r_i) = \frac{V_{ij}}{V_F} = \frac{(R_i + R_j)^3}{R_F^3}. \tag{6.7.6}$$

As we increase the value of r_i, the volume V_{ij} will eventually be only partially contained in the freeze-out volume as shown in figure 6.6. The probability that fragments i and j overlap is then simply given by the part of V_{ij} *which lies inside the freeze-out volume* divided by the freeze-out volume itself. This observation leads us to generalize equation (6.7.6) by writing

$$\Phi_{ij}(r_i) = \frac{V_{ij}^{(o)}(r_i)}{V_F}, \tag{6.7.7}$$

where $V_{ij}^{(o)}(r_i)$ is the part of the volume V_{ij} which overlaps the freeze-out volume. As defined by equation (6.7.7), $\Phi_{ij}(r_i)$ can be interpreted as a probability even in cases where V_{ij} encloses V_F because the overlap requirement then entails $V_{ij}^{(o)} = V_F$ and thus guarantees an upper bound of unity for $\Phi_{ij}(r_i)$ and also for the average value of this ratio taken over the freeze-out volume.

The probability that fragment i does not overlap with any other fragment is the product

$$p_i(r_i) = \prod_{j\neq i}^{N}[1 - \Phi_{ij}(r_i)] \tag{6.7.8}$$

However, this quantity refers to a particular radial position of fragment i. We seek the global survival probability, χ, which is the probability that each and every fragment is not overlapped by one or more fragments. The construction of this quantity is not unique. We adopt a method which is consistent with our two body approximation, i.e., we take the product of all terms, each of which corresponds to the *spatially averaged* non-overlap probability of a distinct pair of fragments (we could have chosen to make a different approximation by taking the average value of the product rather than the product of average values). Thus, the survival probability is expressed using equation (6.7.8) as

$$\chi(R_F) = \prod_{i=1}^{N-1}\prod_{j>i}^{N}(1 - \langle\Phi_{ij}\rangle). \tag{6.7.9}$$

Let us denote the spatially averaged value of the overlap volume for the pair of fragments i and j as $\langle V_{ij}^{(o)}\rangle$. The calculation of this average taken over the position, in the freeze-out volume, of fragment i, involves a simple contribution from the region where equation (6.7.6) is valid. This region may be specified by writing

$$r_i + R_{ij} \leq R_F, \tag{6.7.10}$$

where R_{ij} is the sum of the radii of fragments i and j. For larger radii, r_i, the common volume is given by the sum of the volumes of two spherical caps which with some simple algebra (the volume of a spherical cap of height h in a sphere of radius r is $\frac{1}{3}\pi h^2(3r-h)$) is found to be

$$V_{ij}^{(o)}(r_i) = \frac{2\pi}{3}(R_{ij}^3+R_F^3)+\frac{\pi r_i^3}{12}-\frac{\pi r_i}{2}(R_{ij}^2+R_F^2)-\frac{\pi}{4r_i}[(R_{ij}^2-R_F^2)^2]. \quad (6.7.11)$$

To write down an expression for the average value it is useful to specify the limiting value of r_i such that the sphere with radius R_{ij} is entirely enclosed within the freeze-out volume or, *alternatively*, that the freeze out volume is entirely enclosed in the sphere with radius R_{ij}. Indeed, if we define $R_{\mathrm{LIM}} = |R_F - R_{ij}|$ and $V_{\min}$ as $\min(V_{ij}, V_F)$ then the required average value is expressed as:

$$\langle V_{ij}^{(o)}\rangle = \frac{V_{\min}\int_0^{R_{\mathrm{LIM}}} r_i^2\,\mathrm{d}r_i + \int_{R_{\mathrm{LIM}}}^{R_F} V_{ij}^{(o)} r_i^2\,\mathrm{d}r_i}{\int_0^{R_F} r_i^2\,\mathrm{d}r_i}. \quad (6.7.12)$$

Of course, equation (6.7.12) and thus (6.7.9), must be evaluated numerically. We may, however, gain further insight into the problem by considering two limiting situations. We first note that equation (6.7.9) can also be expressed by writing the logarithm of $\chi(R_F)$ as a sum over distinct pairs.

$$\ln[\chi(R_F)] = \sum_{i=1}^{N-1}\sum_{j>i}^{N} \ln[1-\langle\Phi_{ij}\rangle]. \quad (6.7.13)$$

This last expression is interesting because, in the limit in which, for all pairs i, j, $\langle\Phi_{ij}\rangle \ll 1$, we obtain using (6.7.7) and (6.7.13)

$$\chi(R_F) = \mathrm{e}^{-\sum\langle V_{ij}^{(o)}\rangle/V_F} \sim 1 - \frac{\sum\langle V_{ij}^{(o)}\rangle}{V_F}, \quad (6.7.14)$$

in which the summation sign is to be understood as a short hand expression of the double summation in equation (6.7.13). Equation (6.7.14) states that, for large enough freeze-out volumes, a first approximation to the survival probability can be found by simply dividing the sum of the average overlap volumes of the interaction spheres (with V_F) by the freeze-out volume itself.

As a second illustration of the application of equation (6.7.9), we examine the correction as a function of the partition multiplicity, N, by partitioning the parent nucleus, with mass number A into N *equal* masses represented by uniform density spheres. Let us denote the radius of each fragment as R_0 and use equation (6.7.13) to write

$$\ln[\chi(R_F)] = \sum_{i=1}^{N-1}\sum_{j>i}^{N} \ln[1-\langle\Phi_{ij}\rangle] = \frac{N(N-1)}{2}\ln[1-\langle\Phi_{ij}\rangle]. \quad (6.7.15)$$

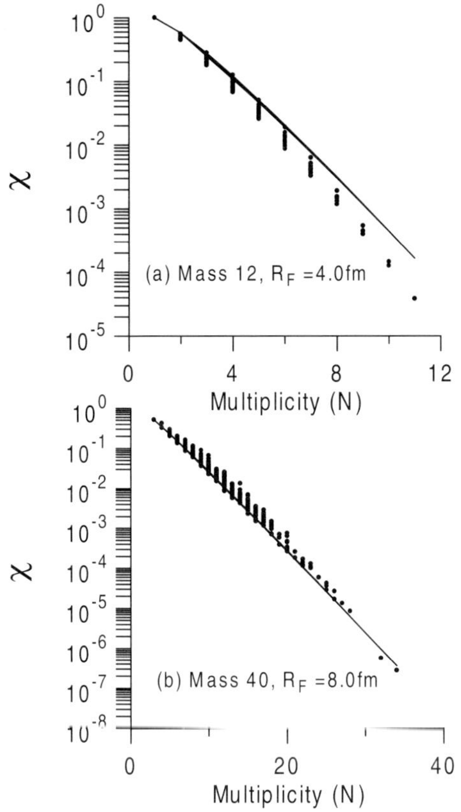

Figure 6.7. Monte Carlo calculations of the survival probability χ [26] The points for given multiplicity are closely grouped so that an approximate estimation is possible by dividing the parent nucleus into N equal mass fragments for multiplicity N . The solid lines show the results obtained using equations (6.7.17) and (6.7.19). (Reprinted from *Phys. Rev.* C, **55**, Cole A J, Heuer D and Charvet M, p 2978. Copyright 1997, by the American Physical Society.)

Neglecting edge effects (i.e., the situation where part of the overlap volume of two fragments lies outside the freeze-out volume), we can approximate $\langle\Phi_{ij}\rangle$ using (6.7.6), i.e.,

$$\langle\Phi_{ij}\rangle \approx \frac{(R_i + R_j)^3}{R_F^3} = \frac{8\rho_F}{N\rho_0}, \tag{6.7.16}$$

where ρ_F is the density at freeze-out and ρ_0 is the density of the fragments (or of the parent nucleus). If N is sufficiently large so that $\langle\Phi_{ij}\rangle < 1$ we can expand the logarithm in equation (6.7.15) to first order to obtain

$$\ln[\chi(R_F)] \approx -\alpha(N-1), \tag{6.7.17}$$

where $\alpha = 4\rho_F/\rho_0$. It turns out that edge effects can be simply included because the average volume overlap of a sphere with radius R_s (in our case $R_s = 2R_0 = 2r_0(A/N)^{1/3}$) and volume V_s, with a sphere of radius R_F ($R_s < R_F$) is quite closely approximated by

$$\frac{\langle V_s^{(o)} \rangle}{V_s} = 1 - 0.53\frac{R_s}{R_F}. \tag{6.7.18}$$

We may simply correct for this effect by *redefining* the slope parameter α of equation (6.7.17) as

$$\alpha = \frac{4\rho_F}{\rho_0}\left[1 - 0.53\frac{2R_0}{R_F}\right]. \tag{6.7.19}$$

This redefined value depends weakly on the multiplicity (via R_0). A comparison of the approximation embodied in (6.7.19) with 'data' generated by a Monte Carlo simulation is shown in figure 6.7 which is taken from reference [26]. The figure includes two case studies which both show a strong, almost exponential, fall of χ with the partition multiplicity. Furthermore, the Monte Carlo calculations indicate that χ depends mainly on the partition multiplicity rather than on the detailed composition of the partition.

It is thus possible to make the following statement: an approximate finite fragment size correction to the freeze-out volume may be included by multiplying the phase space integral by a factor $e^{-\alpha(N-1)}$ where α is given by equation (6.7.19). This is equivalent to multiplying the density of states by e^{α} (of no consequence for branching ratios) and 'renormalizing' the freeze-out volume by writing

$$V_F' = V_F\, e^{-\alpha}. \tag{6.7.20}$$

Finite size corrections are only important for calculations which actually evaluate the statistical weight. In simulations where spatial configurations are explicitly constructed using Monte Carlo techniques the corrections may be included explicitly by directly eliminating spatial configurations involving overlapping fragments.

Finally, it is important to remember that the above corrections form part of a more general problem due to the fact that overlaps (and, *a fortiori*, fragment interactions) may occur during the dynamical evolution following a multifragmentation process. If the freeze-out configuration represents a true scission state (section 6.3) these interactions, apart from Coulomb repulsion, should not be too important. However, it is difficult to make definitive statements without actually carrying out the dynamical simulations. This problem has been considered in the context of the transition state model for multifragmentation, to be discussed in the following section.

6.8 The Berkeley transition state model

In this final section, we discuss the transition state method which was adapted specifically for multifragmentation by Lopez and Randrup [27]. Their theory is conceptually closer to the Bohr–Wheeler theory than either the Copenhagen or Berlin models. It will be recalled that the Bohr–Wheeler theory introduces a specific fission coordinate q, with conjugate momentum p. Increase of the fission coordinate corresponds to increasing the quadrupole deformation of the system (i.e., to the progressive division of the system into two fragments).

Lopez and Randrup formulate their model in terms of mass partitions of a nucleus with mass m_0 and mass number A and refer to the mass number of the nth fragment as A_n. The fragment mass, m_n, is taken to be equal to $A_n m_N$ where m_N is the nucleon mass. They propose that multifragmentation resulting in the production of a partition with multiplicity N, is induced by movement along a 'disassembly' coordinate q given by

$$q^2 = \frac{1}{m_0} \sum_{n=1}^{N} m_n \boldsymbol{r}_n \cdot \boldsymbol{r}_n, \tag{6.8.1}$$

where $\boldsymbol{r}_n$ is the position (with respect to the centre of mass) of the nth fragment. Clearly q is a measure of the spatial extent of the system (analogous to the root mean square radius of a continuous distribution). For any set of fragment positions the interaction energy is $V(\boldsymbol{r}_1, \boldsymbol{r}_2, \ldots \boldsymbol{r}_n)$. As in fission, small values of q imply strongly overlapping fragments so the range over which the definition can be used to describe a specific partition is somewhat limited. The momentum, p, conjugate to q, can be straightforwardly obtained. Denoting the velocity, $\mathrm{d}q/\mathrm{d}t$, as q', the Lagrangian $\mathcal{L} = \frac{1}{2} m_0 (q')^2 - V(\boldsymbol{r}_1, \boldsymbol{r}_2, \ldots \boldsymbol{r}_n)$. It follows that $p = \mathrm{d}\mathcal{L}/dq'$ depends only on the kinetic energy term (changing the velocities of fragments without changing their positions produces no change in the potential energy) so that we find $p = m_0 q'$ which, from (6.8.1), leads to

$$p = \frac{1}{q} \sum_{n=1}^{N} \boldsymbol{p}_n \cdot \boldsymbol{r}_n. \tag{6.8.2}$$

For fission, the mass asymmetry is considered to be 'frozen' at the saddle point, at least in the generalization of the Bohr–Wheeler theory (to include the mass asymmetry coordinate) proposed by Moretto [28]. This, however, is an approximation. We have seen previously that fragments may interact significantly beyond the maximum of the potential energy, $V(q)$, so that the final mass asymmetry is influenced by saddle to scission dynamics. A part of the Lopez–Randrup theory which will be briefly discussed at the end of this section is explicitly concerned with the dynamical evolution beyond the multifragmentation barrier and thus with changes which may be wrought in primary mass partitions by fragment interactions. The major distinction between the earlier 'freeze-out' approaches and the Lopez–Randrup transition state theory is thus that the former theories assume equilibrium at scission with possible modifications arising only from post scission evaporation, whereas the latter assumes equilibrium at the multifragmentation saddle point and includes modifications induced by saddle to scission dynamics as well as post-scission evaporation. The position of the saddle point falls naturally out of the theory so that it is not necessary to introduce the freeze-out volume as a parameter.

A more subtle difference concerns the characteristic scission time, τ_S, introduced in section 6.3 in the discussion of multifragmentation. The Lopez–Randrup theory results in an estimation of the absolute decay rate (as opposed to branching ratios). The characteristic scission time (which is not specifically introduced in the theory) is thus implicitly partition dependent.

We will shortly give explicit expressions for the density of transition states and the decay rate (equivalently, the width) corresponding to any given mass partition. Before doing so, it is useful to recall the essential structure of the Bohr–Wheeler theory for fission. The microstate space is divided into internal (sometimes called 'intrinsic') and collective subspaces. The collective (fission) position and momentum coordinates are q and p and the inertial (mass) parameter associated with movement along the fission coordinate is μ. The corresponding kinetic energy is $\varepsilon = p^2/2\mu$. The number of states in a small energy interval ΔE corresponding to internal excitation of the fissioning system at the barrier is $\rho(E_F, p_F, q_F)\Delta E$ where $E_F = (E_C^* - V(q_F) - \varepsilon_F)$. The energy E_C^* is the original excitation energy of the compound nucleus. The notation $\rho(E_F, p_F, q_F)$ is that introduced in section 1.2. The number of phase space states associated with the fission coordinates which are to be found in the interval $\Delta p \Delta q$, and which are associated with outwardly directed velocities along the fission coordinate, is simply $\Delta p \Delta q/h$ so that the total number of transition states (evaluated at the barrier) corresponding to this interval is the product $N(E_F, p_F, q_F) = \rho(E_F, p_F, q_F)\Delta E \times \Delta p_F \Delta q_F/h$. The number of states which, in a time interval Δt, can escape from a small region Δq_F (are close enough to the outer edge of Δq_F) at the barrier is thus $N(E_F, p_F, q_F)v_F \Delta t/\Delta q_F = \Delta t \times \rho(E_F, p_F, q_F)\Delta E \times v_F \, \mathrm{d}p_F/h$. The fraction of all compound nucleus states which decay in Δt is obtained by dividing by the total number of compound nuclear states, $\rho_C(E_C^*) \times \Delta E$, so that the contribution to the specific decay rate, for values of the collective momentum, p_F, in the interval $(p_F, p_F + \mathrm{d}p_F)$ is $[\rho(E_F, p_F, q_F)/\rho_C(E_C^*)](p_F \, \mathrm{d}p_F/mh)$.

The multifragmentation problem is formulated in the same way. The primary aim is thus, in close analogy with fission, to calculate the density of states for any mass partition at the multifragmentation barrier. The corresponding rate of decay then follows straightforwardly. The energy, to be shared between internal and collective degrees of freedom for a mass partition, $\boldsymbol{n}$, is $E_{\boldsymbol{n}} = E_C^* + Q_{\boldsymbol{n}}$. The Coulomb interaction energy is not included in $E_{\boldsymbol{n}}$ because it is taken into account explicitly. The energy corresponding to the three main groups of degrees of freedom can thus be written in the form

$$E_n = E_r + E_p + E_\mathrm{I} = V(\boldsymbol{r}_1, \boldsymbol{r}_2, \ldots \boldsymbol{r}_N) + \sum_{n=1}^{N} \frac{p_n^2}{2m_n} + \sum_{i=1}^{N} E_i^*, \qquad (6.8.3)$$

where on the right-hand side, the first term, E_r is the total interaction (spatial) energy, the second term is the sum of the kinetic energies and the last term, E_I, represents the sum of the internal excitation energies of the fragments. The density of states corresponding to all degrees of freedom is the convolution product,

$$\rho_n(E_n) = \int_0^{E_n} \int_0^{E_n} \int_0^{E_n} \rho_r(E_r)\rho_p(E_p)\rho_\mathrm{I}(E_\mathrm{I})\delta[(E_r + E_p + E_\mathrm{I}) - E_n]\,\mathrm{d}E_r\,\mathrm{d}E_p\,\mathrm{d}E_\mathrm{I}. \qquad (6.8.4)$$

Lopez and Randrup improve on (6.8.4) by adopting a spatial reference system for all partitions such that the position of centre of mass of the fragments, $\boldsymbol{R} = \sum_{n=1}^{N} m_n \boldsymbol{r}_n / \sum_{n=1}^{N} m_n$, is constant ($\boldsymbol{R}_0$). They also constrain the microstate densities by setting the total momentum, $\boldsymbol{P} = \sum_{n=1}^{N} \boldsymbol{p}_n$, to a fixed value ($\boldsymbol{P}_0$). By a suitable choice of coordinate systems $\boldsymbol{R}_0$ and $\boldsymbol{P}_0$ can both be set to zero (see section 1.2). In freeze-out theories, the fragment positions are further constrained to lie within the freeze-out volume. For the moment, we impose no such constraint so the equivalent of (6.8.4) is

$$\rho(E_n, \boldsymbol{P}_0, \boldsymbol{R}_0) = \left\{\prod_{i=1}^{N} \int \rho_i(\mathrm{d}E_i^*)\,\mathrm{d}E_i^*\right\}\left(\prod_{n=1}^{N} \int \frac{\mathrm{d}\boldsymbol{r}_n\,\mathrm{d}\boldsymbol{p}_n}{h^3}\right) \times \delta[(E_r + E_p + E_\mathrm{I}) - E_n]h^3\delta(\boldsymbol{P} - \boldsymbol{P}_0)\delta(\boldsymbol{R} - \boldsymbol{R}_0) \times \delta\left(\sum E_i^* - E_\mathrm{I}\right)\mathrm{d}E_\mathrm{I}, \qquad (6.8.5)$$

in which the product notation $\left(\prod_{n=1}^{N} \int \mathrm{d}\boldsymbol{r}_n\,\mathrm{d}\boldsymbol{p}_n/h^3\right)$ refers to the N-dimensional integral $\frac{1}{h^{3N}} \int \mathrm{d}\boldsymbol{r}_1\,\mathrm{d}\boldsymbol{p}_1 \int \mathrm{d}\boldsymbol{r}_2\,\mathrm{d}\boldsymbol{p}_2 \ldots \int \mathrm{d}\boldsymbol{r}_N\,\mathrm{d}\boldsymbol{p}_N$ and the first delta function restricts the states to have energy E_n. As we have seen in sections 6.4 and 6.6, the N-fold convolution (in curly brackets), which represents the density of microstates corresponding to internal excitation of fragments, is not too difficult to calculate (at least approximately) for any (summed) energy E_I. The integral of this convolution will be referred to henceforth as $\int \rho_\mathrm{I}(E_\mathrm{I})\,\mathrm{d}E_\mathrm{I}$. Multiplication by h^3 in the second factor (in square brackets) compensates for the fact that the position of the centre of mass and the total momentum are fixed. Consequently, the number of position–momentum degrees of freedom is reduced by a factor of six. Equation (6.8.5) thus represents a density of states in the usually accepted sense (number of microstates per unit energy).

Whereas the energy constraint provides upper limits for the momenta, $\boldsymbol{p}_n$, the distances of fragments with respect to the centre of mass are unconstrained so that the integration in (6.8.5) leads to an infinity. Of course, we could allow all position integrals to run only up to a fixed (freeze-out) radius. However, Lopez and Randrup prefer to base the size constraint on equation (6.8.1). They thus introduce explicitly the 'disassembly' coordinates p and q ((6.8.1) and (6.8.2)) and evaluate the density of states at the saddle point (p_F, q_F).

$$\rho(E_{\boldsymbol{n}}, \boldsymbol{P}_0, \boldsymbol{R}_0, p_F, q_F) = \int \rho_{\rm I}(E_{\rm I}) \times \left(\prod_{n=1}^{N} \int \frac{\mathrm{d}\boldsymbol{r}_n\, \mathrm{d}\boldsymbol{p}_n}{h^3}\right) \times \delta([(E_{\boldsymbol{r}} + E_{\boldsymbol{p}} + E_{\rm I}) - E_{\boldsymbol{n}}][h\delta(p - p_F)\delta(q - q_F)] \times h^3\delta(\boldsymbol{P} - \boldsymbol{P}_0)\delta(\boldsymbol{R} - \boldsymbol{R}_0)\, \mathrm{d}E_{\rm I}. \tag{6.8.6}$$

Simplification of (6.8.6) is possible by integration over the momentum coordinates, $\boldsymbol{p}_n$, although this integration is a little complicated because of the three constraints $(\delta[(E_{\boldsymbol{r}} + E_{\boldsymbol{p}} + E_{\rm I}) - E_{\boldsymbol{n}}]\delta(\boldsymbol{P} - \boldsymbol{P}_0)\delta(p - p_F))$. The result (see appendix D in reference [21]) is

$$\rho(E_{\boldsymbol{n}}, \boldsymbol{P}_0, \boldsymbol{R}_0, p_F, q_F) = \frac{(2\pi)^{3N/2-2}}{m_0^2 h^{3N-4}} \frac{1}{\Gamma\left(\frac{3N}{2} - 2\right)} \times \int \rho_{\rm I}(E_{\rm I}) \times \prod_{n=1}^{N}\left[m_n^{3/2} \int \mathrm{d}\boldsymbol{r}_n\right] \times \left(E_{\boldsymbol{n}} - E_{\boldsymbol{r}} - \frac{p_F^2}{2m_0} - E_{\rm I}\right)^{3N/2-3} \delta(q - q_F)\delta(\boldsymbol{R} - \boldsymbol{R}_0)\, \mathrm{d}E_{\rm I}, \tag{6.8.7}$$

in which the energy contained in the momentum degrees of freedom corresponding to random motion of the fragments (as opposed to the expansion energy $p_F^2/2m_0$) is $\left(E_{\boldsymbol{n}} - E_{\boldsymbol{r}} - p_F^2/2m_0 - E_{\rm I}\right)$. Further integration of (6.8.7) is not possible because the interaction energy $E_{\boldsymbol{r}} = V(\boldsymbol{r}_1, \boldsymbol{r}_2, \ldots \boldsymbol{r}_n)$ depends explicitly on the positions of all the fragments. Under these circumstances, one must either resort to numerical methods or to further approximations (e.g., one could replace $V(\boldsymbol{r}_1, \boldsymbol{r}_2, \ldots \boldsymbol{r}_n)$ by an average value $V(q)$, the average being taken over all sets $\{\boldsymbol{r}_n\}$ which lead to the same value of q).

In analogy with the Bohr–Wheeler theory, the quantity $\rho(E_{\boldsymbol{n}}, \boldsymbol{P}_0, \boldsymbol{R}_0, p_F, q_F)$ $\Delta E(\Delta p_F \Delta q_F / h)$ represents the number of transition states with p and q in the respective ranges $(p_F, p_F + \Delta p_F)$, $(q_F, q_F + \Delta q_F)$. The decay rate, which is evaluated at the minimum value (as a function of q_F) of the density of transition states, follows directly. Setting the vectors $\boldsymbol{P}_0$ and $\boldsymbol{R}_0$ to zero, and removing them from the list of arguments of ρ, we write

$$\frac{\mathrm{d}W(E_{\boldsymbol{n}}, p_F, q_F)}{\mathrm{d}p_F} \Delta p_F \Delta t = \frac{\rho(E_{\boldsymbol{n}}, p_F, q_F)}{\rho_C(E_C^*)} \frac{\Delta p_F \Delta q_F}{h} \frac{v_F \Delta t}{\Delta q_F}, \tag{6.8.8}$$

so that finally, with $p_F = m_0 v_F$,

$$W(E_{\boldsymbol{n}}, q_F) = \frac{1}{h\rho_C(E_C^*)} \int \rho(E_{\boldsymbol{n}}, p_F, q_F) \frac{p_F}{m_0}\, \mathrm{d}p_F. \tag{6.8.9}$$

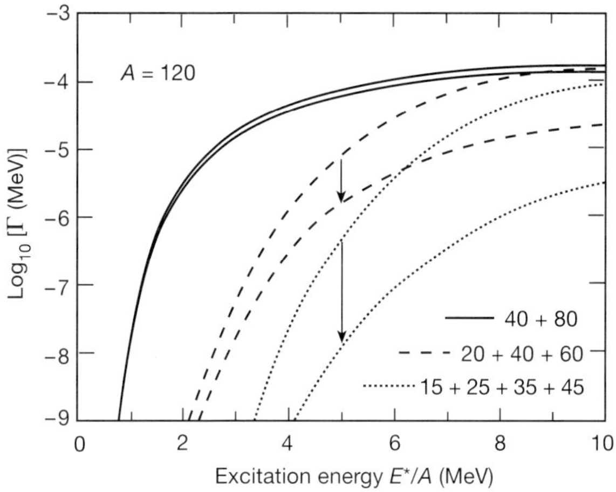

Figure 6.8. Lopez–Randrup model predictions [29] for the multifragmentation of a nucleus with mass number 120 into partitions 40 + 80, 20 + 40 + 60 and 15 + 25 + 35 + 45. The upper curves for each partition are calculated with no frictional forces. The lower curves show the effect of interfragment frictional forces on the partial widths. Fragment–fragment interactions reduce the widths for higher multiplicity partitions and thus have a considerable effect on the observed multiplicity distributions. (Reprinted from *Nucl. Phys.* A, **512**, Lopez J A and Randrup J, p 345. Copyright 1990, with permission from Elsevier Science.)

Equations (6.8.7) and its consequence, (6.8.9), can be considered to represent the main result of the Lopez–Randrup theory of multifragmentation. Several approximations are discussed in reference [27] which are used to simplify practical computations. The most important is concerned with averaging over position coordinates.

In a second paper [29], Lopez and Randrup reported on simulations carried out in order to investigate the effects of post-fragmentation interactions. One body frictional forces resulting from the exchange of nucleons between any pair of fragments as well as Coulomb repulsive forces were considered. The frictional forces convert kinetic energy of relative motion into internal excitation of fragments. This work seems to be the only serious attempt to take such interactions into account. Possibly because the initial configurations are rather compact, the study of reference [29] revealed a strong tendency for pairs of fragments to coalesce, thus reducing the effective multiplicity (see figure 6.8). The calculations show that there is little chance that a partition defined at the barrier survives the passage to complete scission. However, the introduction of a specific radial collective motion (expansion) corresponding to an explosive multifragmentation mechanism tended to reduce this effect. Furthermore, with radial collective motion, the

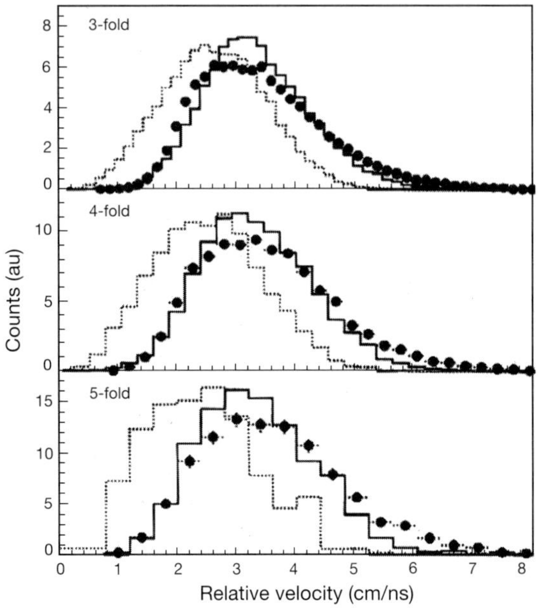

Figure 6.9. Spectra of relative velocities of fragments in the Kr + Au reaction at 60 Mev/nucleon [30] for three values of the complex fragment multiplicity. The data are shown as points. The solid line histogram was obtained using the transition state theory [27]. The broken line represents the result of a sequential decay model calculation. (Reproduced with kind permission of O Lopez.)

observed fragment kinetic energies are similar to those calculated at the barrier. These observations suggest that, if multifragmentation involves a collective radial expansion, the measurement of the asymptotic properties of the fragments can be used to provide information concerning the transition configurations.

The model also seems capable of providing accurate predictions of some dynamical aspects of multifragmentation. Thus, in a study by Lopez *et al* [30], a comparison of the spectrum of relative velocities of complex fragments with predictions using a sequential decay model and the Lopez–Randrup model (figure 6.9) provided strong support in favour of the multifragmentation decay mode. A similar conclusion was obtained using the spectrum of relative angles.

These remarks bring us to the end of our discussion of the main statistical models for multifragmentation. As with fission we have been able to distinguish between scission (freeze-out) models and transition-state approaches. This competition seems to be a recurring feature of attempts to describe nuclear decay in terms of equilibrium statistical mechanics. We may also remark that the theories described above contrast with the Hauser–Feshbach approach because they are based on internal (rather than asymptotic) properties of the decaying system.

References

[1] Bondorf J P 1982 *Nucl. Phys.* A **387** 25c

[2] Gross D H E and Ta-chung M 1981 *4th Nordic Meeting on Intermediate and High Energy Physics* 29 (Norway: Geilo Sportell)

[3] Stöcker H and Greiner W 1986 *Phys. Rep.* **137** 277; Danielewicz P 1979 *Nucl. Phys.* A **314** 465

[4] Siemens P J 1983 *Nature* **305** 410

[5] Jaqaman H R, Mekjian A Z and Zamick L 1984 *Phys. Rev.* C **29** 2067

[6] Bertsch G and Siemens P J 1983 *Phys. Lett.* B **126** 9

[7] Peilert G, Konopka J, Stöcker H, Griener W, Blann M and Mustafa M G 1992 *Phys. Rev.* C **46** 1457

[8] Colonna M, Chomaz Ph and Guarnera A 1997 *Nucl. Phys.* A **613** 165

[9] Konopka J and Stöcker H 1996 *Nuclear Decay Modes* ed D N Poenaru (Bristol: Institute of Physics Publishing)

[10] Pochodzalla J 1997 *Prog. Nucl. Part. Phys.* **39** 443

[11] Hardy G H and Ramanujan S 1917 *Proc. London Math. Soc.* 2 **16** 112; Hardy G H and Ramanujan S 1917 *Proc. London Math. Soc.* 2 **17** 75

[12] Bondorf J P, Donangelo R, Mishustin I N, Pethick C J, Schulz H and Sneppen K 1985 *Nucl. Phys.* A **443** 321; Bondorf J P, Donangelo R, Mishustin I N and Schulz H 1985 *Nucl. Phys.* A **444** 460; Bondorf J P, Botvina A S, Iljinov A S, Mishustin I N and Sneppen K 1995 *Phys. Rep.* **257** 133

[13] Zhang X Z, Gross D H E, Xu S Y and Zheng Y M 1987 *Nucl. Phys.* A **461** 641; Zhang X Z, Gross D H E, Xu S Y and Zheng Y M 1987 *Nucl. Phys.* A **461** 668; Gross D H E 1990 *Rep. Prog. Phys.* **53** 605

[14] Mekjian A J 1977 *Phys. Rev. Lett.* **38** 640; Mekjian A J 1978 *Phys. Rev.* C **17** 1051

[15] Fai G and Randrup J 1983 *Nucl. Phys.* A **404** 551

[16] Randrup J and Koonin S 1987 *Nucl. Phys.* A **474** 173; *Nucl. Phys.* A **471** 355c

[17] Chung K C 1993 *J. Phys. G: Nucl. Part. Phys.* **19** 1373

[18] Cerruti C, Debois J, Boisgard R, Ngô C, Natowitz J and Nemeth J 1988 *Nucl. Phys.* A **476** 74

[19] Das Gupta S, Gale C and Haglin K 1993 *Phys. Lett.* B **302** 372

[20] Brack M, Guet C and Hakansson H-B 1985 *Phys. Rep.* **5** 275

[21] Barz H W, Bauer W, Bondorf J P, Botvina A S, Donangelo R, Schulz H and Sneppen K 1993 *Nucl. Phys.* A **561** 466

[22] Cole A J *et al* 1996 *Z. Phys.* A **356** 181

[23] Metropolis N, Rosenbluth A W, Rosenbluth M N, Teller A H and Teller E 1953 *J. Chem. Phys.* **21** 1087; Bhanot G 1988 *Rep. Prog. Phys.* **51** 429

[24] Cox D R and Miller H D 1972 *The Theory of Stochastic Processes* (London: Chapman and Hall)

[25] Messiah A 1961 *Quantum Mechanics* (Amsterdam: North-Holland)

[26] Cole A J, Heuer D and Charvet M 1997 *Phys. Rev.* C **55** 2978

[27] Lopez J A and Randrup J 1989 *Nucl. Phys.* A **503** 183

[28] Moretto L G 1975 *Nucl. Phys.* A **247** 211

[29] Lopez J A and Randrup J 1990 *Nucl. Phys.* A **512** 345

[30] Lopez O 1993 *PhD Thesis* University of Caen, France; Lopez O *et al* 1993 *Phys. Lett.* B **315** 34

Chapter 7

Percolation, caloric curves and vaporization

That which we call a rose,
by any other name would smell as sweet.

Romeo and Julliet, William Shakespeare

7.1 Introduction

We are approaching the end of our presentation of models of statistical decay of excited atomic nuclei. In this chapter, we discuss topics which are related to multifragmentation viewed as a phase transition. We also examine the logical end point of the multifragmentation mechanism (at least in the nucleonic regime) which is known as vaporization.

Change of phase in a macroscopic system is usually discussed using the language of thermodynamics. The essential 'ingredients' of this discussion are naturally the pressure, P, volume, V (or density, ρ), and the temperature, T, of the system under study. Given that these quantities are related by an equation of state, the behaviour of the system may be examined in any one of the three projection planes (P–T, P–V, V–T). The most useful for our purposes is the P–V (or P–ρ) plane. It is also possible to take advantage of the fact that the entropy, S, is a function of state so that, for example, the equation of state may be examined in the P–S or T–S planes. In fact, these representations are particularly useful because a small increase of entropy, ΔS, considered to take place at constant temperature, T, corresponds to an input of heat energy $\Delta Q = T\Delta S$. The specific volume, v (volume per mole), is also a quantity which is useful for discussing phase transitions. At a given pressure it is easy to see that the specific volume is very different for different phases of a substance (e.g. water and steam).

Phase transitions are often categorized by specifying the order. The order of a transition is related to singularities which may occur as a function of some external parameters which control the system. Thus, for example, the familiar transition (viewed in the T–S plane) between ice and water phases can be viewed as a series of curves $T(S)$ for different pressures (isobars). The change of phase is said to be first order because, at the transition, the specific heat at constant pressure which is given by $c_p = T(\partial S/\partial T)_p$ diverges. Starting from the ice phase, a succession of small inputs of heat induce corresponding increases in temperature (and, of course, entropy). However, at the temperature of the transition, the system responds to each small input of heat energy not by a further increase of temperature, but by changing the relative amounts of ice and water. This situation is maintained until all the ice has melted. The divergence of c_p thus corresponds to the latent heat δQ and to a corresponding 'jump' in entropy $\delta S = \delta Q/T_{\mathrm{M}}$ at the melting point temperature T_{M}.

A liquid–vapour transition is also normally first order. Let us use the P–V plane (see figure 1.8) and induce the transition by compressing the vapour at constant temperature (such a process involves transfer of heat to the heat bath). As the system is compressed the pressure rises. However, at the transition, the system responds to a decrease in volume not by increasing the pressure but by converting vapour to liquid. The liquid and vapour phases thus coexist, the pressure remains constant and there is a jump in specific volume or density in passing from vapour to liquid. Thus, the isothermal compressibility, which is expressed in terms of the specific volume, v, as $\kappa_T = -v^{-1}(\partial v/\partial P)_T$, diverges.

With increasing (heat bath) temperature, the volume at which the vapour to liquid transition takes place diminishes up to the so-called critical point. At this point (along the 'critical' isotherm), the distinction between the two phases disappears, there is no discontinuous decrease in specific volume and the transition is then referred to as second order (because the *second* derivative of the volume is discontinuous). It is therefore important from the outset to distinguish between phase changes taking place below the critical temperature and those which take place at or above the critical point. The critical temperature for the liquid–gas phase change in nuclear matter is estimated to lie between 15 and 20 MeV and the critical density at about one third that of normal density [1]. Some reduction of T_{C} is expected for finite systems. Thus in the Copenhagen liquid drop model, discussed in section 6.4, the critical temperature was set to 12 MeV.

As noted by Moretto, Phair and Wozniak [2], there is nothing particularly remarkable about the phase change which takes place at subcritical temperatures. These authors consider isothermal density variations in a nuclear fluid and point out that, to the extent that the most probable macrostate (corresponding to the maximum number of microstates) may contain both large fragments (resembling the liquid phase) and light particles (gaseous phase), phase coexistence is certainly observed in multifragmentation. If we are justified in using the canonical ensemble, the most probable macrostate is that state which minimizes the free energy ($F = -T\ln[Z]$). For high densities, it corresponds to a liquid and for low densities to a vapour; in between, to some mixture. Thus, phase coexistence comes as no surprise. Of course, it remains an interesting and challenging task to attempt to 'map-out' the liquid–vapour phase co-existence region as a function of temperature and density. This task may be attempted by fitting multifragmentation data with statistical models such as those discussed in chapter 6.

At the critical point in a liquid–vapour transition, the fluctuations in the droplet size (or the density) would be expected to maximize (this gives rise to the critical opalescence observed by scattering light from a 'critical' fluid). It would probably be considered a major achievement to provide experimental evidence for the existence of corresponding fluctuations in nuclear decay. However, the possibility of observing finite nuclei at the critical point of the liquid–vapour phase diagram seems to many authors (including this one) to be remote for at least two reasons. The first is that it is hard to see how one is to simultaneously control both the density and the temperature in the nuclear fluid. The second is that significant fluctuations are, in any case, characteristic of finite systems even below the critical point (the most probable macrostate is accompanied by others which are 'almost as probable'). Furthermore, the 'critical' fluctuations themselves are obviously limited by the finite size of the system.

The phase change observed in percolation studies is rather different because, in the simplest cases, the system is controlled by a single parameter which may, for example, be the average fraction of broken bonds in a finite lattice (analogous to the temperature). Nevertheless, the transition is characterized by a maximum in the fragment size fluctuations and continuity in the entropy. The percolation transition is thus second order and the corresponding bond-breaking probability is often referred to as the critical point.

Let us now turn to the specific organization of this chapter. In the following section we will summarize the arguments which show why the probability distribution of multifragmentation partitions may be expected to mirror quite closely those observed in percolation [3]. These developments provide interesting insights into the nuclear problem. A possible motivation for this study obviously concerns the characterization of the nuclear phase change. It may well be that the phase change in nuclei bears a strong resemblance to the percolation transition because the underlying geometrical aspects of the two problems have a lot in common. On the other hand, the percolation–multifragmentation analogy cannot be developed completely because of the absence of dynamics (motion of fragments and light particles) in the percolation simulations.

Aside from percolation, there are essentially two distinct paths of investigation which have to do with the multifragmentation phase change. The first makes use of the form of the fragment size distributions. It has been shown by Fisher [4] that the formation of droplets from a vapour at the critical point would be expected to be characterized by a mass (or size) probability distribution which follows the simple power law $P(A) = A^{-\tau}$. The experimental observation of such characteristic behaviour may therefore be interpreted as a manifestation of the phase change in nuclei taking place at the critical temperature or, indeed, of a percolation-like transition. Following Campi [5], certain moments of the mass (or charge) distributions would be expected to maximize at the transition (i.e., for the power law distribution) and thus constitute 'signals' which may be sought for in data. This topic is discussed at the end of section 7.2.

The second major line of research which is directly concerned with phase transitions falls under the general heading of 'nuclear calorimetry'. This, somewhat controversial, area of investigation, which is the subject of section 7.3, concerns measurements of the so-called caloric curve (the variation of temperature with excitation energy) which may be derived from studies of multifragmentation. Controversy arises mainly due to the fact that measured temperatures depend on the observables and the methods used to extract them. We will try to make a fair presentation of the current (1998) situation and also discuss some theoretical models.

The point of view concerning phase change which we think is appropriate, given the current state of experimental and theoretical research, can be simply expressed using arguments based on the microcanonical ensemble. At low excitation energies, the density of microstates in a single nucleus increases rapidly and regularly with energy. The first derivative of $\ln[\rho(E^*)]$ is, of course, the inverse of the microcanonical temperature and would be expected to vary approximately as $\mathrm{d}(\sqrt{2aE^*})/\mathrm{d}E^*$ so that the temperature would go as $\sqrt{E^*/a}$ (see equation (1.4.4)). With increasing multiplicity, i.e., as more and more light particles appear in the partitions, the effective level density parameter, a, would be expected to diminish. At the same time, the density of states corresponding to motion of fragments would be expected to play an increasingly important role. At high energies, if the nucleus vaporizes into N light particles, one would expect the state density (considered as a classical gas) to vary approximately as $(E^*)^{3N/2-1}$ so that the temperature would go roughly as E^*/N. The transition between these two variations is certainly not abrupt. Nevertheless, interesting physics may be indicated by any irregular behaviour in one or more of the derivatives of the logarithm of $\rho(E^*)$. Such behaviour is worth investigating whether or not we use the language of thermodynamics to discuss it.

In section 7.4 we turn to investigations at excitation energies which are well in excess of the total binding energy of a nuclear system. In this 'vaporization' regime the reaction products are simply light particles (mainly nucleons and alpha particles). We shall describe the main experimental results and present a quantum statistical model which has been applied to this problem. As indicated in the preceding paragraph, the particular feature that we may expect to be important is that all the available energy must appear essentially as kinetic energy of the fragments. This should provide a severe test for the freeze-out scenario which, at lower energies, is dominated by the density of states corresponding to internal excitation of complex fragments.

We emphasize that a good deal of the material used in preparation of this penultimate chapter has been published in the last few years and, therefore, may be modified by subsequent research. For this reason, we have tried to concentrate on illustrative results (including some which are questionable). Hopefully, the uncertainties will gradually give way to a more consensual appreciation of the subject. In the meantime, one can at least hope that presently unresolved difficulties will stimulate the interest and the creative talents of confirmed researchers and students alike.

7.2 Multifragmentation, percolation and fragment size distributions

Let us begin with a brief description of percolation simulations. Figure 7.1(*a*) shows a two-dimensional (square) lattice in which a set of sites (filled circles in the figure) are connected by bonds. Let us suppose that the bonds are, in fact, resistive wires and that we apply a potential difference between the top and bottom edges of the lattice. We now set up an experiment in which bonds are broken randomly in a large number of similar systems and we record the average current as a function of the fraction of broken bonds. As we increase this fraction, the transmitted current falls slowly until, at some critical fraction, the average current drops rapidly to zero (figure 7.1(*b*)). This 'percolation transition' signals the disappearance of a large cluster (set of connected or linked sites) which spans the lattice from top to bottom. In an infinite lattice this 'percolating' cluster is, itself, infinite.

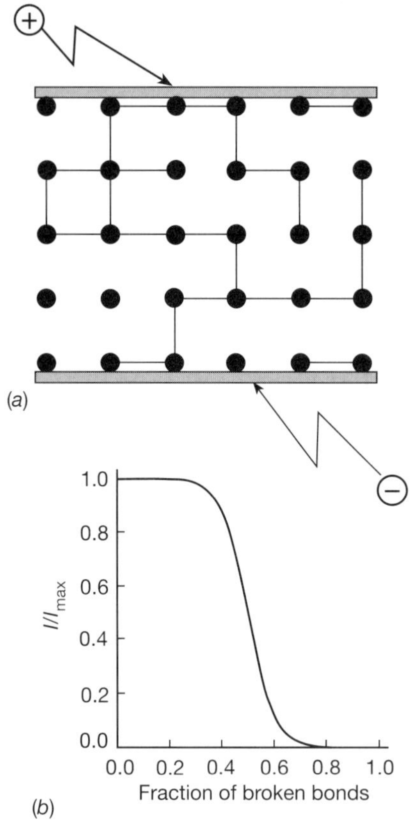

Figure 7.1. (*a*) Two-dimensional square lattice of resistive wires (bonds). (*b*) Fractional current as a function of the fraction of broken bonds.

Our objective in the first part of this section will be to develop, as far as possible, the analogy between a three-dimensional percolation lattice and an excited nucleus. We will then be in a position to comment on the possibility of detecting a percolation-like phase transition in studies of nuclear decay. The primary reason for investigating this analogy is that charge partition probabilities, measured as a function of excitation energy in nuclear decay, closely resemble those generated using a percolation simulation (see below) in which the number of broken bonds or the average value of this quantity plays the role of excitation energy [6,7]. This similarity may be related to universal behaviour, in which case the distribution of partition weights in nuclear multifragmentation may be largely independent of the precise forces involved. That is, of course the point of view which characterizes most 'freeze-out' models.

Before proceeding with the formulation, we should mention that there are a whole set of processes and procedures which may, either accurately or loosely, be labelled as percolation models. It may be that one or other of these procedures is more appropriate for discussing the analogy with nuclear decay. The physics of these models may be somewhat 'lattice dependent' and we therefore feel that it is difficult to make a general discussion. One can note, in this context, the work of Cook and collaborators [8] who have succeeded in building percolation models based on a face centred cubic lattice whose properties exhibit a remarkable similarity to known aspects of nuclear structure. The analogy with nuclear multifragmentation, however, can be quite well developed using the simple bond breaking process on a cubic lattice. One should see no particular restriction here. A similar analogy could be readily established using other percolation lattices.

A word of caution concerning the concept of a broken bond. An object for percolation studies is usually defined as a set of sites linked by bonds. It is normally pictured as being embedded in a lattice which exists only to define the particular geometry of the problem. The lattice itself contains no sites or bonds whereas the percolation structure is made up of occupied or unoccupied sites together with bonds which may be broken or unbroken. A broken bond is thus quite distinct from a line in the lattice structure. We will be later dealing with simple cubic percolation lattices, but, for the sake of explaining the principles involved, we may safely continue to work in two dimensions. Figure 7.2(*a*) shows a two-dimensional lattice in which a set of $4 \times 4 = 16$ sites are connected by $4(4-1)$ 'horizontal' bonds and by the same number of 'vertical' bonds. The structure illustrated (i.e. a square) maximizes the number of bonds for 16 sites in a two-dimensional square lattice. Indeed, if we were to associate an energy with each bond, we could say that the square lattice illustrated in figure 7.2(*a*) represents the ground state for 16 sites because the binding (bonding?) energy is maximized. We are then led naturally to the idea that breaking bonds is equivalent to creating excited states. Based on this remark we can establish a table of correspondence [3] (table 7.1).

If we break just one bond in figure 7.2(*a*) the lattice is always fully connected in the sense that each site is connected to the lattice by at least one bond.

Table 7.1.

Nucleus	Percolation structure
Mass number (A) or charge number (Z)	Number of sites s
Binding energy ($\mathcal{B}(A, Z)$)	$\mathcal{B} = B_s$ = maximum number of bonds for a given number of sites s
Mass excess, $\Delta = 7.29Z + 8.07(A-Z) - \mathcal{B}$. Defined such that mass excess $(^{12}C) = 0$.	Mass excess $= -B_s$. Defined so that the mass excess $= 0$ for $s = 1$.
Excitation energy: ground state binding energy minus the binding energy of the excited state. Described as superposition of single particle excitations or in terms of collective coordinates	Number of broken bonds in a fragment originally in ground state (B_s – number of unbroken bonds). Can arise due to holes in the structure or non-compact shapes. (e.g. sites arranged to lie on a straight line)
Ground state Q-value: difference in mass excesses of parent and fragments	Ground state Q-value: difference in mass excesses of parent and fragments (same definition)

However, if the number of broken bonds, $N_b \geq 2$, the situation changes. Thus, in figures 7.2(*b*) and 7.2(*c*), breaking 5 bonds has the effect of creating fragments which are not connected to the main lattice. In figure 7.2(*d*) all sites are still connected despite the five broken bonds. Here again we have a nice analogy. For small excitations (in this two-dimensional case $N_b < 2$), the lattice is 'below the particle emission threshold'. For 'excitations' above the threshold, there exists a fraction of configurations corresponding to the separation ('emission') of one single site, two single sites, two linked sites etc, but also 'stable' configurations. Let the number of (linked or unlinked) sites which are not connected with the main lattice be n and the number of associated configurations be $\Omega(N_b; n)$. Then the branching ratio for 'n site decay' is simply,

$$P(N_b; n) = \frac{\Omega(N_b; n)}{\sum_{n=0}^{N_b} \Omega(N_b; n)}. \tag{7.2.1}$$

Notice that the upper limit of N_b in the denominator is too conservative. In practice, $\Omega(N_b; n)$ would be expected to fall to zero for $n \ll N_b$. Another interesting observation is that excitation energy may appear in 'collective' or single particle degrees of freedom. Thus, in figure 7.2(*e*), the lattice is 'deformed' but all sites are saturated in the sense that for the shape considered, no further bonds can be accommodated.

The analogy can even be pursued when we consider time evolution. Excited nuclei are dynamical systems which means that they evolve with time. Each nucleus has a precise particle emission threshold. Furthermore, one can establish

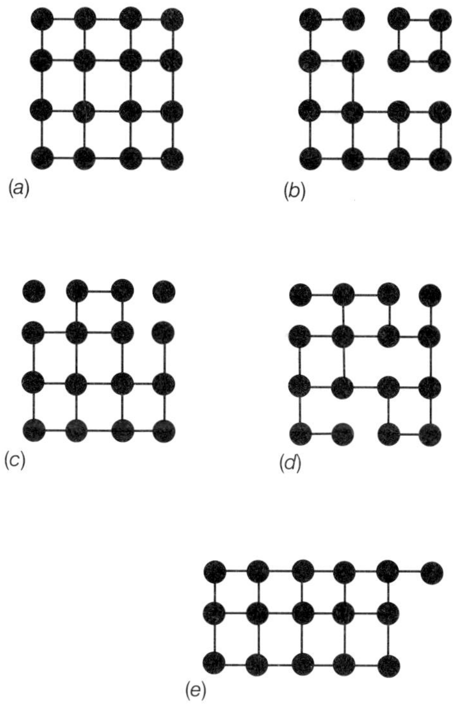

Figure 7.2. Sixteen sites arranged on a square lattice. (*a*) The most stable structure with 24 bonds (no bonds broken). In (*b*–*d*) five bonds have been broken. In (*b*) a cluster of four linked sites is liberated. In (*c*) two clusters with only one site are liberated. In (*d*) no cluster is liberated. In (*e*) there is a 'collective' deformation and the number of unbroken bonds is 23 rather than 24.

a natural discrete timescale based on the mean time between nucleon–nucleon collisions (recall the treatment of pre-equilibrium emission in section 3.7). For excitation energies above the threshold, particle emission is characterized by an exponential decay lifetime, τ, which generally decreases with increasing excitation energy. A correspondence with percolation is established if we imagine a simulation in which a percolation structure is initially produced from the ground state by breaking a fixed number of bonds. At each subsequent time step, two bonds chosen at random (whether broken on not) are interchanged and the lattice is scanned for unconnected sites. If one or more such sites are found, the lattice is considered to have decayed. If the number of broken bonds is less than the 'particle emission threshold' all configurations obtained in this way will be stable.

It is easy to see that the decay process is a finite time step equivalent of exponential decay. Thus, if we consider a small number of bond exchanges, Δ_b, then for a large ensemble of similarly prepared initial systems, the number,

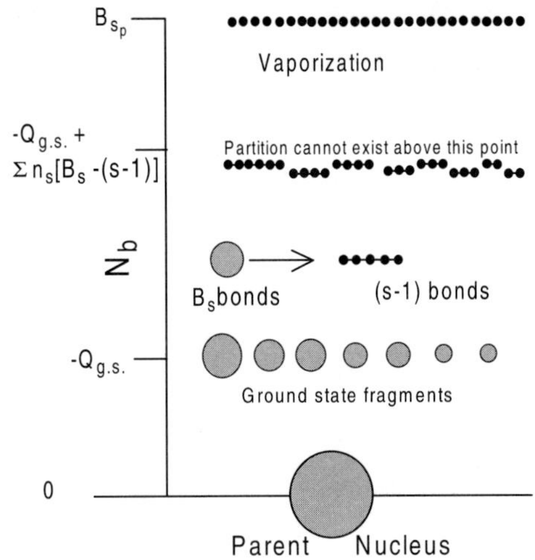

Figure 7.3. 'Energy' thresholds in percolation (see equation (7.2.4)). By breaking a number of bonds corresponding to the Q-value it is possible to form a set of ground state fragments. Above this energy the partition can accomodate broken bonds in the form of excitation energy of the fragments. However, above the second threshold the partition can no longer exist. The vaporization threshold is indicated at the top of the figure.

ΔN_S, of decaying systems in the 'interval' Δ_b, is equal to the current number of remaining systems, N_S, times the (average) probability that interchange of two bonds will produce instability.

$$\Delta N_S = -\gamma N_S \Delta_b. \tag{7.2.2}$$

The 'lifetime' (measured in units of the number of exchanges) is thus

$$\tau = \frac{1}{\gamma}, \tag{7.2.3}$$

in close analogy with radioactive decay. The 'width', γ, for a given disposition of sites, can also be directly obtained from standard percolation simulations. A large number of ground state systems are each modified by breaking N_b bonds chosen at random. The fraction of unstable systems is equal to γ.

Consideration of the widths, γ, leads to the discovery of a limitation on the number of broken bonds which can give rise to a particular partition. This limitation is a consequence of the fact that a fragment of s sites must have a minimum of $s - 1$ bonds (one-dimensional chain) to exist (see figure 7.3). If, in any fragment, which we can label using the number of sites, s, the number of broken bonds $N_b > (s - 1)$, the fragment cannot exist ($\gamma = 1$). Let us now specify a partition by the number of fragments of size s, i.e., by the partition

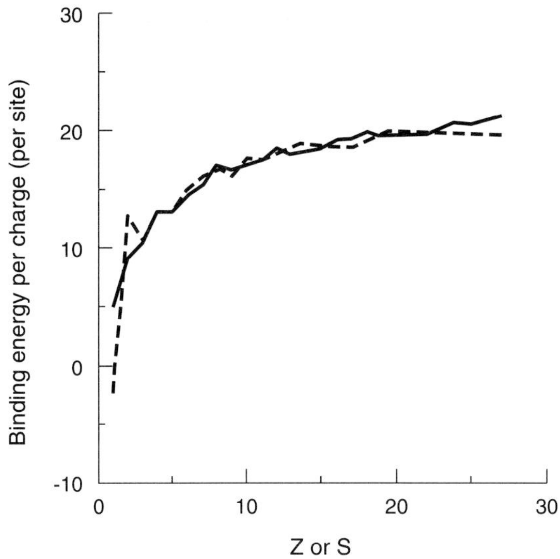

Figure 7.4. Binding energy per charge for most abundant isotopes in nuclei (broken curve) compared with a linear transformation of the binding energy per site for cubic percolation structures (solid curve). The constant in the linear transformation appears because of the the Coulomb contribution to the nuclear binding energy. The similarity of the two curves is remarkable.

vector $\boldsymbol{n} = (n_1, n_2, \ldots n_s \ldots)$. In order to produce this partition, the number of broken bonds must be at least equal to the negative of the Q-value. We can therefore conclude that any given partition can exist only in the range of broken bonds, N_b, for which

$$-Q \leq N_\text{b} \leq B_{s_\text{p}} - \sum_s n_s(s-1), \tag{7.2.4}$$

or

$$B_{s_\text{p}} - \sum n_s B_s \leq N_\text{b} \leq B_{s_\text{p}} - s_\text{p} + N, \tag{7.2.5}$$

where s_p and B_{s_p} are respectively the number of sites and the binding energy in the parent, B_s is the binding energy corresponding to a fragment with s sites, and $N = \sum n_s$ is the partition multiplicity.

Let us first discuss the lower limit. The Q-value is a difference in binding energies. The binding energy for exact cubes ($s = L^3 = 8, 27, 64\ldots$) is easily found to be $3L^2(L-1)$ or $3s - 3s^{2/3}$. We can identify a volume term and a surface term and remark that both terms have the same coefficient '3' which is similar to the nuclear case discussed in section 2.2 ($a_\text{V} = 15.6$, $a_S = 17.2$). For numbers of sites which are not exact cubes, the binding energy is not easy to express as a simple formula but can be obtained numerically. The nuclear case and the percolation case are compared in figure 7.4. The quantity plotted is the binding energy per charge in the nuclear case and a linear transformation (see figure caption) of the binding energy per site in the percolation case. The similarity between the two curves is remarkable. This correlation alone is largely responsible for the analogy between percolation and nuclear multifragmentation. Moreover, the analogy may be further improved by consideration of the Coulomb barrier. The threshold in excitation energy for nuclear decay is given, in principle, by the Q-value but in practice is modified by the Coulomb barrier which imparts a Coulomb kinetic energy to the reaction products. Thus, we have often written the 'available' energy as $E_C^* + Q_n - B_n$ (see after (6.3.10)). The effective Coulomb interaction energy was obtained (6.4.17) in the Wigner–Seitz approximation. We reproduce this equation ('p' for 'parent') writing the second term as a sum over all fragments. Thus

$$B_n = \left(\frac{\rho_F}{\rho_0}\right)^{1/3} \left[\frac{3}{5}\frac{Z_\mathrm{p}^2 \mathrm{e}^2}{R_\mathrm{p}} - \sum_k \frac{3}{5}\frac{Z_k^2 \mathrm{e}^2}{R_k}\right]. \tag{7.2.6}$$

If we write down $Q_n - B_n$ using a simplified liquid drop model which includes only volume, surface and Coulomb terms (2.2.27), we find, writing the Coulomb term in the form $\frac{3Z^2\mathrm{e}^2}{5R}$ rather than as $\frac{a_C Z^2}{A^{1/3}}$, that

$$\begin{aligned} Q_n - B_n &= \text{Mass (Parent)} - \text{Mass(Fragments)} - B_n \\ &= -\mathcal{B}(\text{Parent}) + \mathcal{B}(\text{Fragments}) - B_n \\ &= -a_\mathrm{V} A_\mathrm{p} + a_S A_\mathrm{p}^{2/3} + \frac{3}{5}\frac{Z_\mathrm{p}^2 \mathrm{e}^2}{R_\mathrm{p}} \\ &\quad - \left(-a_\mathrm{V} \sum_k A_k + a_S \sum_k A_k^{2/3} + \sum_k \frac{3}{5}\frac{Z_k^2 \mathrm{e}^2}{R_k}\right) \\ &\quad - \left(\frac{\rho_F}{\rho_0}\right)^{1/3} \left[\frac{3}{5}\frac{Z_\mathrm{p}^2 \mathrm{e}^2}{R_\mathrm{p}} - \sum_k \frac{3}{5}\frac{Z_k^2 \mathrm{e}^2}{R_k}\right]. \end{aligned} \tag{7.2.7}$$

The volume term obviously cancels ($A_\mathrm{p} = \sum_k A_k$) and we are left with a difference in surface terms together with a Coulomb correction.

$$Q_n - B_n = a_S \left[A_\mathrm{p}^{2/3} - \sum_k A_k^{2/3}\right] + \left[1 - \left(\frac{\rho_F}{\rho_0}\right)^{1/3}\right] \left[\frac{3}{5}\frac{Z_\mathrm{p}^2 \mathrm{e}^2}{R_\mathrm{p}} - \sum_k \frac{3}{5}\frac{Z_k^2 \mathrm{e}^2}{R_k}\right]. \tag{7.2.8}$$

If the freeze out density, ρ_F, is 'not too different' from the normal nuclear density, ρ_0, the Coulomb correction can be expected to be small so that the effective threshold, $-(Q_n - B_n)$, for the production of a partition in nuclear multifragmentation is, approximately, a difference in surface energies. This is, of course, precisely the situation in percolation which has no Coulomb interaction. In figure 7.5

we show the correspondence between nuclear and percolation *thresholds* with $R_F/R_p = 2$. The Coulomb correction increases the energy per bond from about 6 MeV (which corresponds to pure surface energies) to approximately 8 MeV but does not destroy the approximately linear relationship between the two quantities.

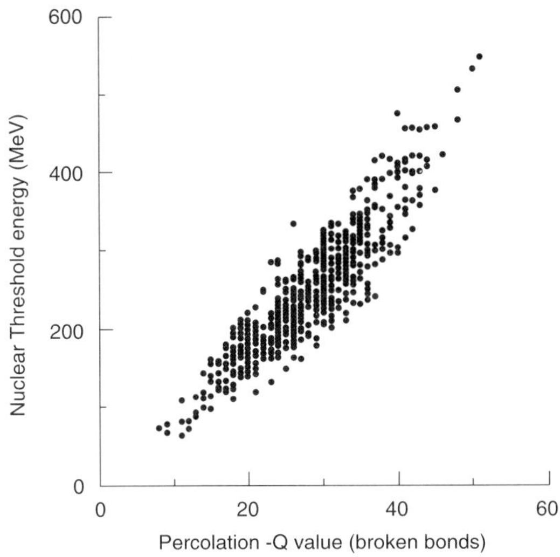

Figure 7.5. Correlation between percolation and nuclear multifragmentation threshold energies [3]. The percolation threshold energies (in broken bonds) were calculated starting from a $3 \times 3 \times 3$ (27 site) cube. The nuclear threshold energies correspond to Q-values for the corresponding charge partitions to which are added the Coulomb interaction energies. This latter contribution was calculated assuming a freeze-out radius which is twice the normal nuclear radius [3].

Having disposed of the lower limit in equation (7.2.5), we now turn to the upper limit. This appears at first sight as something new. Transposed to the multifragmentation situation it says that a partition will not be produced if the original excitation energy is too high. To understand the parallel, let us first suppose that partition probabilities depend only on the convolution of microstate densities corresponding to internal excitation of the fragments. In any given fragment, the lifetime of excited states falls with increasing excitation energy and will eventually become so short that the fragment will be unable to exist as a member of a partition. It is in this sense that the analogy may be established. Each complex fragment is characterized by a 'saturation' energy, just as in percolation there is saturation with respect to the number of broken bonds. However, in nuclear multifragmentation, the fragments also possess kinetic energy so that the partition may continue to be produced provided that 'excess' excitation energy appears as kinetic energy. Here the analogy with percolation must break down. Standard percolation models do not provide for motion of clusters. In multifragmentation,

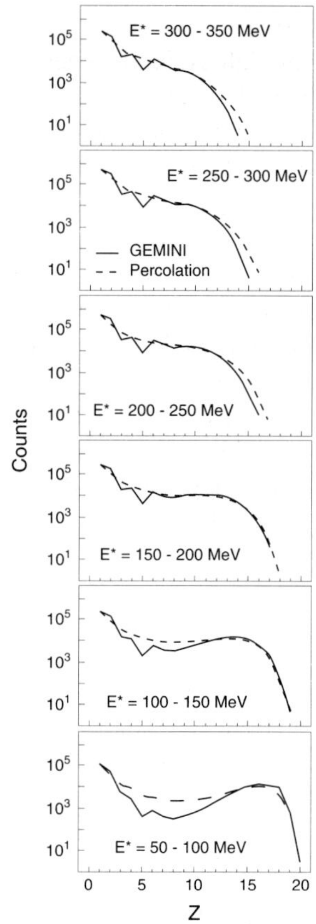

Figure 7.6. Typical evolution of the form of the charge distribution with excitation energy, for excited Calcium nuclei [7], obtained using the GEMINI code as in figure 5.9 but with the detector filter surpressed. The figure also shows that this evolution is quite well reproduced by simple percolation simulations (continuous trace) provided that the distribution of (charge) partition multiplicities at each excitation energy are the same (this condition was imposed on the percolation simulations). (Reprinted from *Phys. Rev.* C, **50**, Lleres *et al*, p 1973. Copyright 1994, by the American Physical Society.)

one would expect a change in the way the density of states, for any given partition, increases with increasing excitation energy. In particular, above the saturation region, the mean kinetic energies (the temperature in thermodynamics) of members of the partition should rise sharply whereas the temperatures which characterize the internal excitation of fragments would be expected to remain roughly constant. Of course, the change would hardly be expected to be abrupt. Even in percolation, the disappearance of a given partition takes place gradually as the upper limit in equation (7.2.5) is approached.

Nuclear level densities at high excitation energies have been investigated by Mustafa *et al* [9]. These authors find that if single particle configurations with unbound nucleons are excluded from the calculation of the density of excited states the density continues to rise with increasing energy although much more slowly than would be expected by the formula, $\rho = e^{2\sqrt{aE^*}}$. It would be interesting to compare these calculations more directly with percolation simulations.

The evolution of charge distributions observed as a function of increasing excitation energy have rather typical features which are reproduced by both binary sequential decay and simultaneous multifragmentation models. An example is shown in figure 7.6. At low excitation energies, the evaporation of light particles from an excited compound nucleus is insufficient to seriously deplete the residue and the charge distribution exhibits two peaks, one at small charges and the other composed of evaporation residues situated close to the compound nucleus charge. At high energies only relatively light fragments remain and the charge distribution is approximately described by an exponential fall off. Between these two extremes one may search for evidence of a phase change. The importance of the form of the fragment size distribution was first emphasized by the Purdue group [10]. They observed a power law distribution as predicted by Fisher [4] for a liquid–gas transition at the critical temperature or for the percolation transition.

An important step was taken by Campi [5] who proposed to characterize charge (or size) distributions for partitions of a 'parent' of size, s_{p}, using the moments

$$m_k(N) = \frac{\sum_{s=1}^{s'} s^k \langle n_s(N)\rangle}{s_{\mathrm{p}}}, \tag{7.2.9}$$

where, for partitions of multiplicity N (or reduced multiplicity $n = N/s_{\mathrm{p}}$), $\langle n_s(N)\rangle$ is the average number of fragments of size s and the sum is carried out by discarding the largest fragment (i.e., s' is the second largest fragment). The reason for this particular construction has to do with similar sums which may be obtained using percolation models. Below the phase transition the largest cluster is essentially infinite so that inclusion of this cluster obscures the information contained in fragments of smaller sizes. We observe that, for $k = 0$, the sum in the numerator of equation (7.2.9) is $N - 1$, and, that $m_1(N)$ is equal to the total system size minus the average size of the largest fragment divided by s_{p} (i.e., the average fraction of the total size contained in fragments other than the largest). Campi remarked that moments of order ≥ 2 diverge in infinite systems for charge distributions of the form $s^{-\tau}$ providing $\tau > 2$. He further showed that, at least in percolation simulations, some remnant of this 'phase change signal' remained even in quite small systems (see figure 7.7). Campi also proposed the ratio

$$\gamma_2 = \frac{m_0 m_2}{m_1^2}, \tag{7.2.10}$$

as a signal for a phase transition. This ratio is exactly equal to two for an exponential distribution, but considerably in excess of two for the 'power law' distribution,

$s^{-\tau}$, when $\tau > 2$.

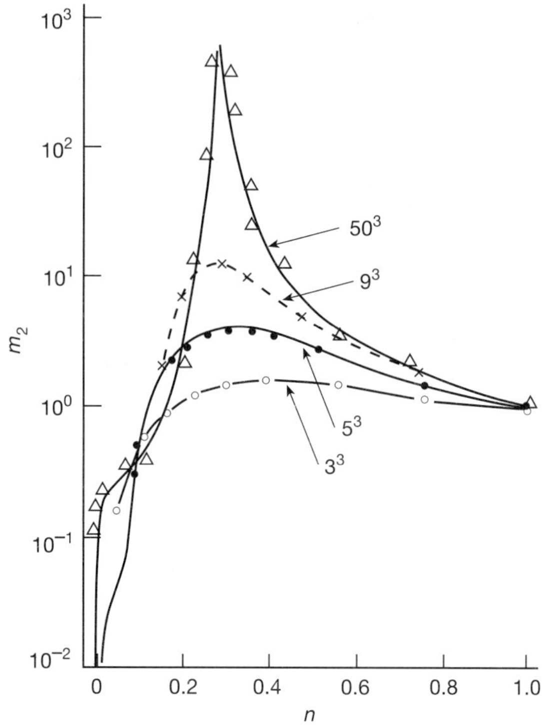

Figure 7.7. The 'Campi' moment m_2 [5] plotted versus n (the multiplicity divided by the number of sites) in cubic lattices of side 3, 5, 9 and 50. Some remnant of the divergence of m_2 is apparent even for the lighter systems. (Reprinted from *Phys. Lett.* B, **208**, Campi X, p 351. Copyright 1988, with permission from Elsevier Science.)

Despite these useful indicators, it is probably fair to say that no strong concensus has emerged regarding the use of fragment size distributions to identify and characterize a possible nuclear phase transition. A large part of the problem is undoubtedly due to the difficulty, in experiments, of isolating appropriate event sets. Thus, for example, molecular dynamics simulations carried out by by Aichelin and coworkers [11] showed that the power law distribution could arise simply due to mixing events with different impact parameters. Another difficulty (mentioned in the introduction) is that the power law behaviour characterizes the liquid–gas transition *at the critical temperature*, T_C. As mentioned earlier, whereas there are manifestly two phases in the nuclear problem it does not appear to be a simple matter to induce the transition exactly at T_C.

One result which remains, however, is that charge partitions in nuclear decay resemble quite closely those provided by percolation simulations even away from the phase transition. This is sometimes considered as a negative finding because, of course, the percolation process is essentially geometrical. The author's guess is that the apparent similarity between percolation and nuclear decay and, indeed the success of statistical models in describing multifragmentation, will turn out to be related to the concept of pseudo-equilibrium. If this is the case, then it still may be possible to find observables which exhibit enhanced sensitivity to the nuclear forces involved.

7.3 Caloric curves

In this section we again discuss the problem of phase change and, in particular, measurements which provide evidence both for and against the experimental demonstration of the existence of this phenomenon in the decay of hot and/or compressed atomic nuclei. The reference to compression is made because the principal means of creating highly excited nuclei is by collision of heavy ions. We have already discussed (in chapter 6) the dynamical studies which indicate not only that small impact parameter heavy ion collisions produce compressed nuclear matter but that the subsequent expansion may drive the system into the spinodal region. In such reactions, one may think that it is the reaction mechanism which is responsible for the phenomenon of multifragmentation. On the other hand, hot nuclei can be produced by using high energy light projectiles in reactions which would not be expected to lead to significant compression. In this case, the dissipation of energy is achieved mainly by thermal heating produced by pions resulting from the decay of Δ resonances. Here again, multifragmentation is observed [12] and is consistent with the formation of an equilibrated source.

The particular objective of studies of caloric curves is to investigate the variation of the density of scission microstates as a function of the excitation energy, E^*, of a system by measuring the variation of the microcanonical temperature $T(E^*)$ with E^* (see figure 7.8(*a*)). In practice, one considers some small part of the system which is thought of as being associated with the 'heat bath' formed by the rest. One may thus begin such a study by considering the microstate density as a convolution of two densities, one of which is considered to act as heat bath. The convolution is

$$\rho(E^*) = \int_0^{E^*} \rho_1(E^* - E)\rho_2(E)\,\mathrm{d}E \approx \rho_1(E^*) \int_0^{\infty} \mathrm{e}^{-\beta E} \rho_2(E)\,\mathrm{d}E, \quad (7.3.1)$$

and is the product of a constant, $\rho_1(E^*)$, and the canonical partition function, $Z_2[\beta(E^*)]$, of the subsystem. This latter quantity may be conveniently expressed as the Laplace transform of the density of states as in (7.3.1) or as a sum over the discrete energy levels, ε_i of the subsystem.

$$Z_2(\beta) = \sum_i g_i\,\mathrm{e}^{-\beta\varepsilon_i}. \quad (7.3.2)$$

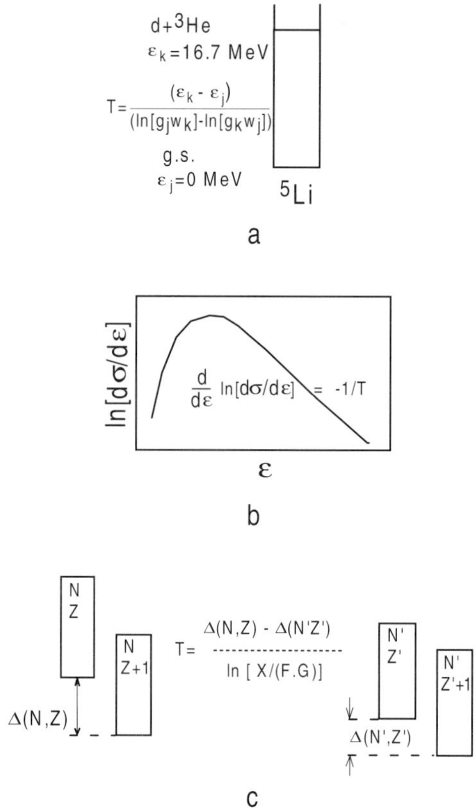

Figure 7.8. This figure was inspired by a similar figure to be found in the work of Pochodzalla [1]. Three different ways of determining a temperature in decay of excited nuclei are represented: (*a*) by comparison of yield of two discrete states of an emitted isotope; (*b*) by measuring the slope of the kinetic energy spectrum of emitted light particles; (*c*) via the double isotope thermometer proposed by Albergo *et al* [13]. In (*c*) X is the ratio of two isotope yield ratios defined in equation (7.3.24), F is a mass ratio factor and G a spin degeneracy factor.

In equation (7.3.2), the factor g_i is normally the spin degeneracy of the (presumably resolved) energy level at ε_i. As an example, we may estimate the (heat bath) temperature by considering the subsystem to be some light fragment such as ^{5}Li (figure 7.8(*a*)). The statistical weight associated with any state is $W_i = g_i \,\mathrm{e}^{-\beta\varepsilon_i}$ that the ratio of the relative probability of observation of any two states with known energies ε_j, ε_k gives the temperature as

$$T(E^*) = \frac{1}{\beta(E^*)} = \frac{(\varepsilon_k - \varepsilon_j)}{\ln[g_j W_k - \ln[g_k W_j]}. \qquad (7.3.3)$$

For a continuous spectrum (figure 7.8(*b*)), the extraction of the temperature requires a rather detailed knowledge of $\rho_2(E)$. Usually this method is applied when $\rho_2(E)$ goes as a small power of E so that the asymptotic slope of the logarithmic energy spectrum is, approximately, $-1/T$. In either case the plot of $T(E^*)$ versus E^* is referred to as a caloric curve. We emphasize that the extraction of T involves the assumption that this quantitity can be considered to be constant at least over the range of energy considered in the subsystem. *A priori*, this condition is harder to enforce when dealing with a continuous spectrum because the energy range is usually more extended. We should also keep in mind that the temperature in equation (7.3.3) refers to the evolution of the logarithmic slope $\mathrm{d}\ln[\rho_1(E^*)]/\mathrm{d}E^*$ for the combined set of degrees of freedom which are considered to constitute the heat bath rather than for the entire system.

It may be thought that the most straightforward application of the above analysis occurs in evaporation from a compound nucleus. In this process the density of states in the residual nucleus plays the role of the heat bath. However, a complication arises from the necessity to include orbital angular momentum in the analysis. We will analyse, as far as possible, how this inclusion may affect extraction of a reliable temperature.

Consider the emission of a light fragment from a compound nucleus with excitation energy, E^*, and angular momentum, J_i (we assume the probability distribution for values of J_i to be known) and let us suppose that accessible states in the residual nucleus may be represented by a state density (the final state is not isolated). We can use the Hauser–Feshbach theory (chapter 4) to write the statistical weight for evaporation of the fragment with relative angular momentum ℓ as a sum over angular momenta (J_f) in the residual nucleus. We recall equations (4.3.10) and (4.3.14) and simplify the treatment of intrinsic spin of the emitted particle in performing the sum over J_f by including a factor $(2s+1)$ where s is the spin of the emitted particle. This is equivalent to assuming that $\rho_f(E_f, J_f)$ can be considered to be constant over small ranges of J_f. The statistical weight then reduces to the difference of two terms. Thus,

$$\begin{aligned}\omega(\varepsilon,\ell) &= T_\ell(\varepsilon)\sum_{J=|J_i-\ell|}^{J_i+\ell}\ \sum_{J_f=|J-s|}^{J+s}\rho_f(E_f,J_f)\\ &\approx T_\ell(\varepsilon)\sum_{J_f=|J_i-\ell|}^{J_i+\ell}(2s+1)\rho_f(E_f,J_f)\\ &= (2s+1)T_\ell(\varepsilon)\sum_{J_f=|J_i-\ell|}^{J_i+\ell}[W(E_f,M=J_f)-W(E_f,M=J_f+1)]\\ &= (2s+1)T_\ell(\varepsilon)[W(E_f,M=|J_i-\ell|)-W(E_f,M=J_i+\ell+1)],\end{aligned} \tag{7.3.4}$$

in which $T_\ell(\varepsilon)$ is the transmission coefficient for emitted fragments with kinetic energy ε, $E_f = E^*+Q-\varepsilon$, and $W(E_f, M)$ is the density of states in the residual nucleus having a z-projection, M, which we take from section 2.5 (equations (2.5.45) and (2.5.49)) as

$$W(E_{\rm f}, M) = \frac{{\rm e}^{2\sqrt{aE_f}}}{\sqrt{48}E_f} \times \frac{1}{\sqrt{2\pi\sigma^2}}\,{\rm e}^{\frac{-M^2}{2\sigma^2}}. \tag{7.3.5}$$

Since W falls rapidly with increasing M we may approximate (7.3.4) by dropping the second term in square brackets. We can now perform the sum over ℓ (which is unobserved) and write the ratio of statistical weights for any two excitation energies in the residual nucleus (attained by varying the kinetic energy of the emitted particle) as

$$\frac{\omega_A(\varepsilon_A)}{\omega_B(\varepsilon_B)} = \frac{(2s_A+1)\sum_{\ell=0}^{\infty} T_\ell(\varepsilon_A) W(E^*+Q-\varepsilon_A, M=|J_i-\ell|)}{(2s_B+1)\sum_{\ell=0}^{\infty} T_\ell(\varepsilon_B) W(E^*+Q-\varepsilon_B, M=|J_i-\ell|)}. \tag{7.3.6}$$

Setting $\varepsilon_A = \varepsilon_B + \Delta\varepsilon$, we expand $W(E^*+Q-\varepsilon_A, M)$ as

$$W(E^*+Q-\varepsilon_A, M) = W(E^*+Q-\varepsilon_B, M)\,{\rm e}^{-\beta\Delta\varepsilon}, \tag{7.3.7}$$

so that we can rewrite equation (7.3.6) as

$$\frac{\omega_A(\varepsilon_A)}{\omega_B(\varepsilon_B)} = {\rm e}^{-\beta\Delta\varepsilon}\frac{(2s_A+1)\sum_{\ell=0}^{\infty} T_\ell(\varepsilon_A) W(E^*+Q-\varepsilon_B, M=|J_i-\ell|)}{(2s_B+1)\sum_{\ell=0}^{\infty} T_\ell(\varepsilon_B) W(E^*+Q-\varepsilon_B, M=|J_i-\ell|)}. \tag{7.3.8}$$

The nature of the difficulty is now clear. Even though we may use equation (7.3.5) to rewrite (7.3.8) as

$$\frac{\omega_A(\varepsilon_A)}{\omega_B(\varepsilon_B)} = {\rm e}^{-\beta\Delta\varepsilon}\frac{(2s_A+1)\sum_{\ell=0}^{\infty} T_\ell(\varepsilon_A)\,{\rm e}^{\frac{-|J_i-\ell|^2}{2\sigma^2}}}{(2s_B+1)\sum_{\ell=0}^{\infty} T_\ell(\varepsilon_B)\,{\rm e}^{\frac{-|J_i-\ell|^2}{2\sigma^2}}}, \tag{7.3.9}$$

there seems to be no simple way of dealing with the energy dependence of the transmission coefficients. The development leading to equation (7.3.3) (in which angular momentum is not considered) can therefore not be clearly justified. Furthermore the first line of equation (7.3.4) shows that the difficulty is not removed by considering decays to isolated final states. In this case only one value of T_ℓ contributes to the sum but its energy dependence cannot be ignored. The one conceivable situation in which equation (7.3.9) may be further simplified is that in which the Gaussian weights are such that all important contributions involve values of T_ℓ close to unity. In this case, indeed, we recover equation (7.3.3).

The application to multistep evaporation is much more complicated. Indeed, in this case, the battle is lost (see the work of Peter described later) because the temperature no longer refers to a particular nucleus. The extracted T value represents some weighted average value (over the various evaporation steps) which would be expected to be rather less than the initial temperature of the compound nucleus. On the positive side one may suppose that, to the extent that multistep evaporation predictions reproduce experimental observations the 'true' nuclear densities do not differ substantial from those assumed in the model calculations.

A rather different 'thermometer' was proposed by Albergo *et al* [13].

Following pioneering work by Mekjian [14], these authors picture light isotopes (H, He, Li...) as being formed from a gas of neutrons and protons in a series of 'chemical' reactions. Relative orbital angular momentum and Coulomb effects are not included in the formulation. The method is based on thermodynamics and assumes thermal and chemical equilibrium. Using the grand canonical ensemble, the total differential of the internal energy (see equation (1.6.11)) is written (setting the average values $\langle N_k \rangle = N_k$)

$$\mathrm{d}U = T\,\mathrm{d}S - P\,\mathrm{d}V + \sum_k \mu_k\,\mathrm{d}N_k, \tag{7.3.10}$$

and chemical equilibrium together with thermal equilibrium implies that

$$\sum_k \mu_k\,\mathrm{d}N_k = 0. \tag{7.3.11}$$

Consider a chemical reaction in which n_1 molecules of a substance S_1 combine with n_2 molecules of S_2 to form n_3 molecules of S_3 and n_4 molecules of S_4. The numbers $-n_1$, $-n_2$, n_3 and n_4 are referred to as the stoichiometric coefficients (note the signs). We find the changes in the numbers of molecules produced in $\mathrm{d}N$ reactions are $\mathrm{d}N_{S_1} = -n_1\,\mathrm{d}N$, $\mathrm{d}N_{S_2} = -n_2\,\mathrm{d}N$, $\mathrm{d}N_{S_3} = n_3\,\mathrm{d}N$ and $\mathrm{d}N_{S_4} = n_4\,\mathrm{d}N$ and from (7.3.11) we have

$$\mu_{S_1}\,\mathrm{d}N_{S_1} + \mu_{S_2}\,\mathrm{d}N_{S_2} + \mu_{S_3}\,\mathrm{d}N_{S_3} + \mu_{S_4}\,\mathrm{d}N_{S_4} = 0, \tag{7.3.12}$$

so that on substituting for $\mathrm{d}N_{S_1}$, $\mathrm{d}N_{S_2}$etc and dividing by $\mathrm{d}N$ we obtain

$$n_1\mu_{S_1} + n_2\mu_{S_2} = n_3\mu_{S_3} + n_4\mu_{S_4}. \tag{7.3.13}$$

Of course, the number of reactants on each side of (7.3.13) may be greater or less than two. One case of particular interest arises in fusion reactions in which a composite system is formed from two or more subunits, i.e.,

$$A_1 + A_2 + A_3 + \ldots \to C, \tag{7.3.14}$$

for which (7.3.13) implies that the chemical potential of C is the sum of the chemical potentials of the constituents. This is particularly relevant if C is a nucleus made up of N neutrons and Z protons. Let us therefore consider such a reaction taking place in a volume V which is sufficiently large for the full volume to be thought of as available to all constituents (neutrons, protons and composites). We then consider each constituent to behave as an ideal gas in which case the canonical (subscript 'C') partition function for N_f particles (fragments) is $Z_c^{N_f}/N_f!$ (equation (1.4.20)). For example, the nucleus composed of N neutrons and Z protons ($A = N + Z$) has the canonical partition function

$$Z_c = \frac{V}{\lambda_T^3} A^{3/2} \left(\sum_i g_i \, e^{-\varepsilon_i/T} \right), \tag{7.3.15}$$

where the first term is the integral over spatial and momentum coordinates, λ_T is the *nucleon* thermal wavelength given by equation (6.4.2), and the second (bracketed) term is the sum over internal excitation energy *levels* (there are g_i microstates at energy ε_i). The Gibbs free energy for N_f such fragments is $F_G = -T \ln[\mathbb{Z}]$ where $\mathbb{Z}$ is the grand canonical sum of states which, from (1.6.22), is

$$\mathbb{Z}(\beta, \mu) = \sum_{N_f} e^{\beta \mu N_f} \frac{Z_c^{N_f}}{N_f!} = \sum_{N_f} \frac{[e^{\beta\mu} Z_c]^{N_f}}{N_f!}. \tag{7.3.16}$$

Setting $q = e^{\beta\mu} Z_c$ we see that (7.3.17) is the expansion of e^q so that the grand potential is

$$\begin{aligned} \Omega^* &= -T \ln[\mathbb{Z}] \\ &= -T\, e^{\beta\mu} Z_c = -T\, e^{\beta\mu} \frac{V}{\lambda_T^3} A^{3/2} \left(\sum_i g_i \, e^{\varepsilon_i/T} \right). \end{aligned} \tag{7.3.17}$$

The mean value of the fragment density $\omega = \langle N_f \rangle / V$ is

$$\omega = \frac{\langle N_f \rangle}{V} = -\frac{1}{V} \frac{\partial \Omega^*}{\partial \mu} = e^{\beta\mu} \frac{A^{3/2}}{\lambda_T^3} \left(\sum_i g_i \, e^{-\varepsilon_i/T} \right), \tag{7.3.18}$$

which can be considered as a relation between the chemical potential and the density. The same result can be obtained immediately by noting that, to within a constant, the terms in the sum in equation (7.3.16) represent a Poisson distribution with mean $\langle N_f \rangle = e^{\beta\mu} Z_c$.

The additivity of the chemical potentials in the fusion reaction can now be used with (7.3.18) to relate the particle density for the fusion product with chemical potential $\mu(N, Z)$ to the neutron and proton densities. This relation is one particular example of the so-called 'law of mass action'. Denoting the neutron and proton chemical potentials respectively as μ_n and μ_p the additivity property is expressed as

$$N\mu_n + Z\mu_p = \mu(N, Z). \tag{7.3.19}$$

For free protons (7.3.18) gives

$$\omega(0,1) = \frac{2}{\lambda_{\mathrm{T}}^3}\,\mathrm{e}^{\mu_{\mathrm{p}}/T}, \tag{7.3.20}$$

where the factor of 2 is the number of microstates $(2s_{\mathrm{p}}+1)$ corresponding to the proton spin. Then from (7.3.20) we obtain

$$\mathrm{e}^{Z\mu_{\mathrm{p}}/T} = \left[\frac{\omega(0,1)\lambda_{\mathrm{T}}^3}{2}\right]^Z, \tag{7.3.21}$$

with a similar expression for neutrons. The particle density for the isotope (N, Z) is given by (7.3.18) with $\mu = \mu(N, Z)$ and $A = N + Z$. This density is now expressed in terms of the nucleon densities, assuming chemical equilibrium (equation (7.3.20)), by first substituting for $\mu(N, Z)$ and then using (7.3.21) (and its analogue for neutrons). Thus

$$\begin{aligned}\omega(N,Z) &= \mathrm{e}^{Z\mu_{\mathrm{p}}/T+N\mu_{\mathrm{n}}/T}\,\frac{A^{3/2}}{\lambda_{\mathrm{T}}^3}\left(\sum_i g_i\,\mathrm{e}^{-\varepsilon_i/T}\right)\\ &= \left[\frac{\omega(1,0)}{2}\right]^N\left[\frac{\omega(0,1)}{2}\right]^Z\left[\frac{1}{\lambda_{\mathrm{T}}^3}\right]^{1-Z-N} A^{3/2}\left(\sum_i g_i\,\mathrm{e}^{-\varepsilon_i/T}\right).\end{aligned} \tag{7.3.22}$$

At this stage, Albergo *et al* [13] make the rather drastic assumption that the nucleus (N, Z) is created only in its ground state so that the sum $\sum_i g_i\,\mathrm{e}^{-\varepsilon_i/T}$ reduces simply to $(2s_{N,Z}+1)\,\mathrm{e}^{\mathcal{B}(N,Z)/T}$ where $s_{N,Z}$ is the ground state spin and $\mathcal{B}(N, Z)$ is the binding energy (remember that the internal energy of a nucleus in its ground state measured from the state of free nucleons is the negative of the binding energy). The justification for this step is that the exponential factor $\mathrm{e}^{-\varepsilon_i/T}$ strongly damps out excited states. With equation (7.3.22) the density ratio for the isotopes (N, Z) and $(N, Z+1)$ is

$$\frac{\omega(N,Z)}{\omega(N,Z+1)} = \frac{A^{3/2}}{(A+1)^{3/2}}\,\frac{2\lambda_{\mathrm{T}}^{-3}}{\omega(0,1)}\,\frac{2s_{N,Z}+1}{2s_{N,Z+1}+1}\,\mathrm{e}^{\frac{\mathcal{B}(N,Z)-\mathcal{B}(N,Z+1)}{T}}. \tag{7.3.23}$$

Equation (7.3.24) can be used to extract a temperature by measuring the ratio of yields on the left-hand side. However, to do so requires knowledge of the proton density $\omega(0, 1)$. This problem can be avoided by constructing a so-called double isotope ratio. Let us denote the ratio expressed in (7.3.23) as $R(N, Z)$ and the corresponding difference in binding energies as $\delta_Z(N, Z)$. Let us further consider a similar ratio for isotopes (N', Z') and $(N', Z'+1)$. Then the ratio of ratios

$$\begin{aligned}X(N,Z,N'Z') &= \frac{R(N,Z)}{R(N',Z')}\\ &= \frac{(A'+1)^{3/2}A^{3/2}}{(A+1)^{3/2}A'^{3/2}}\,\frac{(2s_{N,Z}+1)(2s_{N',Z'+1}+1)}{(2s_{N,Z+1}+1)(2s_{N',Z'}+1)}\,\mathrm{e}^{\frac{\delta_Z(N,Z)-\delta_Z(N',Z')}{T}}.\end{aligned} \tag{7.3.24}$$

It will be noticed that X is independent of the proton density and that the temperature T can be easily extracted from a measurement of X using (7.3.24). Thus, if we denote the spin factor as $G(N, Z, N', Z')$ and the mass ratio as $F(A, A')$, we find

$$T = \frac{\delta_Z(N, Z) - \delta_Z(N', Z')}{\ln[X/(F \times G)]}. \tag{7.3.25}$$

Finally, we may remark that an equation, equivalent to (7.3.25), can easily be derived for isotopes which differ by a single neutron rather than by a single proton. As a numerical example, we consider the double ratio of $[d/t]/[^3\mathrm{He}/^4\mathrm{He}]$. The mass factor F is $[(2/3)/(3/4)]^{3/2} = 0.838$. The spin factor is $[(3/2)/(2/1)] = 0.75$ so that $F \times G = 0.6285$. The binding energies obtained from tabulated mass excesses (equation (2.2.26)) are for d, t, ^{3}He and ^{4}He, 2.224 MeV, 8.481 MeV, 7.718 MeV and 28.295 MeV respectively. Equation (7.3.25) then yields

$$T = \frac{14.32}{\ln[1.59X]}. \tag{7.3.26}$$

Other examples of combinations of isotopes used in the construction of 'isotope ratio thermometers' are the He–Li isotopes (^{4}He, ^{5}Li; ^{5}He, ^{6}Li) and, for nuclei differing by one neutron, the combination (^{6}Li, ^{7}Li; ^{3}He, ^{4}He).

Before considering the application of equation (7.3.25) to experimental data, we should perhaps summarize possible problems with the above analysis. One may, of course, raise doubts as to the applicability of the notion of chemical equilibrium to small isolated systems. In statistical mechanics, equation (7.3.12) (or (7.3.11)) is just the condition which defines the maximum of a convolution integral and, in large (thermodynamic) systems, contributions to the convolution integral away from the maximum may be safely neglected. In small systems, this is usually not the case. Furthermore, it is to be regretted that only ground states are included in the analysis. Indeed, one suspects that if the theory is to be taken seriously we need to consider full chemical equilibrium between all isotopes H, He, Li…etc (including excited states) and not just between a particular isotope and the corresponding set of dissociated nucleons. A possible way of including corrections in an *ad hoc* manner is of course to 'calibrate' the thermometers using statistical decay codes. However, one then encounters again the problem that the extracted temperatures may be somewhat model dependent.

The logical end point of consideration of full thermal and chemical equilibrium would, of course, engender a particular statistical decay theory. Such a theory has already been considered by Mekjian [14]. We remind the reader that it is not necessary to imagine the system existing for long times (although one could argue, using the ergodic hypothesis, that this should be a sufficient condition). We need only the condition that an ensemble of similarly prepared systems provides an approximate representation of the full microcanonical ensemble for the purposes of determining macrostate probabilities. This being said, the approach suffers from the absence of an identifiable heat bath so that the influence of fluctuations about mean values may not be negligible.

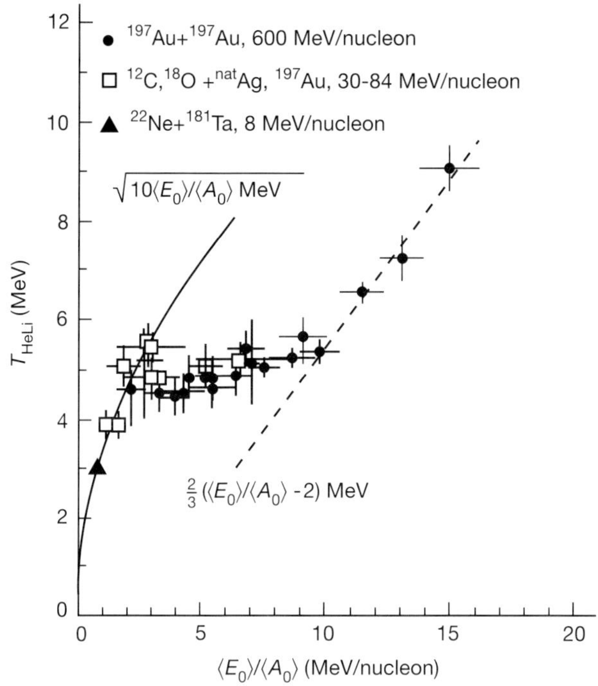

Figure 7.9. He/Li double isotope ratio temperatures versus excitation energy [15]. The plot contains data from several systems but the most interesting feature, which was extracted using the Au + Au data, is the flat portion between about 3 and 10 MeV/nucleon which may be interpreted as evidence for a first order (liquid–vapour) phase change. (Reprinted from *Phys. Rev. Lett.*, **75**, Pochodzalla J *et al*, p 1040. Copyright 1995, by the American Physical Society.)

Let us now turn to a consideration of experimental results. Examination of published reports obliges one to conclude that, at least in the form presented by Albergo *et al*, the operation of equation (7.3.25) is fraught with difficulties. The most spectacular result was published by Pochodzalla *et al* [15] and was obtained mainly from examination of the decay of projectile-like fragments in Au + Au collisions at an incident energy of 600 MeV/nucleon. This result is shown in figure 7.9. For excitation energies from about 3–10 MeV/nucleon, the observed (He–Li) temperatures remain roughly constant at a value of 5 MeV. At first, this observation was taken as evidence of a phase change. However, further work, notably by Peter and collaborators [16], showed that the overall form of the observed 'caloric' curve could result from the side feeding from the decay of (neglected) excited states especially, of course, at high temperature. In reference [16], it was also shown that extracted 'double isotope ratio' temperatures show a strong dependence on the combination of isotopes. Some results obtained

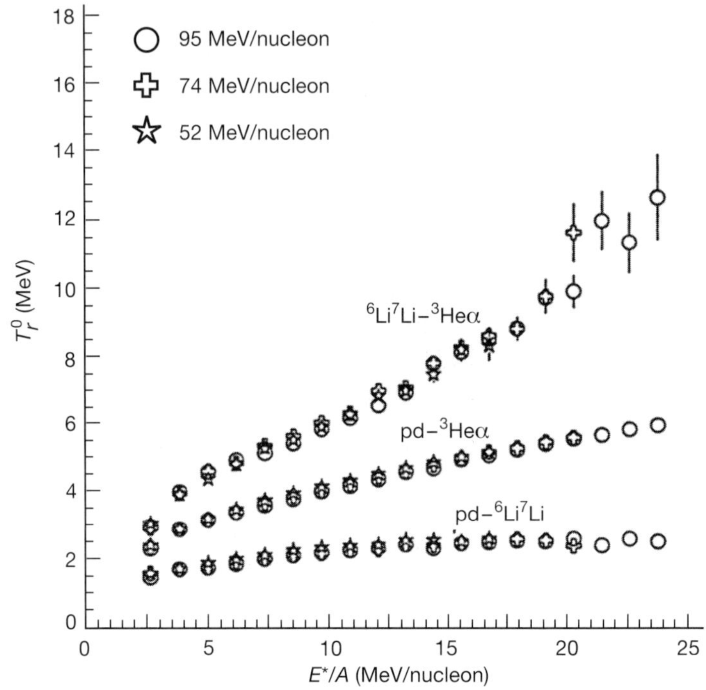

Figure 7.10. Double isotope ratio temperatures versus excitation energy (per nucleon) measured in collisions of ^{36}Ar + ^{58}Ni at several incident Ar energies [16]. Two points are important. The first is that the extracted temperatures depend strongly on the type of thermometer. The second is the absence of any flat portion of the curves which could signal the liquid–gas phase change.

from collisions of ^{36}Ar + ^{58}Ni [17] at three incident Ar energies are shown in figure 7.10. The observations contrast with those of Pochodzalla *et al* in that the extracted temperatures increase rather smoothly with excitation energy and cannot be interpreted as an indication of a phase change.

An interesting result was obtained by Peter and collaborators based on sequential evaporation calculations in which the initial temperatures were calculated directly from the density of compound nuclear states ($T = \sqrt{E^*/a}$). The authors showed that the three principal thermometers (slopes of kinetic energy spectra, double isotope ratios and ratios of excited state populations) gave correct results only if the evaporation sequence was interrupted after just one evaporation. Allowing the cascade to develop fully and then measuring temperatures as if from a unique source produced the results shown in the lower part of figure 7.11. The best estimate of the intial temperature was obtained from the slope of the kinetic energy spectra (7.11(*d*)). The extraction of the temperature using double isotope ratios was rather inaccurate. Furthermore, in the case of the He–Li thermometer,

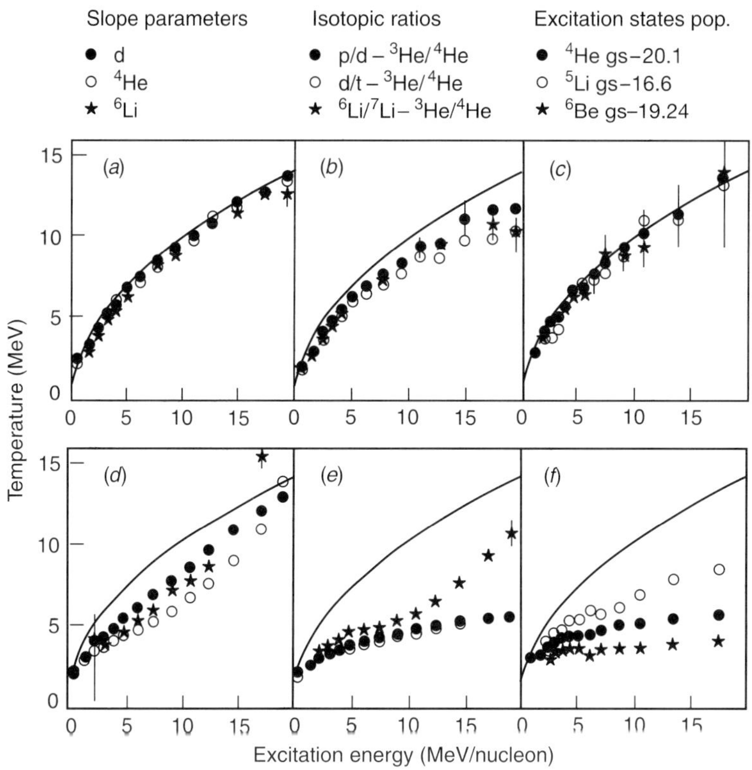

Figure 7.11. Apparent temperatures extracted from 'data' generated by sequential evaporation statistical model calculations [16]. The three methods shown in figure 7.8 were applied first to calculations in which the evaporation process was artificially stopped after evaporation of only one particle. This procedure produced the temperatures shown in the upper panel. The lower panel shows the results extracted by allowing the full evaporation cascade to develop 'normally'. In this case, the extracted temperatures differ considerably and have little in common with the 'true' compound nucleus temperature (continuous lines in all figures).

the variation of the extracted temperature with energy could be mistaken for evidence of a rapid phase change.

The works cited above tend to show that there is little evidence for strong structure in caloric curves and thus no obvious signal of a *sudden* phase change. This finding seems to be confirmed [17] by the observation of temperatures extracted from slopes of kinetic energy spectra which show overall agreement with the Fermi gas model and thus no convincing evidence for a phase change. Despite this setback, there is continuing interest in the caloric curve as witnessed by several papers devoted to this subject in recent conferences (see, e.g., [18]).

Regrettably, at the present time, it seems impossible to draw a firm conclusion concerning the usefulness, in concrete applications, of double isotope thermometers.

As far as the phase change itself is concerned, it is manifestly difficult to characterize the transition using the most probable macrostate as in thermodynamics. The obvious alternative would be to consider a large number of macrostates (partitions), because no one such state is overwhelmingly 'the most probable'. Thus, a more global 'order parameter', such as the *average* fraction of the total charge contained in complex fragments, may be appropriate to characterize the state of the system. This parameter would be expected to fall smoothly from almost one to zero as the excitation energy of the system is increased. A calculation of this kind has in fact been made using the Berlin multifragmentation code (see figure 21 in [1]). However, the observed transition, which is largely independent of the nucleus under consideration, is very broad.

7.4 Vaporization

At excitation energies above approximately 10 MeV/nucleon, experimental measurements reveal a rapidly rising cross-section for the disintegration of the excited parent nucleus into light particles with $Z \leq 2$. This process is known as vaporization. The most obvious question for experiment is to decide whether the production yields of the various light isotopes are consistent with a prompt production mechanism or whether a substantial fraction of light particles results from longer timescale evaporation processes in which the residue is, itself, a fragment with $Z \leq 2$.

The observation of prompt vaporization provides, in principle, an important test of the 'multifragmentation scenario' because practically all the excitation energy appears as fragment kinetic energy. One would, therefore, expect to see the $\mathrm{d}\ln[\rho(E^*)]/\mathrm{d}E^* \sim N/E^*$ proportionality discussed at the beginning of this chapter.

Let us begin with experiment. One of the first indications for the onset of vaporization was obtained in measurements on the $^{197}\mathrm{Au} + {}^{197}\mathrm{Au}$ system at 100, 250 and 400 MeV/nucleon incident energy [19]. The measurements were carried out at the SIS facility at GSI in Darmstadt, Germany, using the small angle ALADIN detectors together with 215 phoswich detectors (MSU Miniball/Miniwall) for fragments emitted at angles $> 14.5°$ and a complementary Si–CsI array for intermediate angles. Indirect evidence for the onset of vaporization was obtained by noting a drop in the complex fragment multiplicity for small impact parameters. However, no direct observation of vaporization events was made. Two years later the INDRA collaboration, working at the GANIL facility in France, reported exclusive measurements of vaporization events in the $^{36}\mathrm{Ar} + {}^{58}\mathrm{Ni}$ system [20] which was studied at incident $^{36}\mathrm{Ar}$ energies between 32 and 95 MeV/nucleon (figure 7.12). These measurements showed that the cross-section for vaporization rises by more than two orders of magnitude above an incident energy of

50 MeV/nucleon. Nevertheless, even at the highest energy studied, it amounts to no more that $1/5000^{\text{th}}$ of the total reaction cross-section. Vaporization is, thus, a rare process for energies below 100 MeV/nucleon.

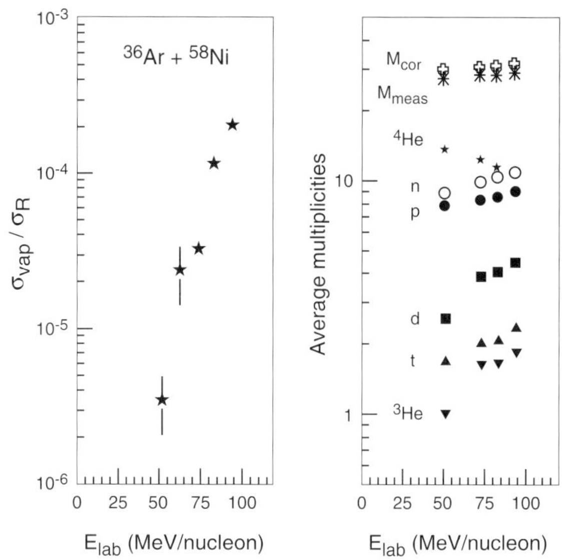

Figure 7.12. Left panel: cross-sections for vaporization (shown divided by the total reaction cross-section) as a function of laboratory energy in the $^{36}Ar + ^{58}Ni$ reaction measured with the INDRA multidetector [20]. Right panel: the corresponding light particle multiplicities. (Reprinted from *Phys. Lett.* B, **353**, Bacri Ch O *et al*, p 27. Copyright 1995, with permission from Elsevier Science.)

As the relative cross-section rises the alpha particle multiplicity falls slowly, giving way to a compensating increase in the multiplicity of nucleons. The summed multiplicity is almost constant over the energy range studied. The authors of reference [20] estimated that the light particles typically have a kinetic energy of 5–6 MeV/nucleon over and above the Coulomb energy characteristic of a single source.

Further work including isotopes with $Z \leq 3$ provided strong indications that the vaporization events observed in the $^{36}Ar + ^{58}Ni$ system arose from quasi-projectile (qp) and quasi-target (qt) fragments produced by deep inelastic scattering [21]. This reaction mechanism gives rise to considerable variation of the mass and excitation energy of the qp and qt sources. However, by directly including these variations in a statistical model for the de-excitation of the qp and qt fragments, it was possible to obtain a reasonable description of the experimental measurements.

The model used in these investigations (extended and applied with success by Gulminelli and Durand in the understanding of calorific curves [23]) is a quantum statistical version of the thermal and chemical equilibrium model proposed by Subramanian *et al* [22]. Similar models (including relativistic treatment of the motion of particles) have been discussed by Das Gupta and Mekjian [24] and by Hahn and Stöcker [25]. The application to vaporization, which is reported in a recent work of Borderie *et al* [26], considers light fragments including nucleons as existing, with uniform densities, in chemical and thermal equilibrium inside a fixed freeze-out volume. Excited bound states of isotopes are considered as distinct *species*. Given the assumption of chemical equilibrium we see that, in order to calculate the mean multiplicities or densities for the various species, we need the grand potential given in equations (1.6.18) and (1.6.21), i.e.,

$$\Omega^* = -T\ln[\mathbb{Z}] = \mp\frac{1}{\beta}\sum_i \ln[1 \pm \mathrm{e}^{-\beta(E_i-\mu)}], \tag{7.4.1}$$

where the upper (lower) signs are respectively appropriate for Fermi–Dirac (Bose–Einstein) statistics. For one particular species, one might first think that there is only one term in the sum. However, we have also to take into account the motion of each isotope in the freeze-out volume. Therefore, for a given species (isotope in ground or excited state) we write the energy of the state under consideration as $E_k = B_k + \varepsilon$ where the kinetic energy $\varepsilon = p^2/2m_k$ (m_k and p denote respectively the mass and the momentum of the isotope) and the energy B_k is the sum of the negative of the binding energy and the energy (with respect to the ground state) of the excited state under consideration.

The sum over states for a particular species can now be written as an integral over momentum and spatial coordinates weighted with the logarithmic factor as in (7.4.1). The integral over spatial coordinates just gives the volume V (the centre of mass is not constrained), so that including a spin degeneracy factor we obtain

$$\Omega_k^* = \mp\frac{V}{h^3\beta}\int_0^\infty \ln\left[1 \pm \mathrm{e}^{-\beta(\varepsilon+B_k-\mu_k)}\right](2s_k+1)4\pi p^2\,\mathrm{d}p, \tag{7.4.2}$$

which with $4\pi p^2\,\mathrm{d}p = 2\pi(2m_k)^{3/2}\varepsilon^{1/2}\,\mathrm{d}\varepsilon$ becomes

$$\Omega_k^* = \mp\frac{2\pi(2s_k+1)(2m_k)^{3/2}V}{h^3\beta}\int_0^\infty \ln[1 \pm \mathrm{e}^{-\beta(\varepsilon+B_k-\mu_k)}]\varepsilon^{1/2}\,\mathrm{d}\varepsilon. \tag{7.4.3}$$

The mean number of particles (the multiplicity) of species k follows by differentiation of Ω_k^*, i.e.,

$$\langle n_k\rangle = -\frac{\Omega_k^*}{\partial\mu_k} = \frac{2\pi(2s_k+1)(2m_k)^{3/2}V}{h^3}\int_0^\infty \frac{\mathrm{e}^{-\beta(\varepsilon+B_k-\mu_k)}}{[1 \pm \mathrm{e}^{-\beta(\varepsilon+B_k-\mu_k)}]}\varepsilon^{1/2}\,\mathrm{d}\varepsilon. \tag{7.4.4}$$

By setting $z = \beta\varepsilon$ and dividing by the volume, V, we obtain the density for each species as

$$\frac{\langle n_k\rangle}{V} = \frac{2\pi(2s_k+1)(2m_kT)^{3/2}}{h^3}\int_0^\infty \frac{z^{1/2}\,\mathrm{d}z}{[\mathrm{e}^{(z+\beta B_k-\beta\mu_k)} \pm 1]}, \tag{7.4.5}$$

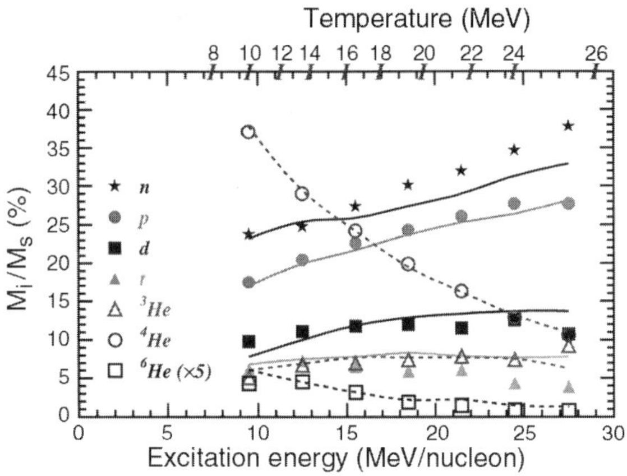

Figure 7.13. Light particle multiplicities (divided by the total vaporization source multiplicity) in the ^{36}Ar + ^{58}Ni reaction at 95 MeV/nucleon [26] plotted as a function of excitation energy. The data are well reproduced by the quantum statistical model of Gulminelli and Durand [23]. The predictions of this model (equation (7.4.5)) are represented by the continuous curves in the figure. (Reprinted from *Eur. Phys. J.* A **6** 197, Borderie B *et al*. Copyright 1999, with permission from Springer Verlag.)

where $T = 1/\beta$ and, as before, the plus and minus signs refer respectively to Fermions and Bosons. The integral in equation (7.4.5) is calculated numerically. As with the model employed by Albergo *et al* (which uses classical statistics) the chemical potential of any isotope is expressed in terms of the neutron and proton chemical potentials.

$$\mu_k(N_k, Z_k) = N_k\mu_{\rm n} + Z_k\mu_{\rm p}. \tag{7.4.6}$$

It follows that the densities for all species can be expressed in terms of $\mu_{\rm n}$ and $\mu_{\rm p}$. Equation (7.4.5) implies that these remaining 'unknowns' are density dependent. However, they may be straightforwardly obtained by application of sum rules to the fragment multiplicities. Thus if $N(Z)$ is the total number of neutrons (protons) then we can use the sums

$$\sum_k N_k\langle n_k\rangle = N,$$
$$\sum_k Z_k\langle n_k\rangle = Z, \tag{7.4.7}$$

($\langle n_k\rangle$ from (7.4.5)) to detemine the neutron and proton chemical potentials and thereby the absolute densities of each species.

In the application of the model by Borderie *et al* [26] to the quasi-projectile vaporization events discussed above, all known discrete levels with lifetimes > 100 fm/c for nuclei up to ^{20}Ne were included in the calculations. Furthermore, following Gulminelli and Durand [23], an excluded volume correction (section 6.7) due to finite fragment size was incorporated into the model. This has the effect of favouring tightly bound species (nucleons and alpha particles) as opposed to loosely bound species such as deuterons and high-lying excited states. The freeze-out density compared with normal nuclear density, ρ_0, was fixed to $\rho = \rho_0/3$ and the temperature was varied at each excitation energy to fit the data. The comparison of the equilibrium model calculations with data for isotopes with $z \leq 3$ is shown in figure 7.13. The description is truly remarkable. Furthermore, the analysis reveals that primary decay is responsible for about 80% of the total multiplicity. Thus, post-vaporization evaporation, while not negligible, does not destroy the picture of a source whose decay is consistent with equilibrium statistical mechanics. On the other hand, as can be seen from the upper temperature scale of the figure, the temperatures required to fit the data vary smoothly with the estimated excitation energy. There is no evidence of any irregular behaviour which could be associated with a rapid change of phase.

This brings us to the end of our discussion of vaporization and, indeed, to the end of our presentation of multifragmentation. It remains, in the final chapter, to offer some remarks which may contribute to a global view of the long road from evaporation to vaporization.

References

[1] Pochodzalla J 1997 *Prog. Part. Nucl. Phys.* **39** 443

[2] Moretto L G, Phair L and Wozniak G 1997 Phase Transitions in Nuclear Physics *5th Int. Workshop on Relativistic Aspects of Nuclear Physics (Rio de Janeiro, Brazil) Int. Report* LBNL-41273

[3] Cole A J, Chabane A, Charvet M, Désequelles P, Giorni A, Heuer D, Lleres A and Viano J B 1995 *Z. Phys.* A **353** 279

[4] Fisher M E 1967 *Physica* **3** 255

[5] Campi X 1988 *Phys. Lett.* B **208** 351

[6] Désesquelles P, Cole A J, Giorni A, Heuer D, Lleres A, Viano J B, Chambon B, Cheynis B, Drain D and Pastor C 1993 *Phys. Rev.* C **48** 1828

[7] Lleres A, Cole A J, Désesquelles P, Giorni A, Heuer D, Viano J B, Chambon B, Cheynis B, Drain D and Pastor C 1994 *Phys. Rev.* C **50** 1973

[8] Cook N D and Hayashi T 1997 *J. Phys. G: Nucl. Part. Phys.* **23** 1109

[9] Mustafa M G, Blann M, Ignatuk A and Grimes S M 1992 *Phys. Rev.* C **45** 1078

[10] Finn E, Argawal S, Bujac A, Chuang J, Gutay L J, Hirsch A S, Porile N T, Sharenberg R P and Stringfellow B C 1882 *Phys. Rev. Lett.* **49** 1321

[11] Aichelin J, Peilert G, Bohnet A, Rosenhauer A, Stöcker H and Greiner W 1988 *Phys. Rev.* C **37** 2451

[12] Kwiatkowski K *et al* 1995 *Phys. Rev. Lett.* **74** 3756

[13] Albergo S, Costa S, Costanzo E and Rubbino A, 1985 *Nuovo Cimento* A **89** 1

[14] Mekjian A Z 1978 *Phys. Rev.* C **17** 1051

[15] Pochodzalla J *et al* 1995 *Phys. Rev. Lett.* **75** 1040

[16] Peter J *et al* 1997 *13th Winter Workshop Meeting on Nuclear Dynamics (Marathon FL Feb. 1997)* and *Internal Report* LPCC97-08 Caen LPC

[17] Ma Y G *et al* 1997 *Phys. Lett.* B **390** 41

[18] *Proc. 6th Int. Conf. Nucleus–Nucleus Collisions* (*Gatlinburg, TN, 1997*) ed M Thoennessen, C K Gelbke, F E Betrand and J D Garrett (New York: Elsevier) (1998 *Nucl. Phys.* A **630**)

[19] Tsang M B *et al* 1993 *Phys. Rev. Lett.* **71** 1502

[20] Bacri Ch O *et al* 1995 *Phys. Lett.* B **353** 27

[21] Rivet M F *et al* 1996 *Phys. Lett.* B **388** 219; Borderie B *et al* 1996 *Phys. Lett.* B **388** 224

[22] Subramanian P R, Csernai L P, Stöcker H, Maruhn J A, Greiner W and Kruse H 1981 *J. Phys. G: Nucl. Phys.* **7** L241

[23] Gulminelli F and Durand D 1997 *Nucl. Phys.* A **615** 117

[24] Das Gupta S and Mekjian A 1981 *Phys. Rep.* **72** 131

[25] Hahn D and Stöcker H 1988 *Nucl. Phys.* A **476** 718

[26] Borderie B *et al* 1999 *Eur. Phys. J.* A **6** 197; Borderie B *et al Proc. 8th Int. Conf. on Nuclear Reaction Mechanisms (Varenna, June 1997)* Ricerca Scientifica ed Educazione Permanente Suppl. no 111 ed E Gadioli p 239

Chapter 8

From evaporation to vaporization

"Oh Oysters," said the Carpenter,
"You've had a pleasant run!
Shall we be trotting home again?"
But answer came there none–
And this was scarcely odd, because
They'd eaten every one.

The Walrus and the Carpenter, Lewis Carrol

8.1 Summing up

This short chapter is mainly devoted to the recapitulation of the main ideas involved in statistical models of nuclear decay. In the course of this summary we also attempt to highlight some topics which, in our opinion, require closer attention. We hope that, in so doing, we provide elements which will lead to an improved global appreciation of the coherence of the various theoretical approaches.

We began this work with an introduction to statistical mechanics and especially to those aspects of the subject which may be applied to the decay of excited atomic nuclei. We therefore particularly emphasized features which distinguish the nuclear decay problem, notably, that observations are made on an ensemble of macrostates long after the equilibrium state envisaged in the statistical models has ceased to exist. We also pointed out difficulties which may occur in applications of thermodynamics to nuclear decay problems.

An important point, which appeared at the end of chapter 1, concerns the *concept* of equilibrium in nuclear physics applications. We argued that the ensemble approach is intrinsically pertinent and, somewhat more tenuously, that the subset of microstates corresponding to measured events is so small compared with the full set that equilibrium may appear to characterize the probabilities associated with macroscopic states, even in the presence of a significant non-uniform population of microstates. This situation was referred to as 'pseudo-equilibrium'.

In the subsequent development of the historical account of the subject, we did not insist on this concept mainly because, as far as we can ascertain, it has not been adequately discussed in the literature. Nonetheless, we believe that it may well be a major contributing factor to the success of statistical models of nuclear decay.

Chapter 2 was devoted to basic nuclear physics including a presentation of equidistant spacing Fermi gas models (ESM) for nuclear level and state densities. The subject matter, although fairly standard, was presented in such a way as to be directly applicable in ensuing chapters. We might just emphasize one interesting particularity. If we assume a constant spacing, $d = 1/g_0$, for single nucleon levels (not distinguishing neutrons and protons), the number of microstates of an N nucleon system, $\Omega_N(E^*)$, at an excitation energy $E^* = Kd$ is equated with $\mathbb{P}_N(K)$ which is the number of ways of partitioning the integer K into not more than N parts. If $K \leq N$ then this quantity is simply $P(K)$, i.e., the total number of partitions of K (because K cannot be partitioned into more than K parts). The asymptotic (large K) form of this quantity, which was given in 1918 by Hardy and Ramanujan [1] (the explicit formula was cited, in a different context in equation (6.2.5)), coincides exactly with the expression derived by Ericson [2] which was obtained by an approximate inverse Laplace transform of the grand canonical partition function. Thus, with the level density parameter $a = g_0\pi^2/6$,

$$\Omega_N(E^*) = \rho_{\mathrm{GC}}(E^*)d = d\frac{\mathrm{e}^{2\sqrt{aE^*}}}{\sqrt{48}E^*} = \frac{\mathrm{e}^{\sqrt{4\pi^2K/6}}}{4K\sqrt{3}}. \tag{8.1.1}$$

The right-hand side of this equation is the leading term in the Hardy–Ramanujan expression for $P(K)$. As far as the author can ascertain, this correspondence has not been exploited particularly with respect to higher order corrections which form part of the Hardy–Ramanujan work.

Let us now turn to chapters 3 and 4 which are mainly concerned with the evaporation of light particles, fusion, pre-equilibrium emission and low energy fission. The treatment of fusion, in chapter 3, was rather schematic. Nevertheless, we did present the basic ideas embodied in a quite successful radial trajectory model referred to as the critical distance model. At the same time, we emphasized that current work considers the fusion mechanism as a rather complex dynamical problem. Consequently, a comprehensive treatment was considered to be somewhat beyond our scope (see Fröbrich and Gontchar [3]).

Looking at studies of pre-equilibrium emission, one is struck by the difficulties associated with this enterprise (even if we consider only nucleon emission). On the one hand, experiments must measure the low statistics high energy tails of energy distributions in order to obtain information on the initial stages of the reaction. On the other, theories based on particle–hole densities are not very satisfactory insofar as they imply that temporal equilibrium is attained between each nucleon–nucleon collision. Indeed, it seems difficult to speak of an equilibrium population of n-exciton states in a single nucleus if the mechanism which is supposed to bring about equilibrium is precisely that which, at each nucleon–nucleon collision, almost always increases the number of excitons. It may be that the concept of pseudo-equilibrium is relevant here. Thus, despite the fact that temporal equilibrium is not attained in the initial stages of the reaction (which are classified in terms of particle and hole numbers), the set of particle–hole microstates sampled by the ensemble of measured events can be considered as a random sample from the full set. The kinetic-equation approach of Harp, Miller and Berne [4] is not affected by this problem since an equilibrium population of (continuum or bound) states is not assumed.

It is convenient to discuss evaporation and fission together, mainly because the Bohr–Wheeler theory [5] represents a coherent approach to the competition between fission and neutron emission. However, there is an evident contrast between the treatment of fission and light particle decay modes at low excitation energies. Researchers in these two disciplines have tended to follow separate highways. It seems probable that this is partly because the first, and perhaps the most difficult, problems in fission involved spontaneous or slow neutron induced fission which are strongly influenced by complicated nuclear structure effects. Furthermore, one has the impression that, whereas the 'fission people' were obliged to worry a lot about the dynamics of the fission mechanism (especially, with respect to the path from saddle to scission), the 'evaporation crowd' were happily gratified by the immediate success of models which simplify, to a rather astonishing degree, the relative motion between light emitted particles and residual nuclei. The Bohr–Wheeler or Weisskopf [6] approaches to evaporation assume randomly directed motion of the light particle in the surface of the nucleus. The Hauser–Feshbach theory [7] makes the same basic assumption. However, in the HF theory, the probability of escape of a particle on a given trajectory is modified by the transmission coefficient, at least at low relative energies. To our knowledge, the possibility that nucleons near the surface of the compound nucleus may have an *a priori* anisotropic distribution of velocities has not been considered. (As an extreme illustrative example, imagine that 'surface' neutrons possess only outwardly pointing radially directed velocities. The corresponding phase space would then be one-dimensional, as in the Bohr–Wheeler theory of fission.) Of course, this is not to decry the resounding success of the Weisskopf and Hauser–Feshbach models for which several examples involving mainly multistep emission were given in chapters 3 and 4, only to suggest that it might be interesting to find out why they work so well.

The latter part of the book was concerned with changes in the decay modes with increasing excitation energy. As far as sequential binary decay is concerned the reader was confronted, in chapter 5, with several approaches which differ significantly in their basic assumptions except in the sense that they rely heavily on the Weisskopf or Hauser–Feshbach models. It will probably have been obvious in the text that the author's personal preference is for the GEMINI model which the author suggested should be understood as a straightforward extension of the Bohr–Wheeler theory. At the very least, it is easy to establish the coherence of this approach with early fission theories. Furthermore, the model is extremely successful in fitting data. For these reasons, the author finds it difficult to escape the idea that the 'democratic' sharing of energy between an emitted large fragment and the residue (which reminds us of Fong's scission model [8]) is essentially correct, especially as this is what would be expected on purely statistical grounds.

The discussion of statistical models of multifragmentation undertaken in chapter 6 highlights a recurrent theme of this work. We are referring to the adoption or rejection of the transition state as the equilibrium point. One of the principal objections to the Fong model for fission, asserts that it is difficult to designate a particular value of the fission coordinate as *the* point where the statistical model is to be applied. We would argue against this objection. Near scission, there exists a minimal separation where, apart from Coulomb acceleration and long timescale evaporation, one may identify 'decaying' states for which further dynamical evolution does not significantly change observables. The same 'point of no return' (beyond which the emission probability is 1 and the final vector momentum of the emitted particle is determined) exists for neutron emission although, in this case, the transition states and states which correspond to 'outwardly directed' scission configurations are indistinguishable.

The 'scission' viewpoint which underlies the earlier multifragmentation models (Copenhagen, Berlin, 'Boltzmann') was discussed at the beginning of chapter 6. A choice made in that discussion was to employ the term 'scission state' to refer to *spatial* configurations which may or may not decay. Thus, in neutron emission, only those scission states corresponding to outwardly pointing trajectories can be considered as decaying states. The most simple multifragmentation theories assume that *all* scission states decay with a constant characteristic break-up time. However, that point of view needs to be corrected to eliminate scission states with overlapping fragments (section 6.7) and more generally (in nice analogy with the neutron problem) those states for which the subsequent motion leads to fragment–fragment collisions. The difficulty of actually assigning a scission point (which becomes the 'freeze-out volume' in multifragmentation) has led to much discussion since it may be that the freeze-out density is relevant for characterizing the nuclear equation of state.

The transition state/scission state problem appears again with the multifragmentation theory of Lopez and Randrup [9], who not only choose the transition state (near the barrier) as the point where equilibrium is established but make an explicit attempt to estimate the effects of fragment interactions on the path from 'saddle' to 'scission'. It is well to propose, as do Lopez and Randrup with characteristic finesse, an expansion coordinate with respect to which one may define a maximum in the potential energy. However, it seems to the author that a reliable treatment of subsequent fragment–fragment interactions which take place beyond the multifragmentation saddle is not only difficult to put into practice but also somehow defeats the object of the exercise. The finally observed partitions (even neglecting the influence of evaporation) may have little resemblance to those formed at the multifragmentation saddle. The same observation may be applied to explicit dynamical models of spinodal break-up. Thus, when the dust has settled, it may be better to adopt the freeze-out approach in which the statistical model is applied to scission macrostates which are normally constituted by a set of non-overlapping fragments contained in a freeze-out volume. Of course, even in this case, one is faced with difficult problems concerning the assumed motion of the fragments at freeze-out (e.g., radial and angular collective, as opposed to random, motion) but these problems do not seem to be insurmountable mainly because they are amenable to experimental investigation. Evaporation corrections which produce long timescale transformations of the freeze-out partitions are a nuisance but enough is known about evaporation to lead us to believe that such corrections do not constitute a major obstacle.

We may retrace our steps somewhat in order to look more closely at angular and radial collective motion. A full discussion of these aspects of multifragmentation was not included in the main text because, at the present time, there are more questions than answers. It will probably be clear that inclusion of angular momentum becomes more and more problematic with increasing complexity of the decay process. Light particle evaporation is assumed to be characterized by intrinsically randomly directed 'outward' motion from the surface of the residual nucleus. This is not the case for binary decay (chapter 5). Thus the GEMINI code, in contrast to the Weisskopf or Hauser–Feshbach models, assumes the angular momentum to reside in a *rigid* rotation of the decay partners perpendicular to an axis through their centres of mass. One immediately anticipates the difficulty of extending this reasoning to the picture of multifragmentation in which the fragments are assumed to be created within a freeze-out volume. If rigid rotation is assumed, the moment of inertia may be rather large so that the rotational energy may be the dominant component of the total energy. Furthermore, the assumption of rigid rotation which fixes the tangential components of the fragment velocities implies a significant modification of the phase space density of microstates (the statistical weight) for any given partition.

The investigation of radial collective motion is a little different because the associated energy is loosely constrained by the conservation of total energy, but not closely constrained by an additional conservation law (as is the case for angular momentum). It may be that the dynamical evolution of the system prior to multifragmentation involves radial expansion. In this case the radial collective (explosive) motion must be imposed in the statistical models. However, it is also possible to understand radial energy (in whole or in part) by again recalling that the calculation of the phase space density for neutron emission assumes outwardly directed motion of the neutron with respect to the surface of the residue. As mentioned above, the same idea applied to 'freeze-out' models of multifragmentation would assert that scission configurations, in which the initial velocities of fragments subsequently lead to fragment–fragment collisions, should be excluded (this restriction was briefly mentioned in chapter 6). In this picture, one would thus expect radial expansion to be an intrinsic feature of realistic statistical models.

The problem of the phase change associated with multifragmentation and vaporization (chapter 7) appears to be rather complex. As a preliminary step, one might be happy to settle for an experimental proof (and not a proof simply based on a simulation code) that the entropy (or density of microstates) shows some non-smooth behaviour as a function of excitation energy of the parent nucleus. Unfortunately, even if the parent nucleus (mass charge, excitation energy, angular momentum) or the distribution of properties of parent nuclei can be identified, experiments can do no more than to provide an estimation of the macroscopic entropy, $S_G = -\Sigma_i p_i \ln[p_i]$ for some set $\{i\}$ of observable macrostates. The set $\{i\}$ may refer simply to the multiplicity probablity distribution, or perhaps the charge distribution. Presumably, as we define macrostates in more and more detail, we approach closer and closer to the microscopic entropy. But where is it safe to work? Charge partitions (possibly with some dynamical information) seem to represent the current limit of what can reasonably be adopted as a macrostate definition with modern multidetectors. However, we do not know, even supposing this degree of detail, how well the microscopic entropy is represented by S_G. This latter aspect of the problem may be amenable to study using simulation models including those currently applied to the analysis of data.

In chapter 7 we discussed the analogy with percolation and the search for critical point characteristics of the fragment size distribution. We also discussed the branch of current research (referred to as calorimetry) concerning the multifragmentation phase change (not necessarily at the critical point), which involves measuring temperatures (as a function of excitation energy) either by using kinetic energy distributions of light particles, or by comparing excited state populations for a single isotope or again by using the double isotope ratio thermometers described in section 7.3. Despite initial difficulties and disagreements (notably with respect to the use of different thermometers), it seems worth pursuing this line of research. The failure to make an uncontested observation of the phase change in hot nuclei should not, in my view, be regarded as a major setback. The smoothing due to finite size fluctuations may well obliterate any sharp signal.

We hope that this recapitulative sketch will provide the reader with *matière à réflexion*. As well as describing successes and difficulties with the statistical description of the decay of excited nuclei, we have attempted to show that there are still many problems to solve not only in multifragmentation, which is a relatively recent research topic, but also in other areas extending backward even to the emission of light particles (i.e., from vaporization to evaporation). Most of them are concerned with the validity of the statistical approach in specific applications, with the detailed assumptions of the corresponding models and with related problems involving the influence of nuclear structure and reaction dynamics on observables.

References

[1] Hardy G H and Ramanujan S 1918 *Proc. London Math. Soc.* 2, **17** 75
[2] Ericson T 1960 *Adv. Phys.* **9** 425
[3] Fröbrich P and Gontchar I I 1998 *Phys. Reps* **292** 131

[4] Harp G D, Miller J M and Berne B J 1968 *Phys. Rev.* **165** 1166
[5] Bohr N and Wheeler J A 1939 *Phys. Rev.* **56** 426
[6] Weisskopf V F 1951 *Phys. Rev.* **83** 1073
[7] Hauser W and Feshbach H 1952 *Phys. Rev.* **87** 366
[8] Fong P 1969 *Statistical Theory of Nuclear Fission* (New York: Gordon and Breach)
[9] Lopez J A and Randrup J 1990 *Nucl. Phys.* A **512** 345

Index